SOCIETA' ITALIANA DI FISICA

RENDICONTI
DELLA
SCUOLA INTERNAZIONALE DI FISICA
«ENRICO FERMI»

LXXIII CORSO
a cura di D. LAL
Direttore del Corso

VARENNA SUL LAGO DI COMO
VILLA MONASTERO
26 GIUGNO - 8 LUGLIO 1978

Sviluppo
del sistema solare primordiale
e sistema solare attuale

1980

SOCIETÀ ITALIANA DI FISICA
BOLOGNA · ITALY

ITALIAN PHYSICAL SOCIETY

PROCEEDINGS

OF THE

INTERNATIONAL SCHOOL OF PHYSICS
« ENRICO FERMI »

Course LXXIII

edited by D. LAL
Director of the Course

VARENNA ON LAKE COMO
VILLA MONASTERO
26th JUNE - 8th JULY 1978

Early Solar System Processes and the Present Solar System

1980

NORTH-HOLLAND PUBLISHING COMPANY, AMSTERDAM · NEW YORK · OXFORD

Library of Congress Cataloging in Publication Data

Varenna, Italy. Scuola internazionale di fisica.
  Early solar system processes and the present
solar system.
  (Proceedings of the International School of Physics
« Enrico Fermi » = Rendiconti della Scuola inter-
nazionale di fisica « Enrico Fermi »; course 73).
  At head of title: Italian Physical Society.
  Added t.p.: Sviluppo del sistema solare primor-
diale e il sistema solare attuale.
  Course held 26th June-8th July 1978.
  Includes bibliographical references.
  1. Solar system—Congresses. I. Lal, Deven-
dra, 1929-   . II. Società italiana di fisica. III. Title.
IV. Title: Sviluppo del sistema solare primordiale
e il sistema solare attuale.
QB500.5.V37 1980   523.2   80-22549
ISBN 0-444-85458-4

Technical Editor

P. PAPALI

INDICE

Preface.

The Seventies have been exciting for scientists in many disciplines, *e.g.* in Chemistry, Biology, Earth and Space Sciences. During this period, with the acquisition of lunar samples into the hands of geologists and meteoriticists, and high-resolution photogeological studies of planets, the field of comparative planetology has fully come of age. The astronomers learnt to recognize signs of birth of stars, a phenomenon which had baffled scientists since the classical work of Jeans. Alongside the development of capability of astrophysicists to herald the stellar event, « A star is born », the meteoriticists began accumulating unmistakable signs associated with the very early history of our solar system, in fact related to the event of birth of our solar system itself. I am referring to the host of isotopic anomalies observed in small grains in certain primitive meteorites using high-precision mass spectrometers. The results can best be understood as grains being *pre-solar* in time, originating from a supernova shock wave. G. J. WASSERBURG has referred to this discovery as Pandora's box, since the anomalies exist in case of several elements.

Our new concepts about the micro- and macro-universe often owe their origin to developments in disparate fields. The independent developments in the fields of Astronomy and Meteoritics provided very good reasons for organizing a communion between scientists who practice these highly specialized fields. This led to the Summer School highlighting the above-mentioned recent developments and their implications, and summarizing the present state of our knowledge about the early-solar-system processes. We also thought it would be good timing to have this particular School, since the last Enrico Fermi Varenna Summer Course dealing with meteorites was held as far back as in 1960.

That the Summer School brought together people from widely different fields was not unexpected considering its motivation, but, happily and importantly, the interactions were smooth and productive. There was a healthy cross-breeding of ideas which stimulated not only the participants but the lecturers too. As an example, I would like to cite P. R. WOODWARD, who was inspired to pursue applying a two-dimensional hydrodynamic approach to the study of protoplanetary disks. On my own part, I was stimulated to extend our investigations of carbonaceous chondrites to see if any further clues can be deciphered about the very early stages of accretion. The problem of delineating the important stages between the proto-Sun state and that of the early Sun with its planets offers a great challenge. It is no doubt a difficult one; but it is solvable.

It is my sincere hope that the readers will also find this Course material to be stimulating as to allow them to take on some detailed investigations of one or the other processes associated with the birth and evolution of the solar system.

Dr. G. CASTAGNOLI and I have greatly enjoyed organizing the Summer Course. It brought us in closer touch with our colleagues and it resulted in new contacts and friendships. In science, one of the greatest rewards is the comradeship, the opportunity to work together with scholars on esoteric problems. For providing the perfect setting, CASTAGNOLI and I are most grateful to the Italian Physical Society, to G. WOLZAK and her colleagues, and to the caretakers of the Villa Monastero.

D. LAL

Star Formation (*).

P. R. WOODWARD

Lawrence Livermore Laboratory, University of California - Livermore, Cal. 94550
Leiden Observatory, Huygens Laboratory - Leiden, The Netherlands

1. – The composition of the interstellar gas.

In recent years our understanding of the star formation process has advanced considerably. This advance has been caused by the development of new means for observing the interstellar gas and also by the increase in size and speed of modern computers. The new observations allow us to test hypotheses about the star formation process by observing gas in regions of star formation efficiently and accurately. The new computer calculations allow us to make ever more realistic experiments which help us to interpret the observational data. In this article, I will concentrate mainly on the computer simulations, but I will indicate where new observations have inspired theoretical work and where they have demonstrated the need for more realistic models.

Our understanding of the structure of the interstellar gas has been closely coupled to developments of new observational techniques. The first observations relating to star formation were of course optical. These observations told us mainly about interstellar gas which was made luminous by ionization or by reflection of nearby starlight. Very dense gas could be observed by its absorption of light. A typical region suited for such optical study is the Orion nebula, shown in fig. 1. A feature of such optical observations is that they pertain to regions where the gas is strongly disturbed by the presence of a newly formed massive star. Thus much information concerning the formation of the first stars in the region has been confused by the disruption of the surrounding gas by these stars. The optical data for these regions are, nevertheless, very valuable in studying the process of formation of second-generation stars.

Optical observations are less suited to a study of the interstellar gas which is not associated with newly formed stars. This gas can be observed optically by determining the reddening of starlight due to dust associated with the gas.

(*) Work performed under the auspices of the U. S. Department of Energy by the Lawrence Livermore Laboratory under contract No. W-7405-ENG-48.

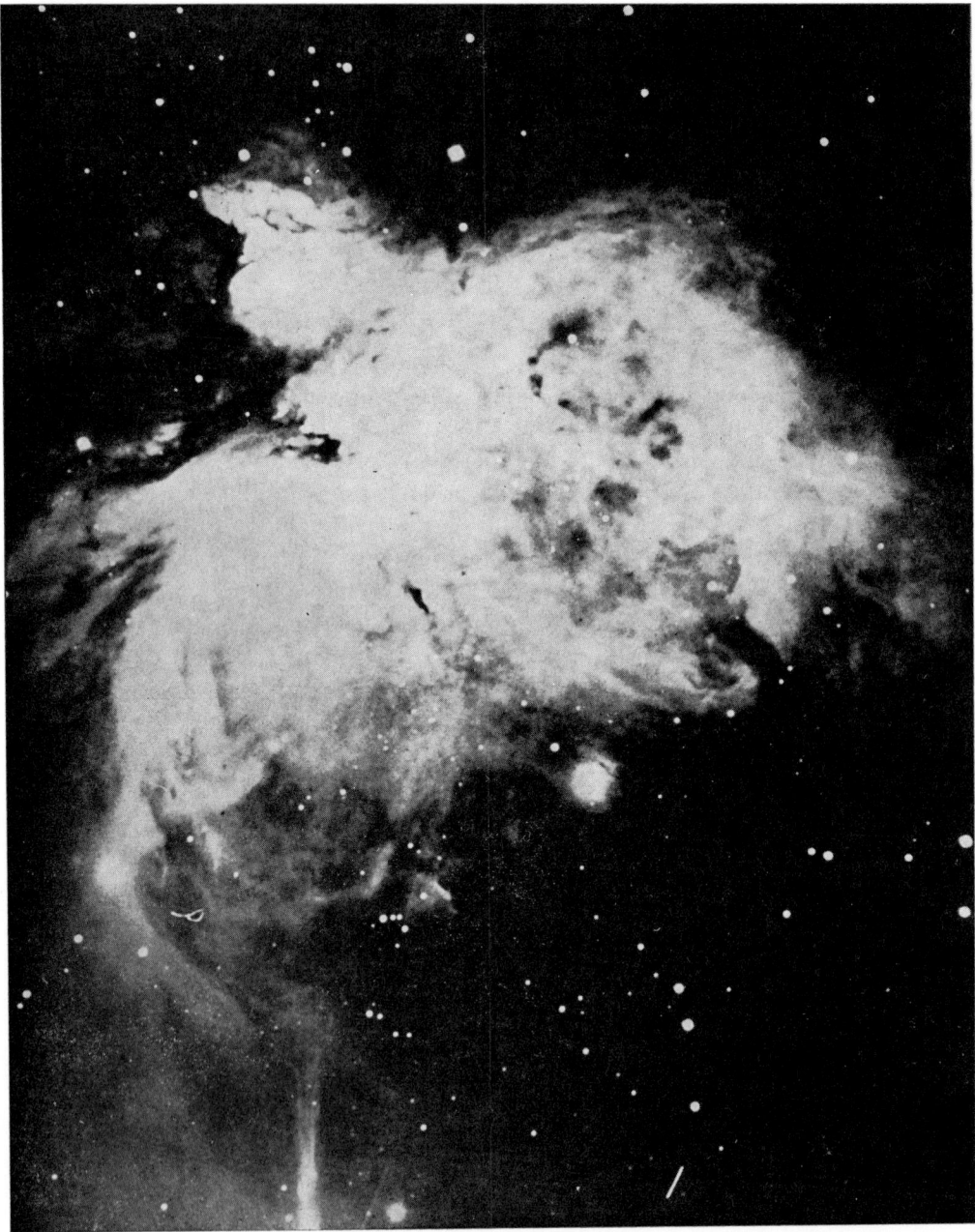

Fig. 1. – An optical photograph of the Orion nebula. The dense gas from which new stars are forming is detected by means of its absorption of starlight. The diameter of the region of bright, ionized gas is about 15 minutes of arc, or about 2.2 pc at the 500 pc distance to the nebula.

It can also be observed by means of the absorption lines of the trace elements sodium and calcium. These latter observations pointed to the existence of concentrations of cold gaseous material, called interstellar clouds. The differential rotation of the galactic disk would smear such an absorption feature by means of the Doppler shift much more than would be observed if the gas were evenly distributed along the line of sight to the background star. The narrow absorption lines, therefore, indicate that the gas is in the form of small clumps, or clouds. An upper limit on the thermal velocities of gas atoms in these clouds is also obtained from the width of the absorption lines. The result is that the clouds are cool in comparison with ionized regions near hot stars, where the gas is seen in emission. Our main interest in these interstellar-gas clouds will be to use them as a starting point in building models of the star formation process.

A fuller understanding of these clouds was made possible by a new observing technique, radio astronomy. Observations of the 21 cm line of neutral hydrogen made it possible to study the interstellar gas out to very great distances from the Sun. A two-phase model of the gas was built to explain these radio observations, and this model will be discussed in detail presently. For now, the important feature of that model is that the 21 cm radiation is produced in about equal amounts by a warm, diffuse intercloud gas at about 10^4 K and $0.3m_H$ cm^{-3} and by cool, dense clouds at about 100 K and $10m_H$ cm^{-3} (m_H is the mass of the hydrogen atom) (*). It is these clouds which we will discuss as starting points for the star formation process. It is by observing radio radiation from neutral hydrogen that we have learned the most about them.

The 21 cm line tells us about warm gas. Hot gas, above about 10^4 K, becomes ionized and does not radiate at this wavelength. Cold gas, at about 10 K, absorbs this radiation. However, cold gas tends to be dense and located in small regions. The absorption it produces is, therefore, difficult to observe with the beam widths of standard radio telescopes. The absorption is also lessened by the formation of molecular hydrogen at the densities characteristic of cold gas. This molecule formation reduces the concentration of absorbing atomic hydrogen in the cloud. A difficulty is now evident. Optical observations show us gas after stars have already formed, while 21 cm observations show us where star formation may only be about to begin.

This observation gap in studying star formation was bridged only recently by observations of radio radiation from trace molecules in the gas, such as CO, OH and H_2CO. Infra-red observations of warm dust have also been helpful,

---

(*) In this paper we will make use of a convenient set of units for interstellar-gas dynamics. We measure densities in hydrogen atom masses per cubic centimeter, distance in parsec, and velocities in kilometer per second. Then the unit of time is equal to $9.778 \cdot 10^5$ y, or nearly 10^6 y. The implied unit of mass is $0.024\,72\,M_\odot$ and typical interstellar pressures range from 1 to 10 units. We will denote the unit of pressure by (km/s)$^2 m_H$ cm^{-3}. In these units the gravitational constant G is $1.062 \cdot 10^{-4}$, and the unit of energy is $4.916 \cdot 10^{41}$ erg.

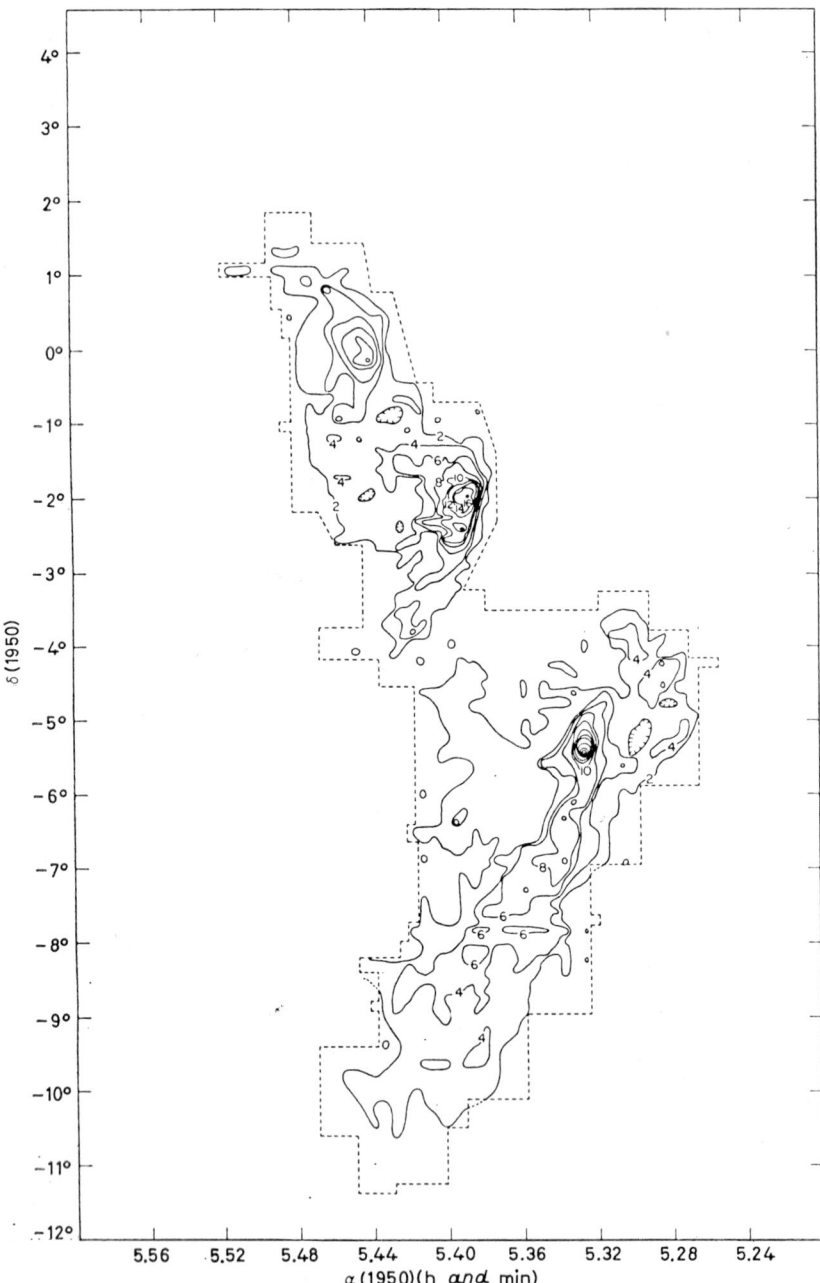

Fig. 2. – Contours of peak CO radiation temperature in the Orion region, taken from [1]. The optical nebula in fig. 1 is located at $\alpha = 5$ h 33 min, $\delta = -5°\,25'$, near the strong CO peak in the lower cloud in this figure. These CO observations reveal a very extensive cloud of dense gas which extends beyond the region of ionized gas for about 45 pc in a direction parallel to the galactic plane.

but mainly in the case where a hot object, such as a star or protostar, is embedded in a dense cloud. This embedded source heats the dust in the gas and makes it emit enough infra-red radiation to be observable. Colder dust is best observed at relatively long wavelengths (around 100 μm), where atmospheric absorption is a difficulty.

The new information provided by molecular observations is strikingly evident in the picture of the region of the Orion nebula obtained from CO observation shown in fig. 2 (cf. [1]). Outside the region seen in visible light and shown in fig. 1 a vast cloud of very cold and dense gas appears, extending for some 45 pc from the region of active star formation. KUTNER *et al.* [1] estimated typical temperatures of $(10 \div 15)$ K and densities of $(400 \div 1000) \, m_H \, \mathrm{cm}^{-3}$ in this gas cloud. They obtain a mass estimate of about $10^5 \, M_\odot$, many times the mass of the stars and gas of the optical nebula. These observations give us fuller information on the gaseous environment in which star formation takes place, and, therefore, molecular and infra-red observations form the bulk of present observational research on star formation. Readers interested in this observational area are referred to the recent reviews of Zuckerman and Palmer [2], Thaddeus [3] and Strom, Strom and Grasdalen [4].

2. – Interstellar clouds.

We will discuss here a mechanism proposed in the 1960's to explain the formation of the interstellar clouds observed in the 21 cm line. This mechanism was also considered a possible means of star formation, but its importance in this regard is now thought to be more indirect. It affects the context in which star formation begins, but it is not likely to drive the process. This mechanism is the thermal instability.

Thermal instability of the interstellar gas can occur because of the particular way in which this gas is heated and cooled. Cooling takes place by radiation in optically thin lines of hydrogen, helium, or of trace elements. The lines are optically thin because of the very low densities in the interstellar medium. The line radiation is produced by collisions, and, therefore, the cooling rate depends upon ϱ^2. Contributions from many lines produce a smooth temperature dependence of the cooling rate below about 10^4 K. Below 100 K the cooling drops rapidly with T because fine-structure lines of trace elements are no longer effective. At about 10^4 K the cooling rate rises rapidly due to the onset of cooling by hydrogen atoms. The cooling rate, Λ, is shown in fig. 3.

Most heating mechanisms proposed for the interstellar gas rely on some pervasive external flux of energy which is partially absorbed by the gas. Thus the heating generally depends upon the first power of the density. Of course, if we consider scales comparable to the absorption length for the external flux, proximity to the source of this flux is also important. Heating sources which

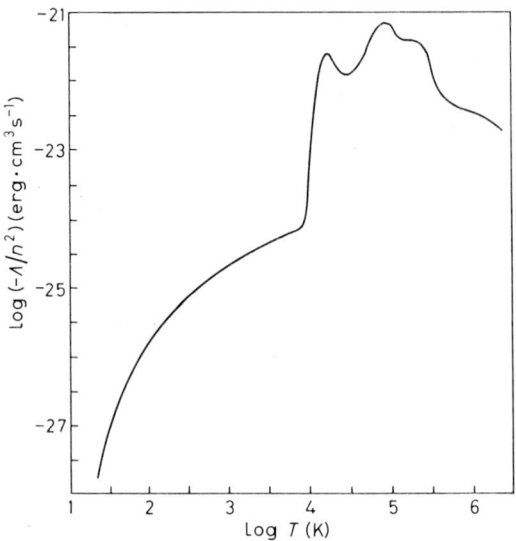

Fig. 3. – The cooling rate Λ divided by the number density, n, squared as a function of temperature. The data below 10^4 K are from D. W. GOLDSMITH (private communication), and those above 10^4 K are from C. B. TARTER (private communication).

have been proposed for the interstellar gas are low-energy cosmic rays, soft X-rays and ultraviolet light from stars (below the hydrogen ionization edge). Of these three, the last probably has the longest absorption length in the gas. It gives heating by knocking photoelectrons from dust grains and by photo-dissociation of molecular hydrogen, where that is present.

Consider a perturbation of uniform interstellar gas where heating and cooling are in balance. The ϱ^2-dependence of the cooling means that it will exceed the heating where the gas is slightly denser. The resulting temperature decrease will reduce the pressure here and cause this gas to be compressed. The compression raises the density here further, and there is a runaway growth of the original perturbation. This process has been treated in detail by FIELD [5]. The proper criterion for instability depends upon the scale of the perturbation. If the scale is very large, the pressure differences set up by the cooling of the gas in denser regions and heating in more diffuse regions have very little dynamical effect. These pressure differences are smoothed out by compression and rarefaction waves, which travel near the speed of sound. If the thermodynamic time scale for the gas is much shorter than the time required for sound to travel from a pressure maximum to a pressure minimum, then the gas must cool off and heat up at approximately constant density. The temperature perturbation will grow if the net cooling rate, $\mathscr{L}$ (erg g^{-1} s^{-1}), increases with decreasing temperature:

$$(1) \qquad\qquad\qquad (\partial \mathscr{L}/\partial T)_\varrho < 0 \ .$$

Now consider a perturbation of very short wavelength, so that the sound crossing time is much less than the thermodynamic time scale. Compressions and rarefactions of very low amplitude will race across the region of the perturbation evening out any pressure differences which arise. The perturbation will, therefore, evolve at nearly constant pressure. Again, instability results when the net cooling rate increases as the gas cools:

$$(2) \qquad (\partial \mathscr{L} / \partial T)_p < 0 \, .$$

There is generally a greater tendency to instability in this case, because of the density dependence of the net cooling rate. For constant-pressure conditions

$$(3) \qquad (\partial \mathscr{L} / \partial T)_p = (\partial \mathscr{L} / \partial T)_\varrho - (\varrho / T)(\partial \mathscr{L} / \partial \varrho)_T \, .$$

$(\partial \mathscr{L} / \partial \varrho)_T$ is generally positive, and hence has a destabilizing effect.

A complete analysis for the general case has been given by FIELD [5], but we can derive the unstable growth rate, n, very easily in the above two limiting cases. The growth rate follows from the internal-energy equation for the gas

$$(4) \qquad \frac{\mathrm{d}\varepsilon}{\mathrm{d}t} = \frac{p}{\varrho^2} \frac{\mathrm{d}\varrho}{\mathrm{d}t} - \mathscr{L} \, .$$

This equation is in the Lagrangian form. The Eulerian equation, for a fixed co-ordinate x, is obtained by replacing $\mathrm{d}/\mathrm{d}t$ by $\partial/\partial t + v \partial/\partial x$. Because the interstellar gas is monatomic, the specific internal energy, ε, is given by

$$(5) \qquad \varepsilon = \frac{3}{2} \frac{p}{\varrho} = \frac{3}{2} \frac{k_\mathrm{B} T}{m_\mathrm{H}} \, ,$$

where k_B is the Boltzmann constant and m_H the mass of the hydrogen atom. If T_1 is the temperature perturbation, we can express $\varrho \mathscr{L}$ near equilibrium as $\varrho_0 T_1$ times the temperature derivative of $\mathscr{L}$ at either constant density or pressure. In either case the other terms in eq. (4) can be expressed in terms of $\mathrm{d}T_1/\mathrm{d}t$ by using eq. (5). The growth rate, n, is then easily obtained:

$$(6a) \qquad n_\varrho = - \frac{T_0}{\varepsilon_0} \left(\frac{\partial \mathscr{L}}{\partial T} \right)_\varrho ,$$

$$(6b) \qquad n_p = - \frac{2}{5} \frac{\varrho_0 T_0}{p_0} \left(\frac{\partial \mathscr{L}}{\partial T} \right)_p .$$

Note that in both cases the growth rate is independent of the wavelength of the perturbation. The wavelength enters only when sound propagation times are comparable to the cooling times. Then we must include in the analysis the continuity and force equations describing these mechanical effects. This has been done by FIELD [5], who also considers effects of thermal conduction. Thermal conduction damps the growth of very small perturbations, so that a minimum unstable wavelength appears. For standard intercloud conditions ($\varrho = 0.08m_{\text{H}}$ cm^{-3}, $T = 10^4$ K), only perturbations of scales larger than about 10 pc can grow [5]. When conduction is absent, the growth rate for modes of scales between the two limits considered above is modified by sound propagation effects. There is a tendency for the shorter-wavelength perturbations to grow more rapidly, because n_p is usually larger than n_ϱ due to the ϱ^2-dependence of the cooling rate.

We have mentioned that the cooling rate increases rapidly with temperature below 100 K and near 10^4 K. This stabilizes the gas in these temperature regimes (see eq. (3)), which, therefore, characterize the two phases of the gas which appear in nearly all models of the interstellar medium. A curve of heating and cooling balance is shown in fig. 4. This curve has been computed assuming an ionization rate, ζ, of $1.2 \cdot 10^{-15}$ s^{-1} due to low-energy cosmic rays. The calculation follows the treatment of Field, Goldsmith and Habing [6].

Similar curves result from other models which invoke a continuous, per-

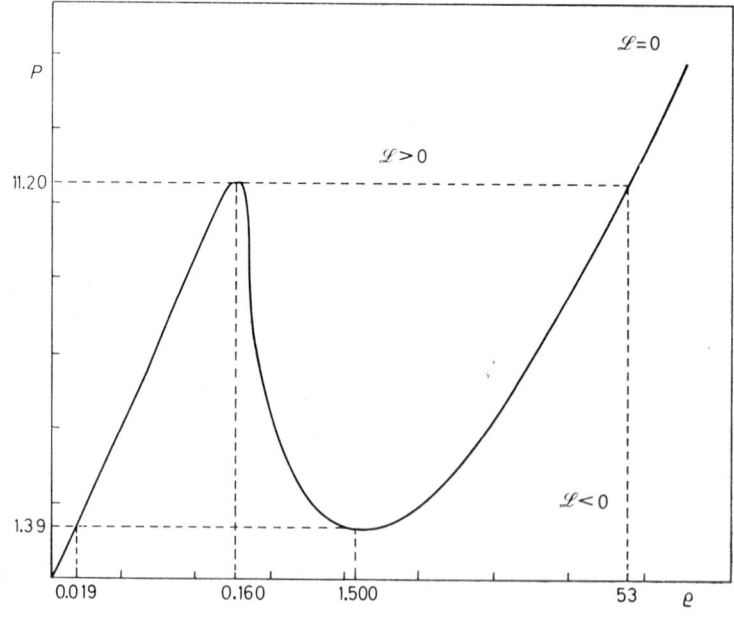

Fig. 4. – The curve of heating and cooling balance. The cooling rate is shown in fig. **3**, and the heating rate is caused by a flux of low-energy cosmic rays with an ionization rate $\zeta = 1.2 \cdot 10^{-15}$ s^{-1}. $\mathscr{L}$ is the net cooling rate. Special values of density and pressure are indicated, although the scales of the plot are logarithmic. Density is measured in H atoms cm^{-3}, and pressure in (km/s)^2 m_{H} cm^{-3}.

vasive, external flux of energy for heating in the gas. In many applications such a curve describes the equation of state for the gas. If changes of density occur slowly compared to cooling time scales, the gas will always adjust to balance heating and cooling. It, therefore, moves along the curve in fig. 4 except for the region where p decreases with increasing ϱ. This is a region of instability, where higher densities tend to set up pressure gradients which further increase the density contrast.

In fig. 4, the stable portion of the curve at low densities represents inter-cloud gas. Here the temperature is nearly constant around 10^4 K. The other stable portion of the curve represents cloud gas. The low-pressure end of the cloud curve is very compressible; that is, the gas must be greatly compressed before a substantial pressure increase will resist further compression. This behavior results because the gas is barely stable in this region. Temperatures here range around 150 K, where the temperature dependence of $\mathscr{L}$ has just become steep enough for stability. This temperature dependence steepens only gradually with decreasing T, hence with increasing ϱ in fig. 4, and, as it does so, the cloud equilibrium curve steepens and the gas becomes less compressible. The high compressibility of warm cloud gas results from the nature of the cooling, and is, therefore, common to most models of the interstellar gas. It will turn out to be a very important consideration for star formation.

The thermal instability provides us with a means of compressing a gaseous region which is too diffuse to collapse under gravity. Once the compression is under way, gravity becomes increasingly important. If the compression driven by pressure forces could proceed far enough, gravitational collapse and star formation would be the ultimate result. Such a process of star formation has been computed by HUNTER [7] and by STEIN, McCRAY and SCHWARZ [8]. However, in the standard picture of thermal instability the result of the pressure-driven compression is the formation of a cloud, not of a star. The reason for this is the development of pressure gradients retarding the collapse once the gas becomes dense and cool enough to be considered cloud phase material. At this point the rapid drop of the cooling rate with further compression stiffens the effective equation of state to a nearly isothermal relation. At this density and temperature the gas can be supported stably against gravitational collapse.

There are two ways in which this stable final state can be avoided, and star formation can be obtained from the thermal instability. The first way is to compress so large a mass of gas that no stable state at cloud density is possible. The second is to remain so far from static equilibrium during the compression that a possible stable cloud state cannot be reached. We now consider the conditions under which a stable state exists. If only pressure forces are available for the support of a cloud, then above some critical cloud mass, M_{crit}, gravity must overwhelm pressure, so that no stable state can be found. An expression for M_{crit} can be obtained by using the virial theorem.

For a static spherical cloud in equilibrium, the virial theorem implies that

$$(7) \qquad 3\int p\,\mathrm{d}V - \int \varrho(\boldsymbol{r}\cdot\nabla\varphi)\,\mathrm{d}V = \int_s p\boldsymbol{r}\cdot\mathrm{d}\boldsymbol{S}\,.$$

The first term on the left is twice the thermal energy of the cloud. Assuming an isothermal cloud, with sound speed $c = (p/\varrho)^{\frac{1}{2}}$ and mass M, we find a thermal energy of $3Mc^2/2$. If no masses outside the cloud contribute to the gravitational potential φ, the second term in eq. (7) is the gravitational energy of the cloud. Taking the result for a sphere of uniform density, we may approximate this energy by $-\frac{3}{5}GM^2/R$. The term on the right in eq. (7) is an integral over the surface of the cloud of the external pressure, p_{ext}, of the surrounding inter-cloud gas. Equation (7) can then be written

$$(8) \qquad 3Mc^2 - \frac{3}{5}\frac{GM^2}{R} = 4\pi R^3 p_{\mathrm{ext}}\,.$$

Consider now a series of clouds of increasing mass M but with the same values of c and p_{ext}. For very small masses, gravity is unimportant and eq. (8) reduces to $p_{\mathrm{ext}} = \varrho c^2$, the condition for pressure balance of the cloud with the external medium. As M increases, the cloud radius R initially increases as $M^{\frac{1}{3}}$, but eventually the gravitational energy must become significant. Then the radius must decrease, and fall to zero at $M = M_{\mathrm{crit}}$:

$$(9) \qquad M^2_{\mathrm{crit}} = 3.15c^8/G^3 p_{\mathrm{ext}}\,.$$

A more detailed treatment for isothermal spheres yields a coefficient of 1.40 in eq. (9) [9]. A discussion of additional supporting forces for clouds using the virial theorem approach is given by SPITZER [10].

Star formation may be obtained from the thermal instability, if we consider the growth of a density perturbation involving a mass greater than M_{crit}. However, we have noted that the shorter-wavelength perturbations are amplified more rapidly. Therefore, it is more likely that small masses will form stable clouds within the region of the large-scale density perturbation before it has a chance to form a single, coherent region of cloud material. Larger clouds may subsequently be built up by collisions and agglomeration of smaller clouds as discussed by OORT [11], FIELD [12], FIELD and SASLAW [13] and FIELD and HUTCHINS [14]. Such an agglomeration process requires about 10^8 y to build up a cloud so massive that gravitational collapse causes star formation.

Before discussing cloud agglomeration further, we should note that the use of an estimate for M_{crit} given by the virial theorem involves the assumption that the process of compression to form a cloud does not take the gas too far from an equilibrium state. This assumption is violated in the work of Hunter [7]

and Stein, McCray and Schwarz [8]. The inward motions toward the center
of a region of enhanced density in this work are driven by pressure, and thus
tend to be approximately sonic. However, these motions, sonic with respect
to the warm intercloud gas, may be quite supersonic with respect to the cool
cloud material being formed. Thus shocks can be produced which serve to
decelerate inflowing warm gas as it strikes the much denser cool gas of the cloud.
The force of this pressure-driven inflow acts as an additional surface pressure
for the cloud. STEIN, McCRAY and SCHWARZ [8] are, therefore, able to obtain
gravitational collapse for a mere $\frac{1}{4}M_\odot$ of material—far less than the value of
M_{crit} given by eq. (9) using for p_{ext} the ambient pressure in the surrounding intercloud gas. The shock these authors obtain at the cloud surface is particularly
violent due to a perfect spherical focusing of the inflow in their one-dimensional
model. Although this perfect focusing is probably unrealistic, we will see that
the enhanced surface pressure caused by gas inflow is an effect of great importance in several other situations.

3. – Star formation as the result of cloud agglomeration, and the motivation for a shock-induced mechanism.

An important feature of star formation which results from the agglomeration
of clouds is that the mass of new stars and gas in a region of star formation
should be approximately equal to M_{crit}. The molecular observations of gas
clouds in which stars are forming indicate that much greater masses than
M_{crit} are involved. This observational evidence argues strongly against a picture where clouds increase gradually in mass by collisions until they reach
M_{crit} and must collapse. To appreciate the strength of this argument, it is
necessary to look at the numbers.

Various observations of the interstellar gas indicate that in galactic spiral
arms, where the gas is most observable, the two phases are close to those given
in fig. 4 at a pressure, p_{max}, at the top of the stable intercloud portion of the
$\mathscr{L} = 0$ curve (cf. [6]). The cloud material then has $\varrho = 50m_H$ cm^{-3}, $p/\varrho =$
$= c^2 = 0.22$ (km/s)^2, and hence $M_{crit} = 400 M_\odot$. The extra compressibility
of this gas resulting from its tendency to cool as it is compressed has been
treated by SHU et al. [15], who obtain an estimate of $M_{crit} = 120 M_\odot$. The
highest estimate has been made by JURA [16]. In his model the gas is heated
by photoelectrons knocked off dust grains by ultraviolet starlight. He obtains
warmer clouds and, therefore, a higher value of M_{crit}. Due to the sensitive
dependence of M_{crit} upon temperature, his warmer clouds give $M_{crit} = 3000 M_\odot$.
It is possible to increase M_{crit} somewhat beyond this value by including support for the cloud due to a frozen-in magnetic field. Equilibria for isothermal
clouds with magnetic fields have been computed by MOUSCHOVIAS [17, 18].
If the field is assumed to have an equipartition value, large increases in M_{max}

apparently do not result. For such a cloud at a temperature of 50 K Mou-schovias [18] finds that M_{crit} is about twice the value for zero field.

We have pointed out that, if cloud collapse and star formation result from a gradual increase of cloud mass, clouds in which stars are forming should have masses near M_{crit}. However, recent CO observations of such clouds indicate that their masses are a great deal larger than this. For the Orion cloud, shown in fig. 2, KUTNER et al. [1] estimated a mass of about $10^5 M_\odot$. For the M17 cloud, shown in fig. 5, ELMEGREEN and LADA [19] obtained a still larger estimate of $10^6 M_\odot$, although more recently ELMEGREEN [20] has given a

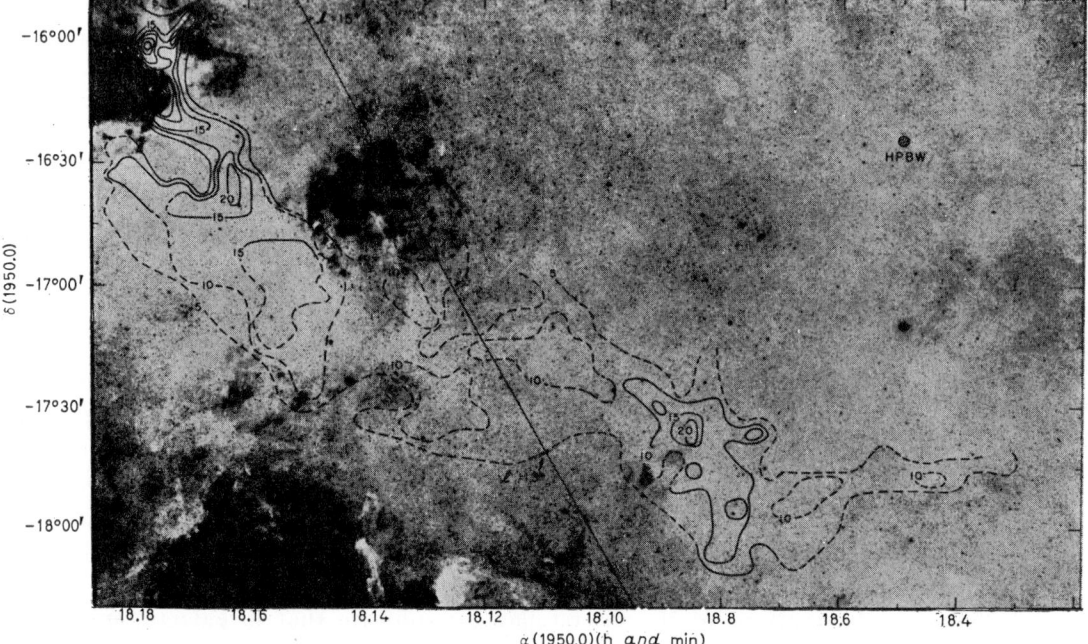

Fig. 5. – Contours of antenna temperature of CO emission at radial velocity (20 ± 2.5) km/s are shown superimposed upon the red Palomar Sky Survey plate of the M17 region. The H II region M17 (distance 2.5 kcp) is at the upper left-hand corner of the figure, near the denser end of a massive molecular cloud. This cloud extends some 85 pc along the direction of the galactic plane, which is indicated by the diagonal straight line in this figure. (From [20].)

revised estimate of about $2 \cdot 10^5 M_\odot$. To obtain values of M_{crit} this large would surely require rather extreme parameters for the interstellar gas. Equation (9) implies that rather high sound speeds, or temperatures, and rather low external pressures would be required. With heating by low-energy cosmic rays, such clouds would lie near the bottom of the dip in the $\mathscr{L} = 0$ curve of fig. 4. For such clouds we would have $M_{crit} = 2 \cdot 10^4 M_\odot$. Extrapolating the results of Jura [16] to this low pressure would give $M_{crit} = 9 \cdot 10^4 M_\odot$. Thus stable clouds might exist with masses comparable to those estimated from CO observations of star-forming regions, but the clouds must be warmer and at lower pressures than are indicated by the various observations of the interstellar medium which are discussed, for example, by FIELD, GOLDSMITH and HABING [6].

We will explain in a later section why such clouds should be expected in the regions between the galactic spiral arms, but for the moment we will merely suppose that such a cloud exists somewhere. If that cloud enters into a region of much higher pressure, such as a galactic spiral arm, gravitational collapse of the cloud will occur. If the cloud has time to get well into the high-pressure region before it collapses, the result will be a cloud far above M_{crit} for the region. If we observed the cloud without realizing that it came very recently from a region of much lower pressure, its existence would puzzle us. We would think that the various pieces of the cloud should have collapsed gravitationally long before they could have been brought together into a single object. The key to resolving this puzzle is that the cloud must be brought into the high-pressure region suddenly. Thus supersonic motions must be involved, and these leave unmistakable signs in the form of shocks. We will defer discussion of these signs to a later section and will first explain how such regions of high and low pressure can arise in the interstellar medium.

4. – The development of shocks along the spiral arms of galaxies.

On scales of a kiloparsec or more, the structure of the interstellar gas can be described in terms of relatively dense and narrow spiral arms separated by much broader regions of lower density. To see this structure most clearly it is necessary to look at external galaxies. A galaxy with particularly clean, symmetrical spiral structure is M81. The radio observations of M81 in the 21 cm line of neutral hydrogen obtained by ROTS and SHANE [21] are shown in fig. 6. The very narrow regions of high density in fig. 6 trace the location of the optical spiral arms of the galaxy. These optical arms are delineated by young, massive stars and the bright regions of ionized gas surrounding them. Thus the spiral-arm regions of high gas density where the newly formed stars are located lie right next to regions of very much more diffuse gas. If we can cause a stable, massive gas cloud to rapidly cross the very narrow compression region into the spiral arm, we can hope to explain the observed masses of CO clouds like those shown in fig. 2 and 5.

In the old theories of spiral structure, spiral features are formed by stretching out condensations of material as a result of differential galactic rotation. In such a theory, a picture of star formation resulting from motion of interarm clouds into spiral arms would be implausible. However, spiral structure is

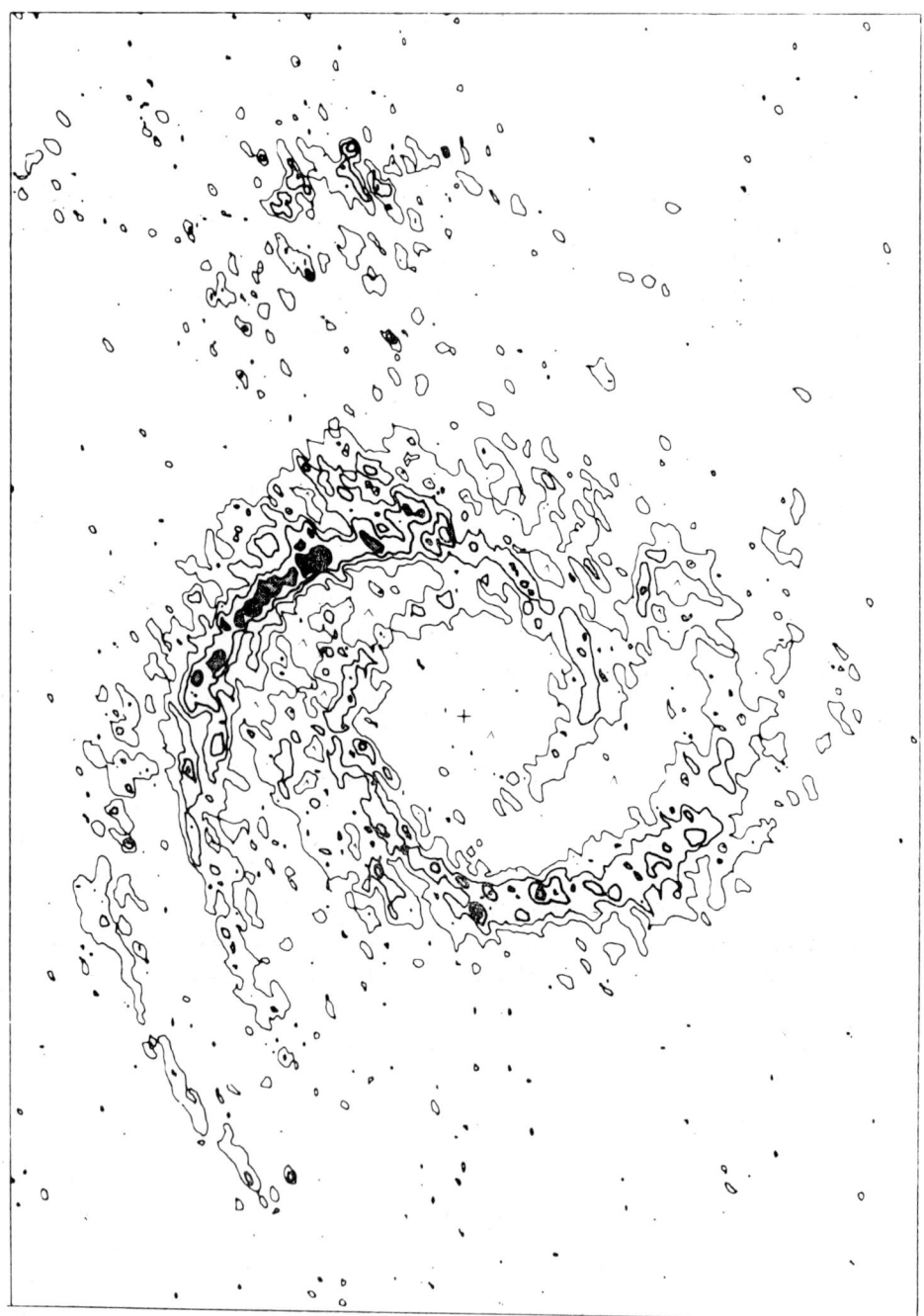

Fig. 6. – Contours of integrated 21 cm emission from neutral hydrogen in the spiral galaxy M81. The galaxy has been deprojected to a face-on orientation. (From [21].)

now believed to be a wave phenomenon, so that gas continually moves from interarm regions to arm regions and back again as the spiral wave propagates in the galactic disk. To obtain collapse of very massive gas clouds in spiral arms, we need the transition from interarm to arm regions to be very sudden.

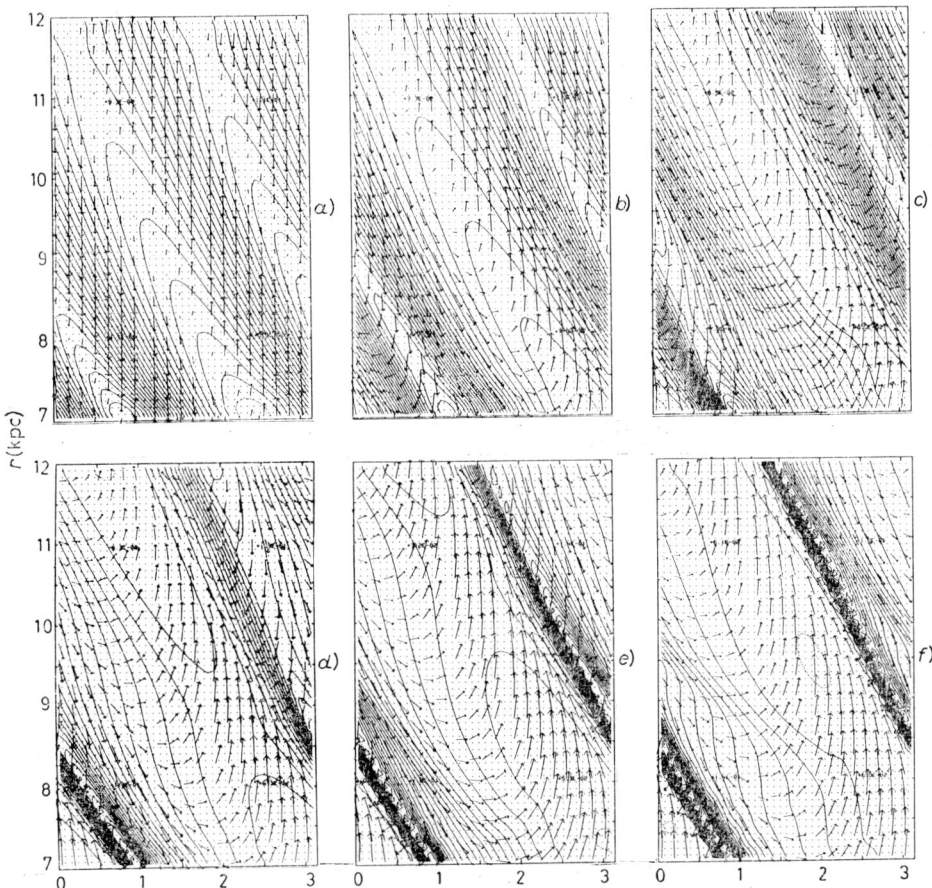

Fig. 7. – The computer response of an annular region within a rotating disk of gas to an imposed spiral perturbation of the gravitational potential: *a)* $t = 5 \cdot 10^7$ y, $\varrho_{min} = 0.0837$, $\varrho_{max} = 0.122$, $v_{max} = 5.44$ km/s; *b)* $t = 10^8$ y, $\varrho_{min} = 0.0642$, $\varrho_{max} = 0.161$, $v_{max} = 6.76$ km/s; *c)* $t = 1.5 \cdot 10^8$ y, $\varrho_{min} = 0.0598$, $\varrho_{max} = 0.174$, $v_{max} = 10.7$ km/s; *d)* $t = 2 \cdot 10^8$ y, $\varrho_{min} = 0.0538$, $\varrho_{max} = 0.282$, $v_{max} = 14.1$ km/s; *e)* $t = 2.5 \cdot 10^8$ y, $\varrho_{min} = 0.0481$, $\varrho_{max} = 0.313$, $v_{max} = 14.5$ km/s; *f)* $t = 3 \cdot 10^8$ y, $\varrho_{min} = 0.0496$, $\varrho_{max} = 0.366$, $v_{max} = 18.9$ km/s. Velocity arrows display only the velocity perturbations to the original circular rotation. Arrows and density contours are plotted in a frame rotating with the spiral perturbation. The 25 density contours in each display are equally spaced between the limits indicated. The half-annulus is displayed at time intervals of $5 \cdot 10^7$ y, beginning $5 \cdot 10^7$ y after the gravitational perturbation was switched on. The steepening of the spiral-wave response of the gas proceeds most rapidly at smaller radii, where the gas flows through the spiral pattern most rapidly. The result is the formation of shocks on the inner edges of the gaseous spiral arms.

Such a sudden transition can occur in a spiral wave as the result of a runaway steepening of the compressive part of the wave. This steepening gives rise to a shock.

In fig. 7, results of a model calculation are given which illustrate shock formation in a spiral wave. The calculation begins with an annular region of a galactic disk, constructed to resemble the galaxy M51, extending from a radius of 7 kpc to 12 kpc. In this region the gas is initially uniform at $\varrho = 0.1 m_H \, \text{cm}^{-3}$ and isothermal, with sound speed $C = 10 \, \text{km/s}$. The gas at all radii rotates at a velocity $v_{\theta 0} = r\Omega = 130 \, \text{km/s}$. A gravitational perturbation φ_1 of spiral form is applied in order to represent the effect of spiral structure in the stellar disk of the galaxy:

$$(10) \qquad \varphi_1 = \frac{1}{2} \alpha F (r\Omega)^2 \cos \left[2\theta + \frac{2}{\alpha} \ln \left(\frac{r}{r_0} \right) \right].$$

In eq. (10), α is the tangent of the angle of inclination of the spiral arms, here chosen to be 15°. The reference radius r_0 was chosen to be 12 kpc. The perturbation φ_1 was constant in a frame of reference rotating with angular velocity Ω_p, the pattern speed. A value of 6.5 m s^{-1} kpc for Ω_p was chosen to match the spiral-wave model of the galaxy M51 constructed by SEGALOWITZ [22]. F is the ratio of radial forces due to φ_1 and the unperturbed potential φ_0. SEGALOWITZ [22] found that $F = 0.03$ gave the best fit to his observations of M51. Here $F = 0.05$ is used to give a more impressive shock.

The spiral-wave theory of Lin and Shu (reviewed by LIN, YUAN and SHU [23]) demands that φ_1 be the consistent potential due to the density perturbation it induces in the stellar disk. This condition relates α, the arm inclination, to the velocity dispersion $c_\star$ of the stars. The stellar spiral arms are broad because $c_\star$ is generally substantial when compared with the velocity $(r\Omega)(\alpha F)^{\frac{1}{2}}$ which a star can gain by falling into a potential well of depth φ_1. Thus the stellar wave is small in amplitude, and only the fundamental harmonics, $\sin 2\theta$ and $\cos 2\theta$, are likely to be noticeable. The distribution of gas in the galactic disk does not significantly affect φ_1, because so much less mass is in gaseous rather than in stellar form. In contrast to the stellar-velocity dispersion $c_\star$, the sound speed c of the gas (10 km/s) is comparable to $(r\Omega)(\alpha F)^{\frac{1}{2}}$ (15 km/s). Therefore, the potential φ_1 has a strong effect on gas motions.

The response of initially uniform gas to the perturbation φ_1 is shown in fig. 7 at intervals of $5 \cdot 10^7$ y. On a grid of r vs. θ, with 60×40 computational zones, density contours, velocity arrows and dots marking zone corners are displayed. The computation was performed by the MUSCL code, written by the author based upon an elaborate, second-order-accurate numerical scheme of van Leer [24]. The particular features of the MUSCL code relating to polar-co-ordinate computations and exact detailed conservation of angular momentum by the method will be described in a subsequent article. An important feature of this code is that it solves perturbation equations explicitly, in which large forces in balance in the unperturbed state have been subtracted out. This immensely increases the accuracy of the computation when we wish to compute small deviations from an unperturbed flow, as is the case here. Because of this formulation, the velocity arrows in fig. 7 represent velocity *perturbations*. Results are displayed in the reference frame of the calculation, which rotates with the perturbation φ_1 at angular velocity $\Omega_p = 6.5$ km s^{-1} kpc^{-1}. Boundary conditions at the inner and outer edges of the annular region displayed simulate a continuation of the computed spiral perturbation into the rest of the disk with the same amplitude and inclination. At $\theta = 0$ and $\theta = \pi$, a periodic boundary condition is applied.

The mechanism for spiral-wave steepening and shock formation is clearly demonstrated by the calculation in fig. 7. The system is shown at $t = 5 \cdot 10^7$ y in fig. 7a). By this time very little radial motion has taken place, but velocities of over 5 km/s have developed from the gravitational attraction of the stellar spiral arms. Because of the small inclination, α, of the arms, this attraction is mainly radial. Thus azimuthal-velocity perturbations, $v_{\theta 1}$, are negligible as yet. The slight radial motion toward the stellar spiral arms has caused small-amplitude gaseous spiral arms to develop. As yet only the fundamental harmonics, $\sin 2\theta$ and $\cos 2\theta$, are present. The amplitude decreases as r increases. This occurs because the constant depth of the potential well φ_1 gives radial acceleration decreasing as $1/r$.

The velocity field in fig. 7a) describes radial motions focussed on the spiral arms. The transport of gas into the arms as a result of these motions increases the arm-interarm density contrast. Where the gas flows radially inward, its azimuthal velocity is least effective in carrying it away from the spiral arm. Therefore, gas piles up in this region. Consequently, in fig. 7b), at $t = 10^8$ y, the density perturbation has grown quite large, and the density peak now occurs at the point of maximum inward motion. The azimuthal-velocity perturbations are now comparable to the radial ones. Gas accelerated radially to give fig. 7a) has now developed azimuthal motions as a result mainly of Coriolis forces. In order to conserve angular momentum, gas moving outward must rotate more slowly. The velocity field of fig. 7b), therefore, describes the epicyclic motion of gas elements as they oscillate about the radii for which centrifugal and gravitational forces are in balance.

The most striking feature of both fig. 7a) and b) is that colliding streams of gas have been set up which converge on the spiral arms from either side. Each stream would like to execute its own epicyclic motion, but these separate motions are not mutually compatible. Instead, the streams collide near the minima of the driving potential φ_1. Adjustments in the flow must be brought about in order to reach a steady state. The forces driving the original, incompatible motions are the gravitational, centrifugal and Coriolis forces. These cannot adjust themselves to produce a steady flow. Instead it is the force of pressure, made strong by the collision of the gas streams, which must determine the form of the final steady flow.

The way in which pressure forces cause the flow to adjust depends upon the ratio of the flow speed to the sound speed. If sound waves originating in the spiral arm can propagate upstream into the interarm region, then these signals can cause changes in the incoming gas which allow a smooth transition from the interarm state to the arm state. However, in fig. 7 the inflow is supersonic. Therefore, sound waves cannot act to soften the collision of the gas streams which occurs at the spiral arms. The area where the streams collide, the compressive part of the flow, therefore, steepens into a shock. This steepening process has been described in detail in an earlier paper [25]. In fig. 7c)-f) shock formation takes place somewhat more rapidly at the inner radii, where the wave amplitude is largest. The subsequent evolution consists of rather minor adjustments to achieve a steady flow.

The arrows in fig. 7 represent velocity perturbations. Therefore, the gas streamlines are not obtained by following the directions of the arrows on the figure. A better idea of the complete flow pattern is given by fig. 8. Here the

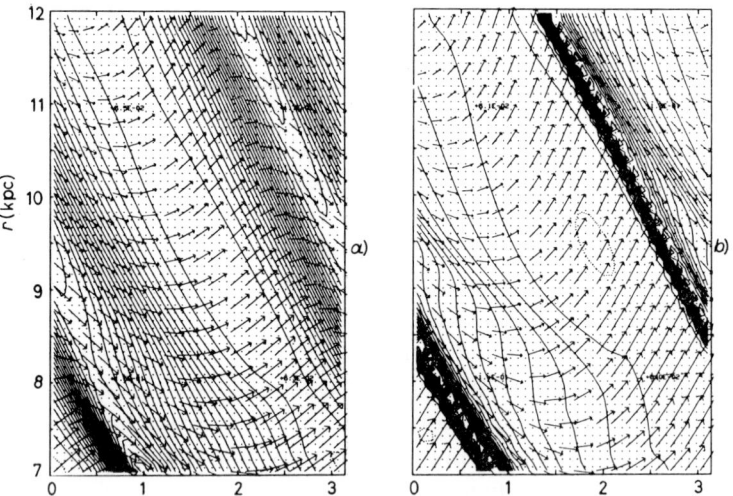

Fig. 8. – The same computation shown in fig. 7, at times $1.5 \cdot 10^8$ y and $3 \cdot 10^8$ y after the spiral perturbation was switched on: a) $t = 1.5 \cdot 10^8$ y, $\varrho_{min} = 0.0598$, $\varrho_{max} = 0.174$, $v_{max} = 16.1$ km/s; b) $t = 3 \cdot 10^8$ y, $\varrho_{min} = 0.0496$, $\varrho_{max} = 0.366$, $v_{max} = 22.0$ km/s. Here full velocity vectors are plotted, so that the gas streamlines can be traced.

same data are displayed as in fig. 7, but the azimuthal component of a velocity arrow is v_θ/r. In fig. 8a) the system is shown at $t = 1.5 \cdot 10^8$ y, the time of fig. 7c). The convergence of the flow upon entering the spiral arm is less striking in this display but no less real. The incompatible epicyclic motions of the gas streams set up in fig. 7a) have created a constriction in the flow at the spiral arm. Sound waves travelling upstream from this region steepen the compressive pressure gradient in order to decelerate the flow into the constriction. If the flow were everywhere subsonic relative to the spiral arms, these sound waves could propagate throughout the system to make the flow adjust smoothly to the squeezing at the spiral arms. However, the flow is only subsonic in the last part of the compression region. Therefore, the sound waves pile up here. No matter how steep they make the pressure gradient in this region, they cannot directly affect the incoming flow. The result is a shock.

Figure 8*b*), at $t = 3 \cdot 10^8$ y, corresponds to fig. 7*f*). Here the gas streamlines have cusps where the gas strikes the spiral arm and immediately turns to the right in the figure. If dense clouds are carried along in the flow, simple common sense tells us that they will not be able to turn this sharp corner. They will instead shoot directly into the spiral arm. As we shall see in the next section, this will enhance the likelihood that stars may form from these clouds.

5. – The implosion of gas clouds by spiral-arm shocks.

The calculation shown in fig. 7 shows the response of a one-component gas to a stellar spiral wave. The sound speed of the gas, 10 km/s, and the isothermal equation of state match the general behavior of the intercloud gas discussed earlier. The additional complication of clouds embedded in this medium was treated by SHU *et al.* [15]. These authors performed one-dimensional computations based on the assumption of a very small spiral-arm inclination α. They pointed out that if the interarm pressure tries to fall below the minimum value $p_{\min}$, at which cloud gas is stable, clouds must evaporate into the intercloud medium. Because evaporation will raise the pressure of the medium, it acts to keep that pressure from falling below $p_{\min}$. This pressure regulator then tells us the state of the interarm gas when it hits a spiral-arm shock, so long as the wave amplitude is sufficiently large.

For star formation, we are interested in the details of the compression of clouds as they flow into the spiral arms. This is not treated by SHU *et al.* [15], but their pressure regulator gives us an idea of the initial state of the clouds as they approach the spiral-arm shock. For the $\mathscr{L} = 0$ curve in fig. 4 we find $\varrho_0 = 1.5 m_{\mathrm{H}}$ cm^{-3}, $p_0 = 1.39$ (km/s)$^2 m_{\mathrm{H}}$ cm^{-3} for such clouds. The detailed interaction of such clouds with the intercloud medium of the spiral arm has been simulated using two rather different numerical techniques. The earlier calculation [26] using the CEL code [27] did not allow radiative cooling in the intercloud gas. A highly distorted Lagrangian grid which developed as the cloud was compressed required the calculation to stop before all the interesting evolution had taken place.

More recent calculations [28] will be reported here. These make use of the BBC code [29]. This code was constructed on the model of an earlier code,

KRAKEN [30], which incorporated several novel features especially suited to the problem we are discussing here. The most important of these is that interfaces between fluids were described as straight lines within any computational zones they crossed. These straight lines were determined only from local information. This method allows much more complicated fluid interface shapes to be treated in the computation. For this reason it was possible to carry the BBC computations far beyond the point where the CEL run was forced to stop. The actual fluid interface definition used in the BBC runs reported here is not that described by DeBar [30] nor the modification discussed by Sutcliffe [29], but instead a method called SLIC [31].

The BBC calculations are carried out on a fixed Eulerian grid of zones. This so simplifies the calculation that very high cloud compressions can be treated. Nevertheless, the most detailed of these computations assumes a spiral-arm shock of only moderate strength. The intercloud gas is treated as isothermal with a sound speed of 8.6 km/s. At the spiral-arm shock its density jumps from 0.02 to $0.11 m_H$ cm^{-3}. An initially spherical cloud is followed as it passes through the shock. The cloud is rather small, with $524 M_\odot$ and a radius of 15 pc. Its density of $1.5 m_H$ cm^{-3} and pressure of 1.39 (km/s)$^2 m_H$ cm^{-3} correspond to the lowest possible cloud pressure given by fig. 4. Results for larger clouds can be obtained from those presented here by multiplying all lengths and times by a common factor.

The implosion of this gas cloud as it enters the spiral arm is depicted by the series of fig. 9a)-i). In each figure the computational grid is indicated by dots. The grid consists of about 34 000 zones. It is finest near the front of the cloud, where eventually most of the material will end up. In fig. 9a) the grid dots in this region blend to make an entirely black rectangle; the dots are resolved on later close-up views of the cloud. Gas velocities are indicated by arrows. These refer to the reference frame of the calculation, which was chosen so that the front of the cloud would be nearly at rest in the region of finest zoning. In this frame the initial cloud moves to the right, while the spiral-arm gas moves more rapidly to the left. The cloud is delineated by contours of average zone density. In zones containing both cloud and intercloud material, this average density is near the cloud density, unless only a very small fraction of the zone volume is cloud gas. Consequently, the contours of average zone density give the cloud boundary a staircase appearance, which masks the much greater accuracy of the internal, multifluid representation of density used in the computation.

In our earlier discussion we stressed the need for the cloud to enter the spiral arm suddenly, so that its collapse and star formation will occur well inside the high-pressure region, where M_{crit} is low. In fig. 9a) we can see that the cloud does indeed enter the arm completely before very much compression is able to occur. The shock at the edge of the spiral arm is delineated by the closely spaced density contours at the far left of the figure. This shock, which struck

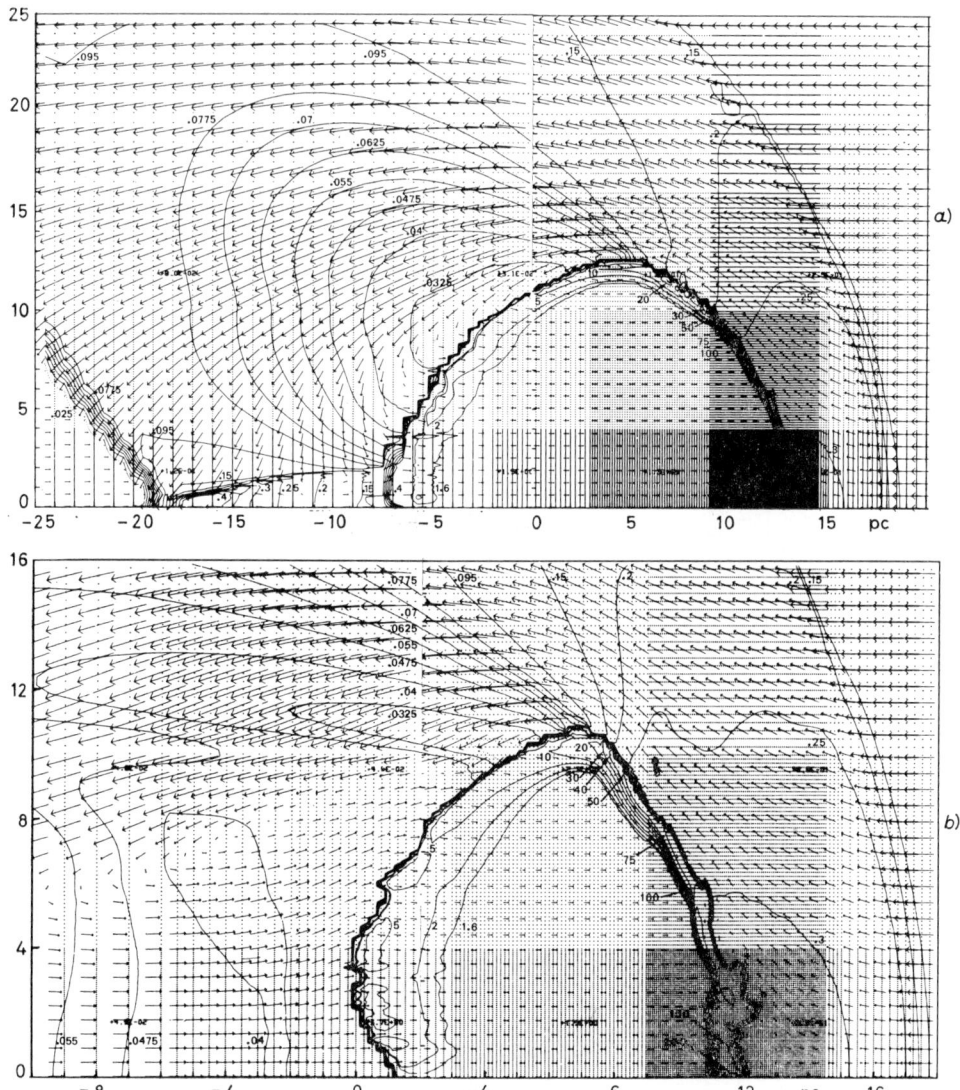

Fig. 9a, b. – A simulation of the passage of a diffuse interstellar cloud through a shock in the intercloud medium and into a galactic spiral arm: a) $t = 3 \cdot 10^5$ y, $v_{max} = 17.9$ km/s, $\varrho_{max} = 137\ m_H\ cm^{-3}$; b) $t = 4.5 \cdot 10^5$ y, $v_{max} = 18.4$ km/s, $\varrho_{max} = 229\ m_H\ cm^{-3}$. Details of the display format are given in the text. The cloud is first flattened into a dense, disklike configuration which is somewhat irregular due to the action of the Rayleigh-Taylor instability. An expanding tail of dense gas then shoots out into the low-pressure region behind the cloud to give it an elongated, « cometary » appearance. Conditions are most favorable for star formation at the front of the cloud. The cloud is shown at times 3, 4.5, 5.5, 6.25, 7, 8, 8.5, 9 and 11 million years after encountering the spiral-arm shock.

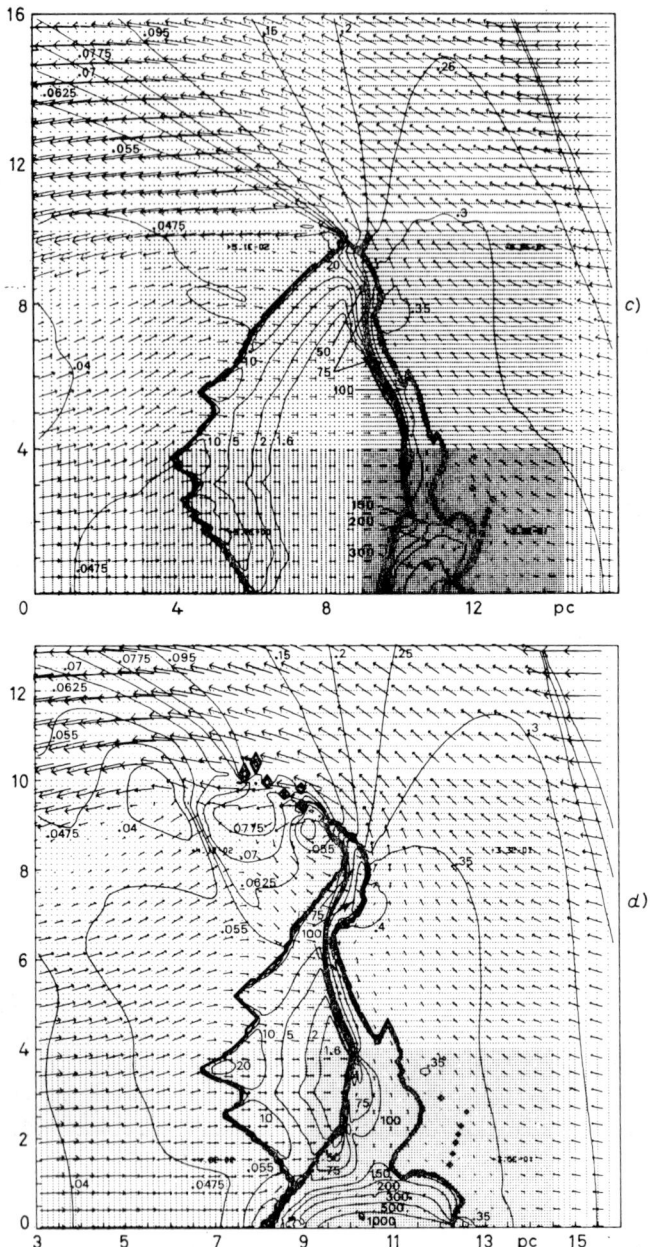

Fig. 9c, d. – A simulation of the passage of a diffuse interstellar cloud through a shock in the intercloud medium and into a galactic spiral arm: c) $t = 5.5 \cdot 10^6$ y, $v_{max} = 17.2$ km/s, $\varrho_{max} = 369$ m_H cm^{-3}; d) $t = 6.25 \cdot 10^6$ y, $v_{max} = 17.2$ km/s, $\varrho_{max} = 15\,246$ m_H cm^{-3}. Details of the display format are given in the text. The cloud is first flattened into a dense, disklike configuration which is somewhat irregular due to the action of the Rayleigh-Taylor instability. An expanding tail of dense gas then shoots out into the low-pressure region behind the cloud to give it an elongated, « cometary » appearance. Conditions are most favorable for star formation at the front of the cloud. The cloud is shown at times 3, 4.5, 5.5, 6.25, 7, 8, 8.5, 9 and 11 million years after encountering the spiral-arm shock.

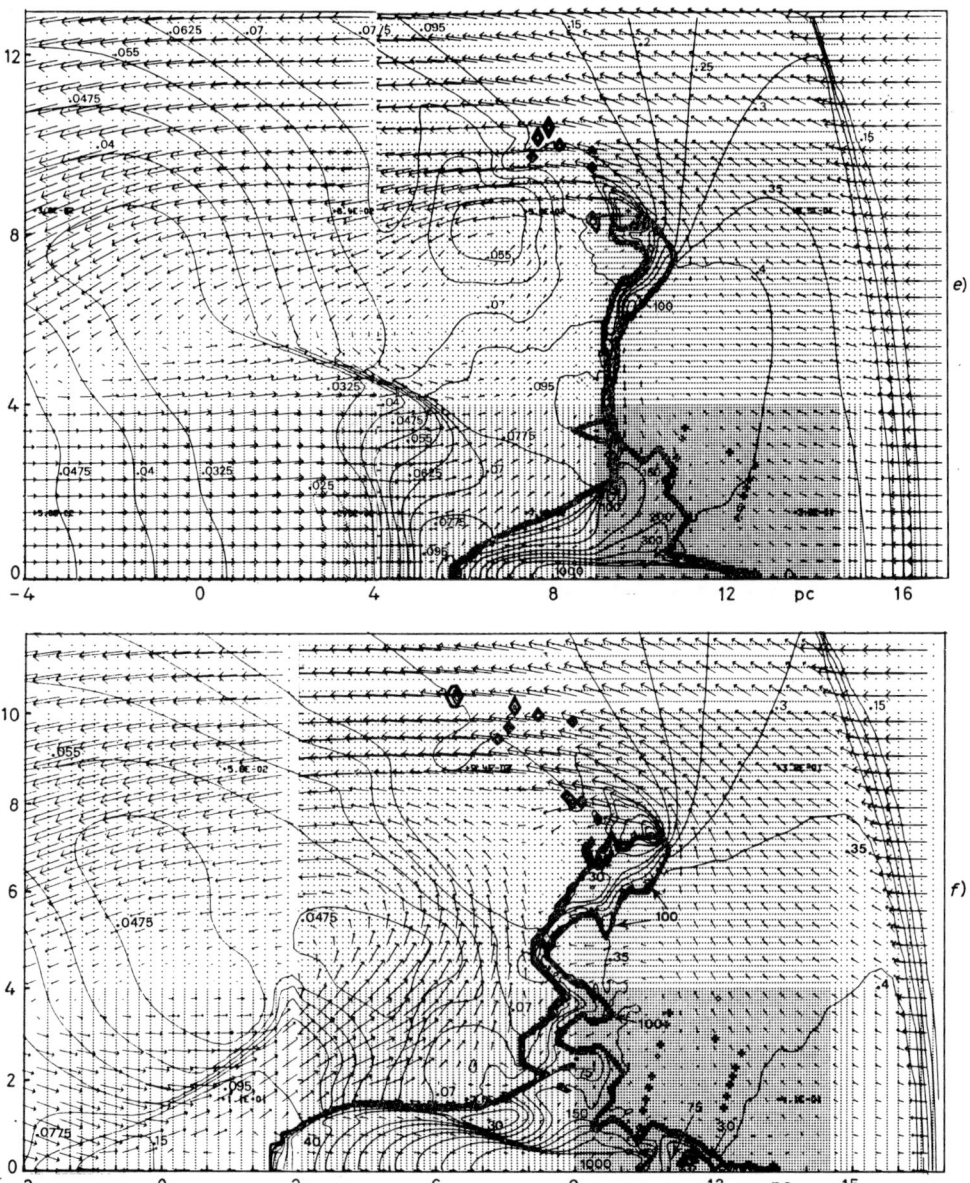

Fig. 9e, f. – A simulation of the passage of a diffuse interstellar cloud through a shock in the intercloud medium and into a galactic spiral arm: e) $t = 7 \cdot 10^6$ y, $v_{max} = 18.4$ km/s, $\varrho_{max} = 9\,963\ m_H\,cm^{-3}$; f) $t = 8 \cdot 10^6$ y, $v_{max} = 17.9$ cm/s, $\varrho_{max} = 16\,275\ m_H\,cm^{-3}$. Details of the display format are given in the text. The cloud is first flattened into a dense, disklike configuration which is somewhat irregular due to the action of the Rayleigh-Taylor instability. An expanding tail of dense gas then shoots out into the low-pressure region behind the cloud to give it an elongated, «cometary» appearance. Conditions are most favorable for star formation at the front of the cloud. The cloud is shown at times 3, 4.5, 5.5, 6.25, 7, 8, 8.5, 9 and 11 million years after encountering the spiral-arm shock.

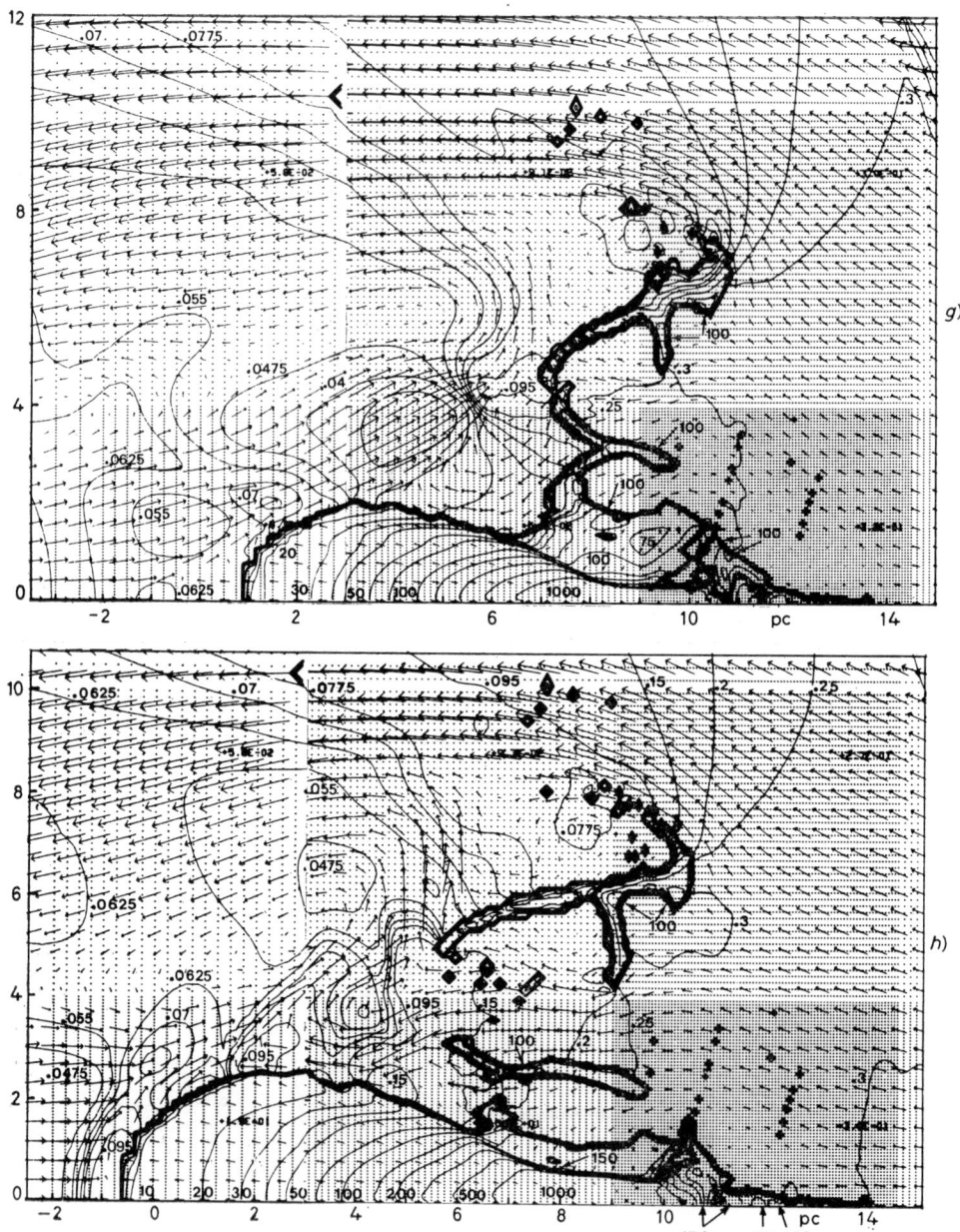

Fig. 9g, h. – A simulation of the passage of a diffuse interstellar cloud through a shock in the intercloud medium and into a galactic spiral arm: g) $t = 8.5 \cdot 10^6$ y, $v_{max} = 17.3$ km/s, $\varrho_{max} = 31\,512\ m_H$ cm^{-3}; h) $t = 9 \cdot 10^6$ y, $v_{max} = 16.9$ km/s, $\varrho_{max} = 4140\ m_H$ cm^{-3}. Details of the display format are given in the text. The cloud is first flattened into a dense, disklike configuration which is somewhat irregular due to the action of the Rayleigh-Taylor instability. An expanding tail of dense gas then shoots out into the low-pressure region behind the cloud to give it an elongated, «cometary» appearance. Conditions are most favorable for star formation at the front of the cloud. The cloud is shown at times 3, 4.5, 5.5, 6.25, 7, 8, 8.5, 9 and 11 million years after encountering the spiral-arm shock.

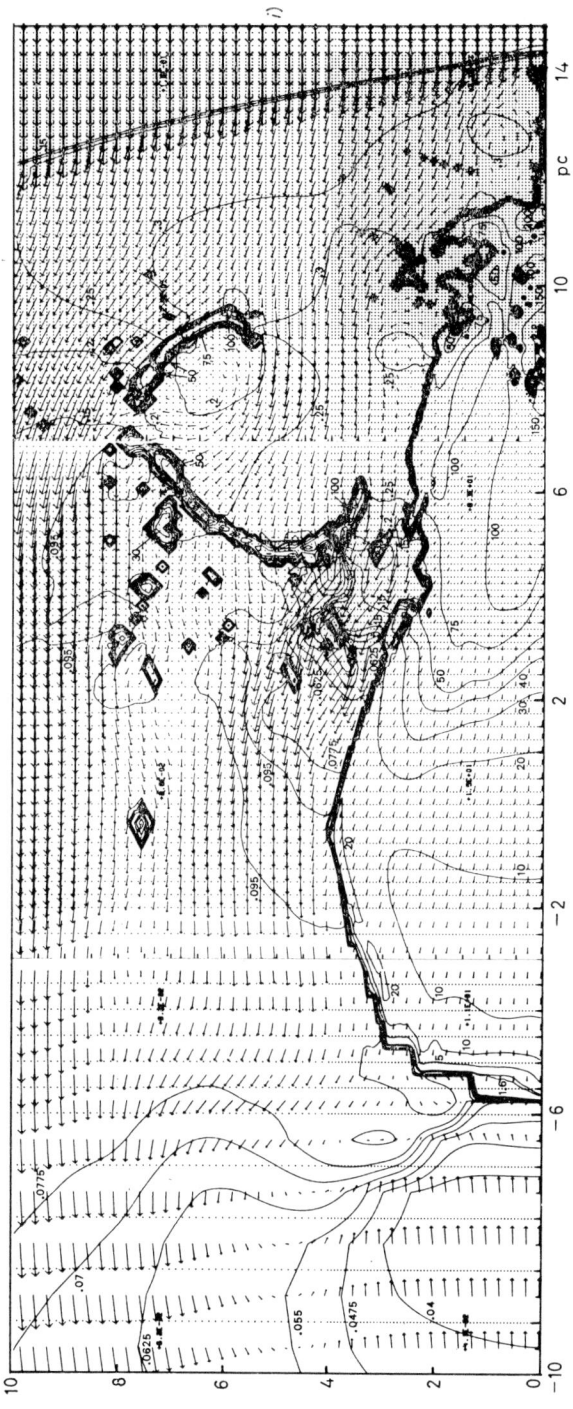

Fig. 9i. – A simulation of the passage of a diffuse interstellar cloud through a shock in the intercloud medium and into a galactic spiral arm: $t = 11 \cdot 10^6$ y, $v_{max} = 16.6$ km/s, $\varrho_{max} = 320\, m_H\, \text{cm}^{-3}$. Details of the display format are given in the text. The cloud is first flattened into a dense, disklike configuration which is somewhat irregular due to the action of the Rayleigh-Taylor instability. An expanding tail of dense gas then shoots out into the low-pressure region behind the cloud to give it an elongated, « cometary » appearance. Conditions are most favorable for star formation at the front of the cloud. The cloud is shown at times 3, 4.5, 5.5, 6.25, 7, 8, 8.5, 9 and 11 million years after encountering the spiral-arm shock.

the front of the cloud about $3 \cdot 10^6$ y before the time of fig. 9a), has by now entirely engulfed the cloud. The cloud is a figure of revolution about the horizontal axis. Thus the back of the cloud is a point of convergence of the various sections of the spiral-arm shock which pass around different sides of the cloud. This convergence has caused a high-pressure region directly behind the cloud in fig. 9a). A reflected shock can be seen which arrests the flow of intercloud gas rushing around the cloud. Because the cloud moves supersonically into the spiral arm, a bow shock has formed in front of it. This shock both compresses and deflects the flow of intercloud gas impinging upon the front of the cloud.

With all these dramatic disturbances which the cloud has caused in the intercloud gas, it is suprising that the cloud itself is not thoroughly disrupted. The reason, of course, is the very high density of the cloud gas. This dense gas is relatively cool, so that the speed of sound is low. Thus signals which communicate the new high pressure at the surface of the cloud to its interior travel very slowly. Therefore, in fig. 9a) the bulk of the cloud volume is entirely undisturbed.

The new, high surface pressure is communicated to the cloud interior by shocks, which compress the surface layers of the cloud to form a thin, dense shell. Near the back and sides of the cloud, the thickness of this shell is artificially enhanced by the necessity to spread the shock over about 3 computational zones (*). Near the front of the cloud enough zones are provided to describe the compressed shell of cloud gas rather well. We have previously noted the extreme compressibility of warm cloud gas. Here it is dramatically evident; the shock at the front of the cloud increases the density from 1.5 to $137 m_H$ cm^{-3}, a compression factor of 90. This very large compression occurs because the gas cools in the shock region by a factor of 4.4 in temperature.

The density contours in fig. 9a) display an asymmetrical distribution of intercloud density, hence pressure, at the cloud surface. The greatest surface pressure is at the front. Here the cloud is subjected to the force of the wind of intercloud gas striking it. This wind force, or ram pressure, is transformed into a gas kinetic pressure by the bow shock which arrests the wind at the front cloud surface. At the sides of the cloud the wind need not be stopped, as gas can flow around the cloud. Therefore, pressures are lower here, and consequently densities are also lower here in the shell of compressed cloud gas. These lower pressures drive the shock more slowly into the cloud in this region. The result is that the cloud becomes flattened.

In fig. 9b), $4.5 \cdot 10^6$ y after the cloud entered the spiral arm, this flattening

---

(*) The calculation does not make use of a Von Neumann-Richtmeyer artificial viscosity, as described by SUTCLIFFE [29]. Instead dissipative truncation errors from first-order remapping of Lagrangian results at each time step serve to spread out shocks and keep the results smooth.

is much more noticeable. A distortion of the cloud surface is also apparent. This is a result of the instability of the cloud surface during the compression. There are two types of instability which can occur at the cloud surface. The first of these is the Kelvin-Helmholtz instability. This is the slip surface instability which causes ripples to form on a pond when the wind blows. We can visualize the driving force for the instability in two ways. The first takes a purely mechanical point of view. When the intercloud wind flows into a depression in the cloud surface and back out again, it exerts a centrifugal force on the surface because it is forced to follow a curved path. This centrifugal force acts to amplify the distortion of the cloud surface.

The driving mechanism for the Kelvin-Helmholtz instability can also be seen by analogy to flow of a gas through a channel with a constriction. If we consider only the top or bottom half of the cannel, the flow is similar to that of gas over a part of the cloud surface. In the limit of very large channel width, the two flows should become identical. It is a well-known feature of steady flow in such a channel that the pressure and velocity profiles change qualitatively when the flow velocity is increased from a subsonic to a supersonic value. For subsonic flow, sound waves cause the flow to adjust so that the velocity at the constriction is highest and the pressure there is lowest. For supersonic flow, the opposite behavior occurs. The subsonic behavior will cause the constriction to grow if, in fact, the walls are made of cloud material which is free to move. The supersonic pressure profile, on the other hand, would push hardest on the movable walls at the constriction point. The constriction would then be relieved.

This analogy to flow in a channel, which we have already encountered in discussing the steepening of spiral waves in galaxies, shows that, when the flow becomes supersonic, its character reverses. Instead of causing growth of the surface perturbation, it causes decay. In the case of cloud and intercloud gas, the dense cloud material with its very low sound speed is relatively immovable, like a channel wall. For more similar isothermal fluids the analogy is less apt, and the transition from wave growth to wave decay occurs when the relative flow velocity exceeds $(c_1^{\frac{2}{3}} + c_2^{\frac{2}{3}})^{\frac{3}{2}}$ [32]. Here c_1 and c_2 are the isothermal sound speeds in the two fluids.

The flow velocity over most of the cloud surface is supersonic, so that the Kelvin-Helmholtz instability does not appear in the calculation of fig. 9. In the earlier CEL calculation [26], where the intercloud gas of the spiral arm was not allowed to cool, the Kelvin-Helmholtz instability distorted the cloud surface considerably. This distortion is evident in fig. 10, at the end of that calculation, some $6 \cdot 10^6$ y after the cloud entered the spiral arm. The flow of intercloud gas over the cloud surface was subsonic in this case, so that surface perturbations were amplified. These grew further at a later stage by a second instability to produce the long tongues of dense material in fig. 10.

In the calculation of fig. 9, the isothermal treatment of the intercloud gas

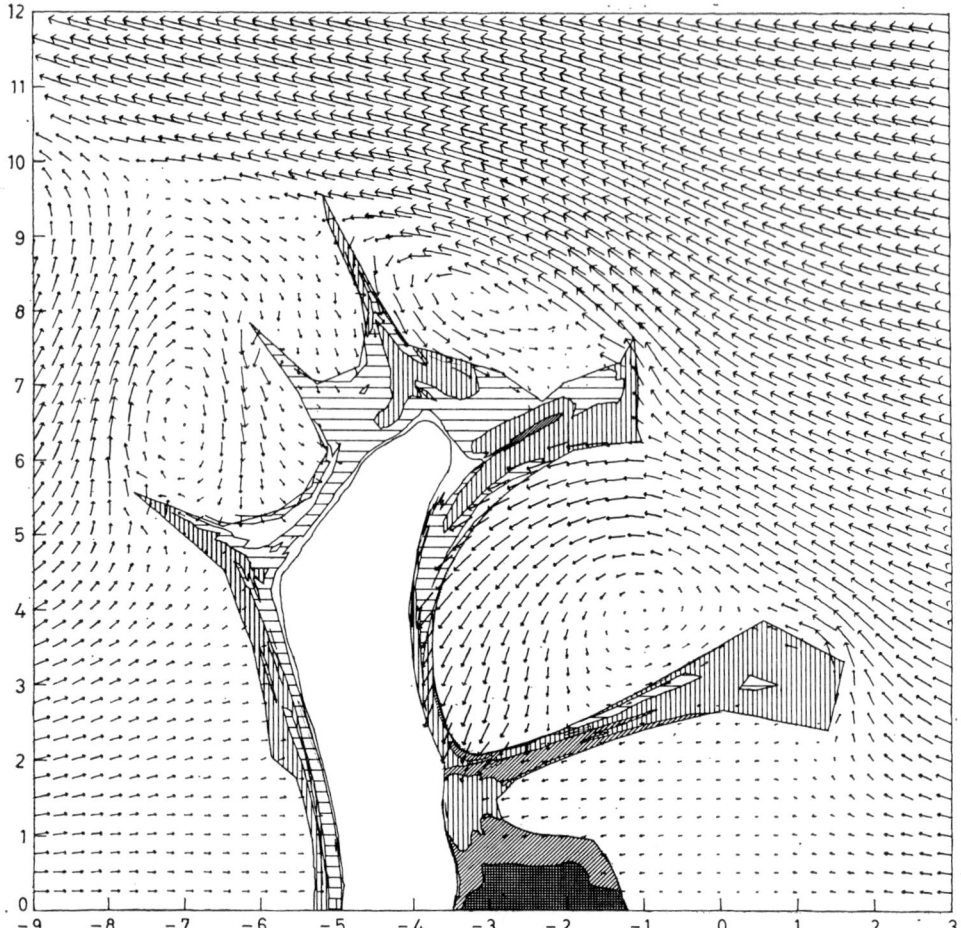

Fig. 10. – The density distribution in a diffuse interstellar cloud which has passed through a shock in an intercloud medium which is not permitted to cool. Because of heating of the intercloud medium in the shock, the cloud motion is subsonic in the shocked gas. Therefore, Kelvin-Helmholtz instabilities not present in the calculation in fig. 9 have added to the distortion of the cloud surface. The cloud is shown about $6 \cdot 10^6$ y after encountering the shock, $v_{max} = 18.4$ km/s, cycle 2228. Density contour levels are 10, 20, 40, 100 and 200 m_H cm^{-3}. This calculation was performed with a coupled Eulerian-Lagrangian method, CEL, in contrast to the multifluid Eulerian method, BBC, used to obtain fig. 9. (From [26].)

is more realistic. It causes the cloud surface to remain smoother, but this effect may be unrealistic. FEJER and MILES [33] have pointed out that oblique, « fluted » modes of the Kelvin-Helmholtz instability may still grow. However, these modes are not permitted by the axial symmetry imposed in the present calculation. Figure 10 is included here as an example of the much greater surface distortions we might expect in a three-dimensional calculation in which the fluted modes might grow.

Although the Kelvin-Helmholtz instability does not occur in the calculation of fig. 9, the cloud surface in fig. 9*b*) is clearly distorted. This distortion is the result of a Rayleigh-Taylor instability. The Rayleigh-Taylor instability occurs when a heavy fluid is superposed on a light fluid in a gravitational field. When a light fluid accelerates a heavy fluid, the instability of the fluid interface is referred to as the Rayleigh-Taylor instability, because the acceleration can be considered as an effective gravity. The presence of shocks in the fluids complicates this simple picture, because shocks by themselves are extremely stable structures. However, we can picture the instability without reference to any sort of gravitational potential. We take instead a more mechanical point of view.

Consider the region of the cloud surface near the symmetry axis. Any local kink in the surface elsewhere will respond to similar mechanical forces to be discussed here, and the amplification of such a kink in the surface will proceed along the same lines. The high pressure of the intercloud gas drives a shock into the cloud which is curved. In the frame moving with this shock, cloud gas rushes into the shock at an oblique angle. Only the normal component of this velocity is reduced by the shock compression. Behind the shock there is hence a tangential motion. The shock curvature thus causes the flow of shocked gas to be focussed onto the symmetry axis. Now, if there were no cloud-intercloud boundary nearby, this focussed flow would be stabilizing. It would increase the pressure at the symmetry axis and drive the shock forward more rapidly there. This would straighten out the shock and reduce the flow of gas toward the symmetry axis. This behavior gives rise to the mechanical stability of shocks in uniform media.

In our problem the curved shock in the cloud gas does not straighten out, but instead grows more curved. This is caused by the presence of the cloud boundary. As the focussed flow behind the shock increases the pressure near the symmetry axis, this pressure excess is very effectively relieved by pressure waves which race out into the intercloud gas. Rarefaction waves travel back into the cloud, keeping the pressure down and causing the cloud gas to squirt out along the symmetry axis. The pressure waves in the intercloud gas quickly transmit the higher pressure near the symmetry axis to the region above, causing the shock to be driven more rapidly into the cloud there. The result of all this is a growth of the curvature of the cloud surface near the symmetry axis and the formation of a dense lump of cloud gas there. This dense lump is visible in fig. 9*b*) and becomes more pronounced in fig. 9*c*) and *d*).

Above the dense lump on the symmetry axis, other distortions of the cloud surface are evident. The original perturbations they grew from were caused by very small numerical errors. Sudden changes in zone size caused small kinks to appear in the cloud boundary. Had the boundary been stable, these kinks would have remained small. It is the Rayleigh-Taylor instability of the boundary which has amplified them to form noticeable lumps. In these lumps, densities are not as high as near the symmetry axis. This is partly due to the much weaker azimuthal convergence of the flow in these regions. In a three-dimensional calculation, these lumps would truly be lumps, not rings, and their densities would be higher. Even so, the symmetry axis would still play a special role because of the general convergence of flow there which results from the shape of the initial cloud. For an aspherical initial cloud, points of stagnation of the incident flow of intercloud gas would presumably become the sites of the strongest cloud compressions.

Figure 9c), at $t = 5.5 \cdot 10^6$ y, corresponds most closely to the final point of the CEL calculation shown in fig. 10. The lack of long tongues of dense material in fig. 9c) is due to the stabilizing effect of the supersonic flow around the cloud in this calculation. Densities in the shell of compressed gas are nevertheless quite similar in the two cases. This is because a weaker spiral-arm shock was chosen for the BBC calculation in order to match the effective surface pressure at the front of the cloud to that in the earlier work. At this point in either calculation gravity would become important for the cloud evolution, if it were included. If we scale up the results of the calculation to higher cloud masses by multiplying all lengths and times by a common factor, gravity becomes relatively more important. As is evident from the density contours in fig. 9c), the lump of dense gas on the symmetry axis is the favored site for gravitational collapse and star formation in such a cloud.

Because gravity was not included in the calculation, the results in fig. 9d)-i) are best interpreted as an indication of the kind of flow which pressure forces will try to establish in such a cloud once it has been compressed into an irregular disk of dense, cool gas. The formation of stars, brought about by gravity, will also affect the pressure forces in the cloud by providing local sources of heating. Therefore, the calculation beyond the time of fig. 9c) can give only a rough picture of the flow pattern which pressure forces will try to set up. Nevertheless, the general features we will now discuss should be expected in a calculation of this type which includes gravity.

Figure 9d) shows the cloud about $6.25 \cdot 10^6$ y after it has entered the spiral arm. At this point the shock driven into the cloud from the front has just reached the back surface. This has occurred on the symmetry axis, where the shock has been driven into the cloud most rapidly as the lump of dense gas in fig. 9c) has moved downward and squashed the cloud gas in this region. In fig. 9d), the density rises to a peak value on the symmetry axis of about $1.5 \cdot 10^4 m_H$ cm^{-3}. This corresponds to a pressure some two orders of magnitude greater than that in the intercloud gas in front of the cloud and some three orders of magnitude greater than that in back. A rapid expansion ensues, which is directed mainly into the low-pressure region behind the cloud. In fig. 9e), at $t = 7 \cdot 10^6$ y, a jet of dense gas is forming at the back of the cloud, driving a shock in the intercloud gas before it.

In fig. 9e), the cloud material above the dense lump on the symmetry axis has been squashed into a very thin disk. This dense gas can now expand into the low-pressure region behind the cloud, and, as it does so, the front surface of the disk is accelerated rapidly to the left in the figure. If we consider this sudden, very large acceleration as an effective gravity, we see that the Rayleigh-Taylor instability should occur with a rapid growth rate. From a mechanical point of view we simply have a slightly bent slab of dense gas with a low pressure on the back and a high pressure on the front. The high intercloud sound speed causes the pressure to remain nearly uniform on each side as the cloud gas is accelerated. Because the resulting pressure gradients bring about motions orthogonal to the slightly curved surfaces of the slab, the curvature is rapidly enhanced. The result is shown in fig. 9f)-h); the thin disk buckles and finally breaks. Of course, the point at which the high-pressure intercloud gas breaks through the disk to flow into the low-pressure region behind the cloud is determined in the calculation by the size of the zones in the region. The effect is nonetheless real.

As the disk of cloud gas buckles, cloud gas also squirts out along the symmetry axis behind the cloud to form an elongated tail. In fig. 9h), at $t = 9 \cdot 10^6$ y, this tail has an orderly velocity and density gradient. Maximum expansion velocities in the tail are about 3 km/s. The maximum density in the cloud is still high, about $4 \cdot 10^3 m_H$ cm^{-3}, due to continued flow focussed onto the symmetry axis. In fig. 9i), at $t = 11 \cdot 10^6$ y, this focussed flow has ceased, the cloud has re-expanded, and the peak density is about $320 m_H$ cm^{-3}. Of course, we might well expect the presence of newly formed stars in the cloud to alter its appearance, but the most important point is that the cloud has been transformed from a flattened disk configuration into an elongated « cometary » shape with systematic gradients of density and velocity along the axis of symmetry, that is in the direction of flow relative to the intercloud gas of the spiral arm. Star formation is always preferred at the front end of the cloud, where the densities and pressures are highest.

It is very tempting to compare the results of the calculation of fig. 9 to the molecular observations of dense, cool clouds such as that near Orion (fig. 2) or that near M17 (fig. 5). Both of these clouds have an orientation and a density and velocity structure which is indicated by the calculation discussed above. In addition, newly formed massive stars are located at the denser ends of these clouds, as the calculation would lead us to expect. The difficulty in drawing a correspondence is the very large mass which the CO observations indicate is contained in these clouds. For such high masses of $10^5 M_\odot$ or more, the clouds would have already collapsed before entering the spiral arms, unless the pressures are lower and temperatures are higher for interarm clouds than those assumed in the calculation. Even if stable clouds of such high masses can be constructed in interarm regions, the calculation presented here is at best suggestive, as gravitational forces have not been included. For such clouds gravity must surely play an essential and no doubt very complicated role in determining the cloud evolution upon entering a spiral arm. Nevertheless, the morphological similarity between clouds like the Orion cloud and the one treated in the calculation is striking; it can hardly be accidental.

6. – The formation of second-generation stars — Star formation chain reactions.

We have seen how the asymmetrical compression which a cloud experiences upon passing through a spiral-arm shock can produce a cloud morphology similar to that shown in fig. 2 and 5. Once gravitational collapse and star formation take place in the cloud, the behavior of the cloud gas may be dramatically altered. Stars form preferentially at the denser end of such a cloud, where they are observed in both Orion (fig. 2) and M17 (fig. 5). Once this denser gas is converted into stellar form, it is no longer affected by the wind of intercloud gas which has helped to compress the cloud. The wind will then compress the gas which is now at the front of the cloud and decelerate its motion into the spiral arm. The action of the wind on newly exposed cloud gas could give rise to a second generation of new stars.

If massive stars are formed in the first generation, the sequence of events can be quite different. Massive stars can affect the cloud gas near them by means of strong stellar winds and by ionizing radiation. Stellar winds compress the surrounding gas by means of their ram pressure. Ionizing radiation first heats the gas. Subsequent expansion of this heated gas compresses the gas lying outside the ionized region. If massive stars form on one end of a cloud, they will cause the cloud gas nearest them to be compressed. In the immediate neighborhood of massive stars this source of cloud compression will certainly overwhelm any effect of the intercloud wind discussed above. If this compression can bring about further star formation, we can have a chain reaction of star formation. The chain reaction is initiated at one end of a cloud as it passes through a spiral-arm shock. A wave of star formation can then be driven down the length of the cloud by the stellar winds and ionizing radiation from massive stars.

Such a model of sequential star formation has been discussed recently by ELMEGREEN and LADA [34]. The idea that star formation may be triggered by the expansion of H II regions has been discussed before (cf. [11]), and compression of globules within H II regions to form stars has been investigated by several authors [35-41]. The new feature of the model of Elmegreen and Lada [34] is that it deals with the compression of a long, thin, dense cloud by ionizing radiation from a cluster of stars formed at one end of the cloud. This geometry was suggested to ELMEGREEN and LADA by the CO observations of regions of star formation such as that near M17 [19]. As we have seen, this geometry can be produced when a cloud passes through a shock in the intercloud medium.

ELMEGREEN and LADA [34] approximate the ionization front and shock driven into a long, dense cloud as planes perpendicular to the long axis of the cloud. By assuming a uniform density in the cloud and a uniform illumination by the massive stars at one end, they are able to construct a simple one-dimensional model of the cloud compression. We will present their model here, making a few more simplifying assumptions than they do, in order to present the essence of their picture with a minimum of complication.

When stars first form out of the gas at one end of the cloud, they ionize the gas near them very rapidly. Ionization fronts move out from the stars so rapidly that the gas becomes ionized, but does not have time to move. In our one-dimensional picture we can imagine these ionization fronts merging fairly early on to produce a roughly planar front. We will treat the front beginning with the stage when it propagates into the cloud relatively slowly. This occurs when the ionizing photons are almost completely used up in maintaining the ionization of the gas between the stars and the cloud. Changes in the density of this gas affect the ionizing flux at the edge of the cloud and cause the ionization front there to move. The motion is rather slow, because density changes can only be produced on time scales comparable to the sound travel time from the stars to the cloud. Because the motion of the ionization front into the cloud is slow, a signal, in the form of a shock, can propagate into the cloud ahead of the ionization front. This shock communicates the high pressure of the hot ionized gas to the cloud interior.

We have the following one-dimensional picture. At the left we have a group of stars which emit a flux $F_\star$ of ionizing photons per unit area per unit time. Between these stars and the edge of the cloud, a distance D away, we have a uniform ionized gas with density ϱ_{II}, pressure p_{II}, velocity v_{II}. We will treat this gas as isothermal with sound speed c_{II}, so that $p_{II} = \varrho_{II} c_{II}^2$. At the edge of the cloud is an ionization front. This front moves slowly into the cloud at velocity v_{IF}. Nearly all the flux $F_\star$ is used up in balancing recombinations in the ionized gas. Thus, if α is the recombination rate to states other than the ground state, we have approximately

$$(11) \qquad\qquad F_\star = n_{II}^2 \alpha D \, .$$

Here n_{II} is the number density of protons in the ionized gas.

Near the edge of the cloud we have a region of dense gas, which we denote by a subscript I. This gas has been compressed by a shock propagating into the cloud at velocity v_s and located a distance δ ahead of the ionization front. We treat this compressed cloud gas as isothermal with sound speed c_I. We treat the undisturbed cloud gas ahead of the shock as isothermal with sound speed c_0. The sound speed c_I might be less than c_0 if the cloud gas cools in the shock compression, or it may be greater than c_0 if the compressed gas is heated by nonionizing radiation from the ionized region beside it. The unshocked cloud has zero velocity and has density ϱ_0 and pressure $p_0 = \varrho_0 c_0^2$. In the reference frame moving with the shock, the mass flux, ϱv, and momentum flux, $p + \varrho v^2$, must be the same on either side of the shock. We will also assume that the shock is strong, so that the compression will be large enough to give interesting results. Then $v_s \gg (c_0, c_I)$, and the momentum flux balance merely equates the incident ram pressure $\varrho_0 v_s^2$ to the kinetic pressure p_I of the shocked gas. Hence approximately

$$(12) \qquad\qquad p_I = \varrho_I c_I^2 = \varrho_0 v_s^2,$$

$$(13) \qquad\qquad \varrho_I(v_s - v_I) = \varrho_0 v_s.$$

We have assumed that the shock is strong. Hence $\varrho_I \gg \varrho_0$, and eq. (13) implies that v_I is nearly equal to v_s. Now $v_{IF} > v_I$, because the ionization front is driven into the cloud by the small part of the flux $F_\star$ which we will assume is not used up in reaching the edge of the cloud. Because the ionization front drives the shock ahead of it, $v_{IF} < v_s$. The near equality of v_I and v_s, therefore, implies that

$$(14) \qquad\qquad v_{IF} \approx v_I \approx v_s.$$

Across the ionization front we must also balance fluxes of mass and momentum. Equation (14) then implies that we can neglect the ram pressure of the neutral gas. Thus

$$(15) \qquad \varrho_0 v_s^2 = \varrho_I c_I^2 = p_I = p_{II} + \varrho_{II}(v_{IF} - v_{II})^2 = \varrho_{II}[c_{II}^2 + (v_{IF} - v_{II})^2],$$

$$(16) \qquad\qquad \varrho_I(v_{IF} - v_I) = \varrho_{II}(v_{IF} - v_{II}) = \mu f_\star.$$

In eq. (16) we have equated the mass flux at the ionization front to that consistent with the incident ionizing flux $f_\star$. The mean mass per positive ion in the ionized gas is denoted by μ. A little algebra yields an expression for v_{II}:

$$(17) \qquad\qquad v_{IF} - v_{II} = \frac{p_I}{2\mu f_\star} \pm \left[\left(\frac{p_I}{2\mu f_\star}\right)^2 - C_{II}^2\right]^{\frac{1}{2}}.$$

The $+$ sign in eq. (17) corresponds to a «strong-D type» ionization front. For such a front the ionized gas streams away from the front at supersonic velocity. The $-$ sign in eq. (17) corresponds to a «weak-D type» ionization front. For such a front, the ionized gas streams away subsonically. This front is called weak because it has a smaller density contrast ϱ_I/ϱ_{II}. In order to avoid a choice between these two types of fronts and to eliminate the small residual flux $f_\star$ from their analysis, ELMEGREEN and LADA choose $f_\star$ to give a «D-critical» front:

$$(18) \qquad\qquad p_I/2\mu f_\star = C_{II}.$$

ELMEGREEN and LADA [34] give no justification for this choice, but they point out that it does not have a large effect on their results. For this choice of $f_\star$, the ionized gas streams away from the front at the speed of sound:

(19) $$v_{IF} - v_{II} = C_{II} \,.$$

The equations we have written down do not describe a steady state. As the ionization front and shock move further from the stars at the side of the cloud, these wave fronts must slow down. If we assume a constant flux $F_\star$, eq. (11) implies that ϱ_{II} must decrease in time as $D^{-\frac{1}{2}}$. Using eq. (19) in eq. (15), we find that

(20) $$\varrho_0 v_s^2 = 2\varrho_{II} c_{II}^2 \,.$$

Therefore, the shock and the ionization front must slow down, with v_s decreasing as $D^{-\frac{1}{4}}$.

The principal adjustable parameters of this one-dimensional model are $F_\star$, ϱ_0 and c_I. Therefore, it is useful to display the dependence of important quantities such as v_s on these parameters as well as on the time-related quantity D. We first define a useful combination of these parameters, a length D_0:

(21) $$D_0 \equiv 4\mu^2 F_\star / \alpha \varrho_0^2 \,.$$

D_0 is 4 times the thickness of a slab of density ϱ_0 which can be kept ionized by the flux $F_\star$. We now use eq. (11) to obtain $\varrho_{II} = \mu n_{II}$, then eq. (20) to obtain v_s, and finally eq. (12) to obtain ϱ_I:

(22a) $$\varrho_{II}^2 = D_0 \varrho_0^2 / 4D \,,$$

(22b) $$v_s^4 = D_0 c_{II}^4 / D \,,$$

(22c) $$\varrho_I^2 = \varrho_0^2 (D_0/D)(c_{II}/c_I)^4 \,.$$

Recognizing that dD/dt is just v_{IF} and that this is nearly v_s, we obtain a differential equation for D, which is easily solved to yield

(22d) $$D^5 = D_0 (\tfrac{5}{4} c_{II} t)^4 \,.$$

Here we see that D depends principally upon $c_{II} t$ and only weakly upon the other parameters. This time dependence of D implies that D is just $\tfrac{5}{4} v_s t$. Hence

(23) $$v_s^5 = \tfrac{4}{5} D_0 c_{II}^4 / t \,.$$

Our main interest in considering this model is to obtain gravitational collapse and star formation as a result of the cloud compression. Precisely when this will occur depends upon the rate of accumulation of mass in the compressed layer between the ionization front and the shock. Mass enters the layer at a rate $\varrho_0 v_s$ and leaves it at a rate $\varrho_{II} c_{II}$. From eq. (20) we see that $\varrho_0 v_s$ exceeds $\varrho_{II} c_{II}$ by a factor $2c_{II}/v_s$. Here it is clear that our assumption of a D-critical ionization front is inconsistent if $v_s \geqslant 2c_{II}$. Then the shock is no longer driven into the cloud ahead of the ionization front. In this case the ionization front would be of weak-D type, and the mass flux across it would be lower. Integrating the mass accumulation rate over time and dividing by ϱ_I, we obtain the thickness, δ, of the compressed layer:

$$(24) \qquad \delta = \frac{5}{6} \frac{c_I^2 t}{c_{II}} \left[\frac{3}{2} \left(\frac{5c_{II} t}{4D_0} \right)^{\frac{1}{3}} - 1 \right].$$

A rough criterion for gravitational collapse and star formation in the layer is that a cubic section of it, a length δ on a side, should contain the critical mass M_{crit} given by eq. (9). In eq. (9) we use the more accurate numerical coefficient of 1.40 instead of 3.15. The external pressure p_{ext} is just $p_I = \varrho_I c_I^2$. Star formation will, therefore, occur when the layer thickness grows to a value $\delta_\bigstar$ given approximately by

$$(25) \qquad \delta_\bigstar^2 = 1.12 c_I^2 / \varrho_I G.$$

Let us assume that at the time of gravitational collapse, $t_\bigstar$, the first term in eq. (24) is dominant. Then eqs. (24) and (25), together with ϱ_I determined from (22c) and (22d), imply that the time $t_\bigstar$ depends only upon the initial density of the cloud ϱ_0:

$$(26) \qquad t_\bigstar^2 = 0.717/G\varrho_0.$$

This amazing result is a consequence of the one-dimensional nature of the model. A one-dimensional compression is an extremely inefficient way to initiate gravitational collapse. In a one-dimensional situation, gravitational forces are not increased in the compression, and pressure forces are unlikely to decrease. In the above model, an increase in the ionizing flux driving the compression causes an increase in the pressure of the dense layer. However, the favorable effects of this pressure increase are cancelled by a decrease in the layer thickness.

We must keep in mind that this model is to be applied to clouds like the one near M17. Without any additional compression due to newly formed stars, such massive clouds are already above the critical mass for gravitational collapse. We are, therefore, interested in the efficiency of the ionization-driven mechanism for initiating star formation as compared to simple collapse in the undisturbed dense cloud.

An estimate of the time required for this latter process is simply the free-fall collapse time, t_{ff}, in the undisturbed gas. Since free-fall collapse assumes that pressure forces are negligible, t_{ff} depends only on the density of the gas (cf. [10]):

$$(27) \qquad\qquad t_{ff0}^2 = 3\pi/32G\varrho_0 \,.$$

We conclude that the ratio of $t_\star$ to t_{ff0} is independent of all model parameters:

$$(28) \qquad\qquad t_\star = 1.56 t_{ff0} \,.$$

This startling result underscores the general inefficiency of one-dimensional compressions in initiating star formation. In order for the ionization-driven mechanism to be the dominant cause of star formation in the cloud, the undisturbed gas must be supported against collapse long enough for an unstable compressed layer to accumulate. Low cloud temperatures and high cloud masses suggested by CO observations imply that pressure cannot support the clouds so long. ELMEGREEN and LADA [34] suggest magnetic fields for this purpose, but the required field values are rather high.

An intriguing possibility suggested by ELMEGREEN and LADA [34] is that low-mass stars form in the dense cloud ahead of the ionization front and shock. The dense layer behind the shock would then be the preferred site of formation of massive stars but not of low-mass stars. A consequence of this picture is that the low-mass star formation ahead of the shock would produce density inhomogeneities there. These inhomogeneities might then be compressed by the shock in three dimensions, as for the case of a cloud encountering a spiral-arm shock. If such departures from strict one-dimensional symmetry are allowed, then the ionization-driven star formation must become more efficient, and the ratio $t_\star/t_{ff0}$ reduced from the value in eq. (28). However, a proper calculation of ionization-driven star formation in two dimensions remains a fairly distant goal for future research.

Although the one-dimensional model of Elmegreen and Lada [34] is too idealized to justify detailed quantitative predictions, it is interesting that choices of the parameters ϱ_0, c_I and $F_\star$ which observations indicate are reasonable yield values for $\delta_\star$ and $t_\star$ in the correct range. ELMEGREEN and LADA [34] choose $\varrho_0 = 10^3 \, m_H \, \mathrm{cm}^{-3}$ and $F_\star = 10^{11}$ photons $\mathrm{cm}^{-2} \, \mathrm{s}^{-1}$, corresponding roughly to 1 main sequence O star per 10 pc². They assume that radiation from the ionized region heats the dense layer to about 100 K, so that $c_I \approx 1$ km/s. With $\mu = 1.4 m_H$, $c_{II} = 10$ km/s and $\alpha = 3.1 \cdot 10^{-13}$ cm³ s⁻¹, we obtain $D_0 = 0.82$ pc. Therefore, at the point of gravitational collapse our simplified model gives

$$t_\star \quad = 2.5 \cdot 10^6 \text{ y},$$

$$D_\star \quad = 16 \text{ pc},$$

$$\delta_\star \quad = 0.46 \text{ pc},$$

$$v_{s\star} \quad = 4.8 \text{ km/s},$$

$$\varrho_{I\star} \quad = 2.3 \cdot 10^4 m_H \text{ cm}^{-3},$$

$$\varrho_{I\star} \delta_\star^3 = 56 M_\odot.$$

ELMEGREEN and LADA [34] stress that these numbers agree fairly well with the observational data on subgroups of $0B$ associations discussed by BLAAUW [42].

LADA, ELMEGREEN, CONG and THADDEUS [43] have argued that the dense clouds and massive young stars located near the H II regions W3, W4 and W5 give us a good example of sequential star formation driven by ionizing radiation and stellar winds from massive stars. The CO observations of this region obtained by these authors are shown in fig. 11 superimposed on red Palomar Sky Survey prints. It appears that the group of massive young stars which have ionized the H II region W4 on the left side of the right-hand cloud have compressed much of this cloud to form the dense rim of gas apparent on the CO contour map. Near the top of this dense rim stars have already formed and have ionized a small spherical region, W3. The dense clouds in fig. 11 are in the Perseus spiral arm near galactic longitude 135°. A picture of star formation initiated by passage of these clouds through a spiral-arm shock would predict that, when projected onto the plane of the sky, the new stars should appear on the side of the clouds at higher galactic longitudes. The galactic plane is indicated in fig. 11, and we can see that indeed the newly formed stars are located on the high-longitude sides of the clouds.

We have been discussing star formation initiated by cloud compressions driven by ionizing radiation from massive stars. Our numerical estimates for the time interval $t_\star$ between generations of massive stars indicate that, before a new generation is formed, some of the most massive stars may explode as supernovae. These stars may only live 10^6 y, while star formation may take place in a large cloud for as long as 10^7 y. Observational evidence that such stars do explode while star formation is still going on comes from the lack of such stars in the older subgroups of $0B$ associations like the one near Orion [42].

More direct evidence of supernova explosions in regions of active star formation is given by the example of the Canis Major R1 region. HERBST and ASSOUSA [44] have argued that a supernova explosion has compressed the gas in this region to form a dense shell in which stars are now forming. The com-

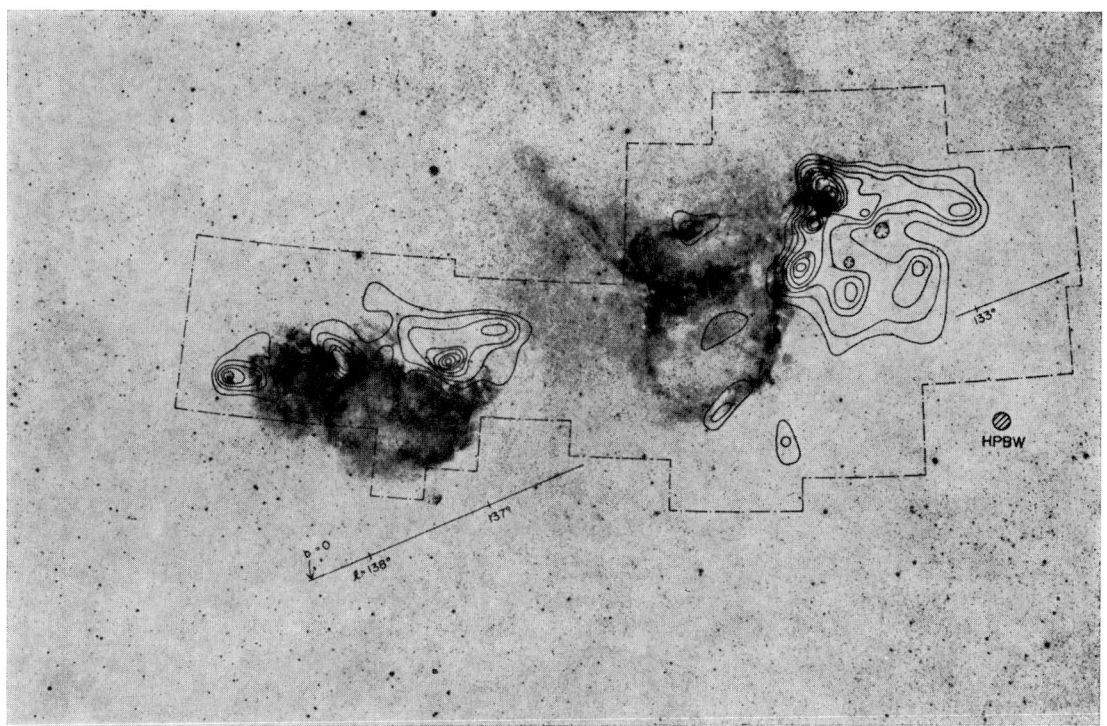

Fig. 11. – Contours of integrated CO emission from the region of the H II regions W3, W4 and W5 (distance 2 kpc) are shown superimposed upon red Palomar Sky Survey prints. The contour interval is 1.5 K·MHz. The galactic plane is indicated by a straight solid line. These dense clouds are possible examples of initiation of star formation by a spiral-arm shock followed by subsequent star formation chain reactions driven by strong stellar winds and ionizing radiation from massive stars. (From [43].)

pression from such an explosion acts in a very similar way to that caused by ionization, so that it is difficult to distinguish the two processes observationally. Because stars which explode are also likely to have ionized the surrounding gas, it is reasonable to expect some combination of the two mechanisms to be responsible for secondary star formation in the Canis Major R1 region.

The dense shell of gas near Canis Major R1 has been observed in CO by L. BLITZ (1977, private communication). His CO map of the region is shown in fig. 12 superimposed on an optical photograph. HERBST and ASSOUSA [44] estimated an age of about $3 \cdot 10^5$ y for this dense shell. They based this estimate on the models of Chevalier [45] of supernova explosions in uniform media. In such models we have again a one-dimensional compression of the gas, and a great deal of material must be swept up before gravitational collapse occurs. To obtain their age estimate of $3 \cdot 10^5$ y, HERBST and ASSOUSA take an explosion of energy $E_0 = 3 \cdot 10^{50}$ erg in a medium of density $\varrho_0 = 1 m_H$ cm^{-3} which forms a shell of radius $R = 30$ pc. This shell would contain about

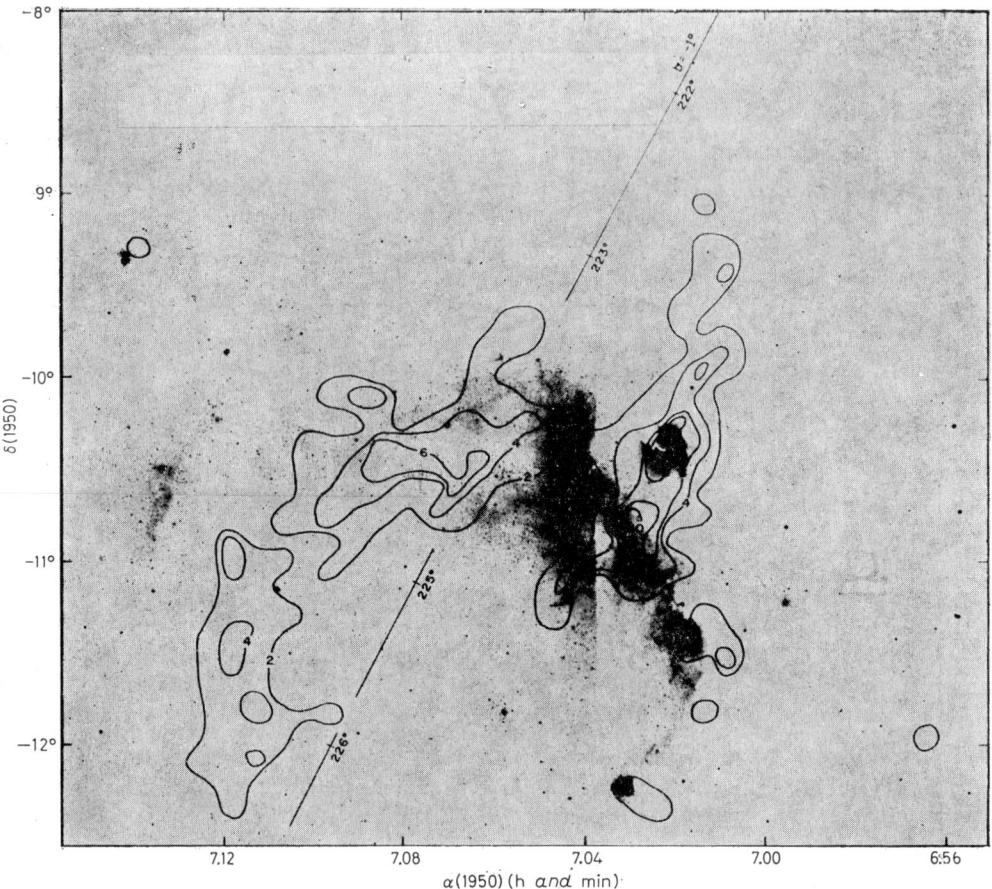

Fig. 12. – Contours of integrated CO emission from the region of Canis Major OB I (distance 1150 pc). This region may be an example of secondary star formation brought about, at least in part, by a supernova explosion. A line at galactic latitude $-1°$ is indicated. This figure was provided by L. BLITZ (private communication).

$2800 M_\odot$. For a compression to density $\varrho_1 = 100 m_H$ cm^{-3}, the shell thickness would be only 1/10 pc. A cubical portion of such a shell would contain only about $1/400 M_\odot$, for which gravitational collapse would be very unlikely.

The observations, both optical and CO, clearly display an asymmetry which suggests that such a simplified model of the shell formation cannot be correct in detail. Of course, HERBST and ASSOUSA did not use this model to derive detailed predictions; they merely suggested that the age estimate it gives is consistent with the ages of the young stars in the region. The extension of dense gas which goes off from the shell in fig. 12 in the direction of decreasing galactic longitude is unlikely to have been swept up by the supernova shock. It may instead be the portion of a cloud pre-dating the explosion which has not yet interacted with the supernova shock. It may not have been an accident that a massive star exploded near this cloud; on the contrary, the star may well have formed from the cloud.

To illustrate the sensitivity of the age estimate for the dense shell to the model adopted, a calculation with the BBC code for a supernova explosion within a dense cloud is presented in fig. 13. A cylindrical cloud of radius 4 pc with a hemispherical end was used to roughly represent the cloud extending from the dense shell in fig. 12. A uniform density of $1000 m_H$ cm^{-3} was chosen for this cloud. Gravity was not included in the calculation, and the dense cloud was contained by a hot intercloud gas of uniform density $1 m_H$ cm^{-3}. This gas was treated as isothermal with sound speed 9.1 km/s. The cloud gas was also isothermal, with sound speed 0.29 km/s. In fig. 13a) we see the model cloud very near the beginning of the calculation. The calculation simulates the effects of a supernova explosion of energy $E_0 = 3 \cdot 10^{50}$ erg at the center of the hemispherical end of the dense cloud.

The early expansion of the supernova shock has not been computed. This early expansion takes place in a uniform medium, so the evolution can be derived from the similarity solutions discussed by SPITZER [10]. The calculation was begun at the point where the supernova shock reaches the cloud boundary. At this point the material swept up by the shock should be expanding at 11 km/s (cf. formulae given by SPITZER [10]). In order to give the numerical method a swept-up shell thick enough to treat on the grid shown by the dots in fig. 13, the shell thickness was set to 1 pc at the outset of the calculation. Inside the expanding shell hot isothermal gas with density $1 m_H$ cm^{-3} and sound speed 9.1 km/s was placed to represent the low-density hole created by the supernova explosion. This gas has practically no dynamical effect upon the cloud evolution.

In fig. 13a) the expanding dense shell is driving a shock before it into the intercloud gas. This shock is weak, because the expansion of the shell is only slightly supersonic. The compression in this shock is so low that the shocked intercloud gas would be unlikely to cool down to the low temperature of the gas in the shell. Therefore, the isothermal treatment of the interloud gas is reasonable and the shock which runs ahead of the dense shell into the intercloud gas should be a realistic phenomenon. The original, arbitrary thickness of the shell is not altered as the calculation proceeds, at least in the region of expansion into the intercloud medium. This is because the shell thickness in this region adjusts so that the pressure in the shell matches that of shocked intercloud gas travelling at the same velocity. A convenient choice of the original shell thickness has made very little adjustment necessary.

Where the dense shell expands into the cloud, it is made thinner by the very large ram pressure of the dense cloud material striking it. In this region the shell thickness is limited by the numerical grid. As the shell is eventually decelerated in this region, this falsification is reduced because the ram pressure of the cloud gas decreases. In fig. 13a) the dense shell is just striking a

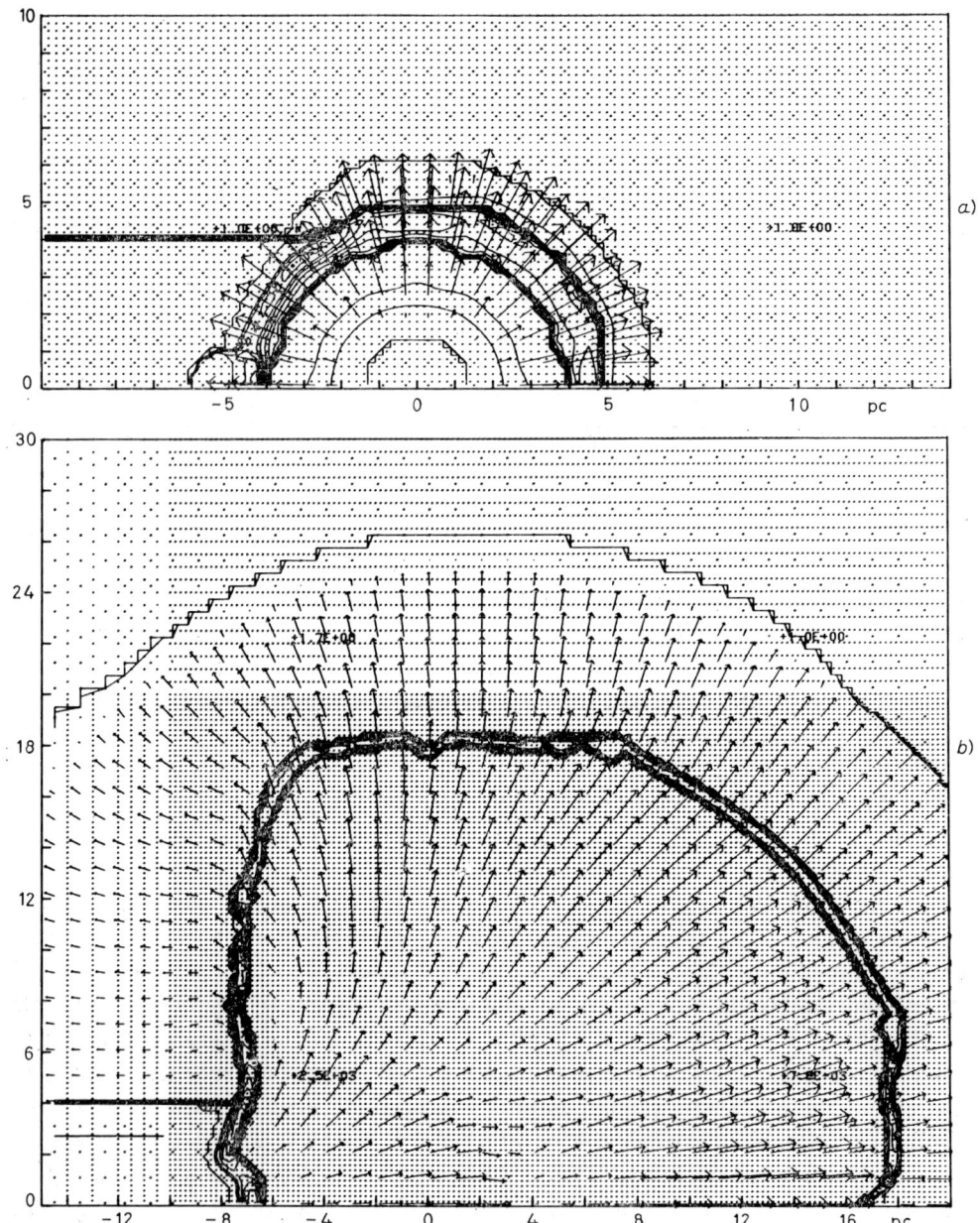

Fig. 13. – A simulation of a supernova explosion within a dense cloud is shown to illustrate the sensitivity of the age estimate for the dense shell in fig. 12 to the location of the supernova explosion. Here an explosion of energy $3 \cdot 10^{50}$ erg takes place in an elogated cloud of density $1000 \, m_{\mathrm{H}} \, \mathrm{cm}^{-3}$. A spherical lump of density $3000 \, m_{\mathrm{H}} \, \mathrm{cm}^{-3}$ and radius 1 pc was situated within the cloud just outside the expanding supernova at the beginning of the computation. The cloud is shown $2.4 \cdot 10^{5}$ y and $1.7 \cdot 10^{6}$ y after the explosion: a) $t = 2.4 \cdot 10^{5}$ y, $v_{\max} = 11.1$ km/s, $\varrho_{\max} = 6058 \, m_{\mathrm{H}} \, \mathrm{cm}^{-3}$; b) $t = 1.7 \cdot 10^{6}$ y, $v_{\max} = 16.3$ km/s, $\varrho_{\max} = 6113 \, m_{\mathrm{H}} \, \mathrm{cm}^{-3}$. The computational grid is marked by dots. Density contour levels shown are 0.2, 0.4, 0.7, 1, 2, 4, 7, 10, 20, 40, 70, 100, 200, 400, 700, 1000, 1500, 2000, 3000, 4000, 5000, 6000 $m_{\mathrm{H}} \, \mathrm{cm}^{-3}$.

spherical region within the cloud which has been given a higher initial density of $3000 m_H$ cm^{-3}. This region is located on the symmetry axis just beside the expanding shell at the outset of the calculation. It has a radius of 1 pc and contains $310 M_\odot$. In fig. 13a) it is evident that this denser region has impeded the expansion of the shell significantly.

The expansion of the supernova shock to the edge of the cloud, at which point the calculation begins, requires $1.4 \cdot 10^5$ y. In fig. 13a) the cloud is shown about $9.5 \cdot 10^4$ y later, at $t = 2.4 \cdot 10^5$ y. Had the explosion taken place in a medium of density $1 m_H$ cm^{-3}, the radius of the dense shell would nearly equal the 30 pc radius of the Canis Major R1 shell. However, the very large density of the cloud gas in which our computed explosion has occurred has kept the shell radius below 5 pc. The part of the shell which expands into the inter-cloud medium is now essentially coasting at 11 km/s. To reach the 60 pc diameter of the Canis Major R1 shell, this model would require about $5 \cdot 10^6$ y. This would give a much greater age for that expanding shell than the estimate of Herbst and Assousa [44]. Such a larger age would allow more time for gravitational collapse and star formation to take place in the compressed cloud gas.

In fig. 13b) the expanding shell is shown at $t = 1.7 \cdot 10^6$ y. The part which expands into the intercloud gas is still coasting along at nearly 10 km/s, but the shell moves into the cloud gas at only about 1 km/s. Although the shell expands quite asymmetrically, it still presents a fairly spherical appearance. Therefore, it seems possible that the shell in Canis Major R1 was produced by a supernova explosion within a dense cloud. Such a model would give an age of several million years for the expanding shell. It would also require a massive star formed in the cloud to evolve to the supernova stage before its ionizing radiation could disperse the cloud material surrounding it. A model in which both ionizing radiation and a supernova shock play a role in the cloud compression would probably be more realistic than the simple illustrative calculation presented here.

The main interest of the Canis Major R1 example for models of the formation of the solar system is that it illustrates the possibility of triggering star formation by a supernova shock. Recently discovered isotope anomalies in meteorites indicate that some such triggering mechanism may have caused the solar system to form [46-49]. Whether or not sufficient quantities of supernova ejecta can enrich the surrounding gas may well depend upon whether the supernova occurs inside or outside of a dense cloud. A good deal of further research will be necessary before such questions can be answered.

7. – Collapse of rotating clouds.

In this section we will discuss the dynamics of a small section of a larger cloud as it collapses under gravity to form a star. The first work on this subject assumed spherical symmetry in order to make the problem more tractable. The results of this work are very useful in illustrating effects of radiation transport and hydrodynamics which must certainly apply to more general situations. However, we will limit our attention to the calculations of the axisymmetric collapse of rotating clouds. Readers interested in the calculations of spherical collapse are referred to the recent reviews of Woodward [50], Larson [51, 52], Bodenheimer [53] and McNally [54].

Our main interest here is to understand how the solar system might have been formed from a collapsing fragment of a larger gas cloud. We would like to compute the gravitational collapse of a rotating-cloud fragment all the way to the point where, presumably, the Sun would form at the center of a rotating disk of gas. Such a final computed state could then be used as an initial configuration for models of planetary accumulation. CAMERON and PINE [55] have constructed such final states, but they have not obtained them from calculations of gravitational collapse. Instead, they have looked for equilibria with specified distributions of specific angular momentum over the variable M_r, the mass contained within a cylinder of radius r. Although the specific angular momentum is a constant of the motion during collapse, it is not clear that its distribution over M_r should be preserved in the collapse. Therefore, we would like to compute the collapse directly.

In this section we will discuss calculations of the collapse of rotating gas clouds which begin with a spherical, uniformly rotating configuration of uniform density. The density will be initially very low, so that radiative cooling of the gas is very efficient. The collapse will, therefore, proceed nearly isothermally at a temperature near 10 K. Few of the calculations we will discuss carry the computation beyond the isothermal stage. There is as yet no simulation which produces a final state which looks like an early solar system—a protosun embedded in a rotating gaseous disk. At an earlier stage in the collapse, the various simulations begin to give conflicting results. For similar cases, some calculations yield rotating disks, while others yield rotating rings.

It has generally been supposed that rings will fragment to produce multiple-star systems [56, 57]. Indeed, such fragmentation has been computed by NORMAN and WILSON [58]. Rotating clouds which collapse to form rings, therefore, are unlikely to produce planetary systems. Instead, those clouds which collapse to form disks are the likely progenitors of planetary systems. Because the various simulations do not agree on the conditions which are necessary for disk, and hence planetary-system formation, this issue has become a focus for recent research. Because of the importance of this question for models of the early solar system, we will discuss it at some length here.

Before considering the various calculations of dynamical collapse of rotating clouds, we can get some idea of what results to expect by looking at equilibrium models of self-gravitating Maclaurin spheroids. These spheroids are in hydrostatic equilibrium with uniform rotation and uniform density.

We can parametrize this sequence of spheroids by the ratio, β, of the total kinetic energy of rotation to the absolute value of the gravitational potential energy. As β is increased, these spheroids become secularly unstable to non-axisymmetric perturbations at $\beta = 0.137$, dynamically unstable to nonaxisymmetric perturbations at $\beta = 0.274$, secularly unstable to axisymmetric perturbations at $\beta = 0.36$ and dynamically unstable to axisymmetric perturbations at $\beta = 0.46$ (cf. the analysis of Bardeen [59]). For sequences of differentially rotating polytropes OSTRIKER and BODENHEIMER [60] found secular and dynamic instability to nonaxisymmetric perturbations at about these same values of β. Presumably, the dynamical instability at $\beta = 0.274$ produces bars, while that at $\beta = 0.46$ produces rings.

If the calculations of axisymmetric collapse of rotating clouds never produced states far from hydrostatic equilibrium, the results quoted above would lead us to expect rings to form when β exceeded about 0.46. For lower β values we would expect to find rotating disks, the presumed progenitors of planetary systems. All the calculations of axisymmetric collapse first yield rotating disks, but there is disagreement between various authors as to whether or not rings form at a later stage in the collapse. The expectations indicated by the study of equilibrium spheroids cannot be used to decide this issue, because all the calculations show that the collapse is nonhomologous, and the collapsing cloud is generally far from equilibrium.

The most detailed published calculations are those of Black and Bodenheimer [57] and of Fricke, Möllenhoff and Tscharnuter [61]. The latter authors have enlarged upon the earlier work of Tscharnuter [62]. All these authors have performed calculations using similar initial conditions and apparently similar grid resolution (FRICKE *et al.* do not clearly indicate their grid resolution), but BLACK and BODENHEIMER [57] find rings, while FRICKE *et al.* [61] do not. Recently, M. L. NORMAN (1978, private communication) has computed the collapse of Black and Bodenheimer's « case 1*B* » ($\beta = 0.08$) with a different numerical scheme and a variety of grid resolutions. His hydrodynamic method differs from that of Black and Bodenheimer [63] in a number of ways to be described in a paper he is now preparing.

NORMAN has computed « case 1*B* » with the grid resolution used by BLACK and BODENHEIMER [57]. He finds that a ring forms which is similar to the one which BLACK and BODENHEIMER obtained. However, when NORMAN refined his grid substantially, no ring was formed. This result is still preliminary, and a good deal of work remains to be done to clearly separate numerical from physical effects in these very difficult collapse calculations. At the moment, the question of whether or not rings form at such low values of β as the value 0.08 in case 1 *B* is not settled. Because of the great importance of this issue for theories of planetary-system formation, a discussion of the processes involved, both numerical and physical, is in order here.

The ring obtained by BLACK and BODENHEIMER for their case 1*B* is shown in fig. 14. Two features of the results shown in this figure should be carefully noted. First, the density minimum which causes the ring occurs over the first 3 or 4 radial zones. Second, near the equatorial plane, and particularly near the origin, density gradients are very large. These features of the computed results indicate that nearly any numerical scheme will have difficulty

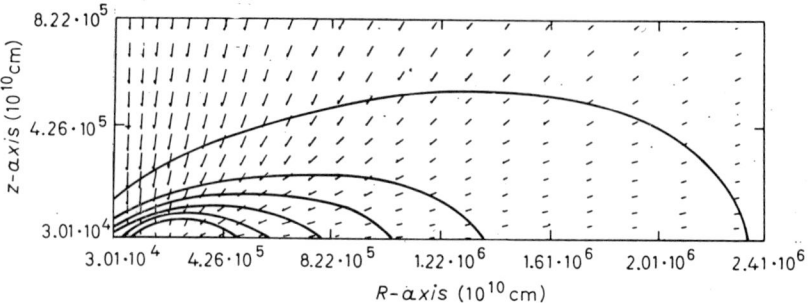

Fig. 14. – The ring obtained in the gravitational collapse of a rotating cloud, case $1B$ of Black and Bodenheimer [57], at $t = 7.78 \cdot 10^4$ y. A velocity vector is plotted for each zone, with the maximum vector length corresponding to 1.17 km/s. Density contours begin at 10^{-17} g cm^{-3} and increase by a factor 2.07 from contour to contour. Initially this cloud of mass $1M_\odot$ was spherical with uniform density $1.38 \cdot 10^{-18}$ g cm^{-3}, temperature 10 K, and was rotating uniformly with a total angular momentum of $1.2 \cdot 10^{54}$ g cm^2 s^{-1}. The ratio β, defined in the text, was 0.08 initially. A constant-pressure boundary condition was applied.

in computing this cloud evolution. This is a familiar feature of gravitational-collapse calculations in general, and BLACK and BODENHEIMER have brought considerable numerical expertise and computing power to bear on the problem. Despite the quality of their computation, it is important to bear in mind the following limitations of any numerical simulation under conditions such as these.

Numerical methods for hydrodynamics in cylindrical co-ordinates generally have difficulties near $r = 0$. This is because the difference equations usually involve the assumption that $\Delta r/r$ is small, either explicitly or implicitly. Consequently, it is normally expected that the results in the first couple of radial zones may be significantly in error. However, in most calculations this makes very little difference. The small volume of the first radial zones means that they usually contain an insignificant amount of mass. Therefore, errors in these zones usually have little dynamical effect. In gravitational-collapse calculations, strong density concentration toward the origin is likely to occur. For these calculations zoning requirements near $r = 0$ are much more stringent than is usually the case. The fact that the ring in fig. 14 is centered on the fifth radial zone means that errors from the approximation $\Delta r/r \ll 1$ may have played a role in its formation.

In replacing the differential equations of hydrodynamics by difference equations it is generally assumed that the change in any variable over a zone is in some sense small. During gravitational collapse, very large density gradients are produced. This makes it very difficult to satisfy the condition $\Delta\varrho/\varrho \ll 1$ even with a fine numerical grid. Near the origin in fig. 14 we have $\Delta\varrho/\varrho$ of order unity in a number of zones. It is not impossible to perform a reasonable computation with this kind of zoning, but it is very difficult, especially for a scheme of only first-order accuracy.

We have just mentioned two causes for concern about the accuracy of the results in fig. 14, or, in fact, for the results of any other published calculation of collapse with rotation. We now list a number of effects, both physical and numerical, which might cause rings to form. We begin with numerical mechanisms for ring formation.

We can think about the treatment of the fluid equations quite generally as a product of two operations, each of which has some difference approximation. The first operation is a calculation of Lagrangian hydrodynamics, that is a set of zones is used which move with the fluid. The second operation is a simple remapping of these Lagrangian zones, which have moved during the time step, back onto the original, fixed Eulerian grid. With each of these operations are associated specific terms in the fluid equations.

The two operations are illustrated in fig. 15. On the left-hand side of the figure, zones are shown for a calculation where the variables depend only upon the cylindrical radius r. At the top, the zones are shown at the beginning of the time step. The radii of the zone interfaces are indicated along with the velocities at these interfaces. The average values of density in the zones are also indicated. Beneath this row, the same zones are shown at the end of the Lagrangian step. Each interface has moved a distance $v_i\Delta t$, so that the new quantities are given by

$$(29a) \qquad \tilde{r}_i = r_i + v_i\Delta t,$$

$$(29b) \qquad \tilde{\varrho}_i(\tilde{r}_i^2 - \tilde{r}_{i-1}^2) = \varrho_i(r_i^2 - r_{i-1}^2).$$

In the third row in fig. 15, the portion of each Lagrangian zone which is mapped into the Eulerian zone originally to the left of it is indicated by a shaded area. Each such area contains a mass per unit length dm_i which we might estimate as

$$(30) \qquad dm_i^{(\text{BBC})} = \tfrac{1}{2}(r_i^2 - \tilde{r}_i^2)\,\tilde{\varrho}_{i+1}.$$

This formula, of first-order accuracy, was used in the BBC calculations described earlier. The calculations of Black and Bodenheimer [57] use a different first-order approximation:

$$(31) \qquad dm_i^{(\text{BB})} = r_i(r_i - \tilde{r}_i)\,\varrho_{i+1}.$$

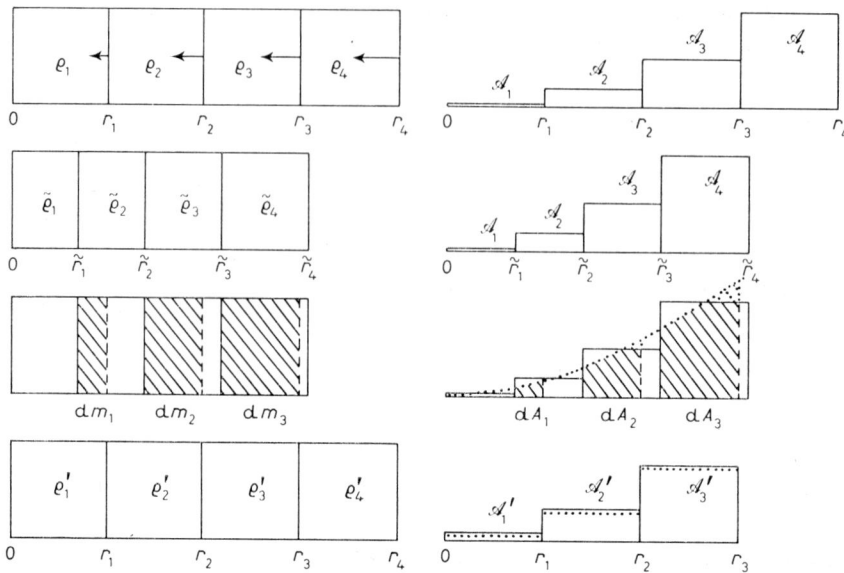

Fig. 15. – A numerical scheme for hydrodynamics is represented here in terms of separate component operations. The zones used for the computation are shown (top) at the beginning of the time step, (second row) at the end of the calculation of Lagrangian hydrodynamics, (third row) during the remapping step and (bottom) at the end of the remapping step. On the left, the zones themselves are indicated, with advected volumes shaded in row 3. On the right, the specific angular momenta are indicated, with two possible determinations of the advected momenta shown in row 3.

The BBC formula makes a bit more sense physically, because it uses the simple approximation of constant density within an Eulerian or a Lagrangian zone. The formula of Black and Bodenheimer has a larger or smaller error than eq. (30) depending on the signs of v and its derivatives.

These first-order remapping schemes cause errors in the solution of the flow equations which can be described as a numerical diffusion. Any sharp features in the density distribution at the end of the Lagrangian step will be smeared out by this crude remapping to the Eulerian grid. Errors of this sort can act in the direction of ring formation in a gravitational-collapse calculation. First we will consider an effect of numerical diffusion of angular momentum.

The right-hand side of fig. 15 illustrates a possible first-order scheme for updating specific angular momentum, $\mathscr{A}$. The value of $\mathscr{A}$ in each zone is represented by the height of the box in the figure. The radially increasing distribution of $\mathscr{A}$ and the flow converging on the origin is characteristic of calculations of gravitational collapse of rotating clouds. At the end of the Lagrangian step, $\mathscr{A}$ is unchanged, as it is a constant of the motion. The last two rows in the figure show how the distribution of $\mathscr{A}$ in Lagrangian zones is falsified in the remapping operation.

In the third row in fig. 15, the advected angular momenta, dA_i, are indicated. A first-order mapping, with $\mathscr{A}$ constant within a Lagrangian zone, would give

$$(32) \qquad\qquad dA_i = \mathscr{A}_{i+1} dm_i .$$

A second-order-accurate mapping would use the linear distributions of $\mathscr{A}$ within Lagrangian zones which are indicated by the dotted lines in the figure. This more accurate mapping displays the error in the first-order method; systematically too much angular momentum is mapped by this scheme into the new Eulerian zones shown in the last row of the figure. Therefore, too large a centrifugal force will be computed for these zones, and the material will not fall inward far enough.

Clearly, this effect is small so long as $\Delta\mathscr{A}/\mathscr{A}$ is small in each zone. But, near $r = 0$, $\Delta\mathscr{A}/\mathscr{A}$ is never small. Also, the errors caused in the general flow by this effect will be small if the mass in the first few zones is small. As fig. 14 and other collapse results indicate, this criterion is difficult to satisfy. Therefore, this mapping error which gives diffusion of angular momentum radially inward might help to form a ring in a rotating-collapse calculation.

Another numerical effect depends upon the large negative density gradient, $\partial\varrho/\partial r$, near $r = 0$, which develops early on in the collapse. Because motion here is generally radially inward, the first-order remapping brings too little mass into the inner zones. This error affects the computation of the gravitational force. The r^{-2}-dependence of gravity means that computing too low a mass in the first couple of radial zones can significantly affect the gravitational force there. Again, this happens because $\Delta r/r$ is not small, and the effect is greatly enhanced when $\Delta\varrho/\varrho$ is not small either. The effect is systematic and cumulative; its result is a tendency of the central material to move outward from its proper position.

In the very first radial zone another effect occurs when $\Delta\varrho/\varrho$ is not small. If $\partial\varrho/\partial r$ in this zone is large and negative, a first-order approximation to the gravitational force will significantly underestimate it. Again this is caused by the r^{-2}-dependence of the force and the fact that $\Delta r/r$ is not small. This effect also causes the material not to fall inward far enough.

All the numerical effects we have been discussing act in the direction of ring formation. They can be greatly reduced in importance by a refinement of the mesh or by using a numerical method of second-order accuracy. We have focussed on the results of Black and Bodenheimer [57], shown in fig. 14, but surely those of Fricke et al. [61] have similar difficulties. These cannot be discussed here because these authors have not clearly specified the resolution of their grid nor have they given full description of their numerical method (which is not of a standard type). A brief discussion between TSCHARNUTER and BODENHEIMER which appears in *I.A.U. Symposium No. 75*, p. 274, indicates that the numerical method of Fricke *et al.* [61] yields rings or disks depending upon the differencing scheme used to approximate the angular-momentum equation.

A physical effect which may cause ring formation has been pointed out by BODENHEIMER (1978, private communication). As the gas falls radially inward, the gravitational acceleration it feels increases very roughly as r^{-2}, while its centrifugal acceleration increases more rapidly, as r^{-3}. Eventually, centrifugal forces will overcome gravity and halt the inward motion. However, there will be some overshoot and oscillation about the equilibrium position of each particular gas element. The innermost gas elements will begin the outward motion of this oscillation first, and, as they move outward, they will strike gas which is still moving inward. The result will be a density concentration in the region where the outward and inward flows meet. If the density here exceeds that at the center, we would have a ring. LYNDS and TOOMRE [64] have used an effect of this nature to explain the appearance of « ring galaxies ».

If the flow is inviscid and not self-gravitating, then a ring formed in the manner just described must dissipate. The specific angular momentum of each fluid element is conserved, so that it must ultimately return to its equilibrium radius. However, self-gravity could preserve such a ring if it is strong enough. An analysis of the formation of rings in thin, self-gravitating, rotating disks can be found as a special case of the work of Goldreich and Lynden-Bell [65]. These authors found that gas can be gathered over short ranges in radius to form rings which will persist if the gas is sufficiently cold. It is more difficult to gather gas over large ranges in radius because of the stabilizing effect of rotation. GOLDREICH and LYNDEN-BELL [65] considered small-amplitude departures from equilibrium, while in the gravitational collapse of rotating clouds we expect shocks where inward and outward flows collide to produce large perturbations. Thus rings may be easier to form in this context than in the near-equilibrium disks treated by GOLDREICH and LYNDEN-BELL [65].

Another effect, which has both physical and numerical causes, would help to preserve a ring formed by the above mechanism. This effect is angular-momentum diffusion. If the flow is turbulent in the region where outward and inward streams collide, mixing of the gas and of its specific angular momenta will take place. Once the inner gas shares some of the higher angular momentum of the outer gas, it cannot return to its original equilibrium radius. This effect would help to preserve a ring once it has formed. Unfortunately, mixing of angular momentum also occurs as a result of numerical errors, so physical and numerical effects of this type are difficult to unravel.

The above physical mechanism which BODENHEIMER has suggested may very well cause rings to form. However, the recent calculations of Norman and the variety of ring-forming numerical effects discussed above cause doubt that this physical mechanism was responsible for the ring shown in fig. 14.

Future calculations with finer grids and lower intrinsic numerical errors will hopefully tell us more clearly under what conditions rings form in the gravitational collapse of rotating clouds.

8. – Summary.

We have presented here a picture of star formation which begins with very diffuse gas clouds between the spiral arms of galaxies and follows their implosion by spiral-arm shocks and their subsequent further compression by ionizing radiation and supernova shocks. We have also discussed the collapse of rotating fragments of such clouds to produce dense rotating disks or rings which are presumably the progenitors of planetary systems or multiple-star systems.

All along the evolutionary path we have described there is abundant need for further research. Magnetic fields have been consistently ignored, although they are likely to play a significant role at each stage of the star formation process. In the later part of the time evolution of clouds entering spiral arms, important effects of self-gravity were not treated. In discussing secondary star formation, the effects of ionizing radiation, strong stellar winds and supernova explosions have not been combined in a single, two-dimensional calculation. In computing gravitational collapse of rotating clouds, the tremendous central condensation and density gradients which develop push present computational techniques to the limits of their power.

All of these faults of the discussion of star formation given here are the attention of present research. These faults exist not because researchers have been unaware of them, but because they have proved extremely difficult to deal with. However, progress in this field has been rapid, and observers of regions of star formation can expect improved theoretical models to continue to appear for comparison with their observational data.

REFERENCES

[1] M. L. KUTNER, K. D. TUCKER, G. CHIN and P. TADDEUS: *Astrophys. J.*, **215**, 521 (1977).
[2] B. ZUCKERMAN and P. PALMER: *Annu. Rev. Astron. Astrophys.*, **12**, 279 (1974).
[3] P. THADDEUS: in *Star formation, IAU Symposium No. 75*, edited by T. DE JONG and A. MAEDER (Boston, Mass., 1977), p. 37.
[4] S. E. STROM, K. M. STROM and G. L. GRASDALEN: *Annu. Rev. Astron. Astrophys.*, **13**, 187 (1975).
[5] G. B. FIELD: *Astrophys. J.*, **142**, 531 (1965).
[6] G. B. FIELD, D. W. GOLDSMITH and H. J. HABING: *Astrophys. J.*, **155**, L149 (1969).
[7] J. H. HUNTER jr.: *Mon. Not. R. Astron. Soc.*, **142**, 473 (1969).

[8]   R. F. STEIN, R. McCRAY and J. SCHWARZ: *Astrophys. J.*, **177**, L125 (1972).
[9]   L. SPITZER jr.: in *Nebulae and Interstellar Matter*, edited by B. M. MIDDLEHURST and L. H. ALLER (Chicago, Ill., 1968), p. 1.
[10] L. SPITZER jr.: *Diffuse Matter in Space* (New York, N. Y., 1968).
[11] J. H. OORT: *Bull Astron. Inst. Neth.*, **12**, 177 (1954).
[12] G. B. FIELD: in *Evolution stellaire avant la sequence principale, VI Colloque International d'Astrophysique, Liège, 1969* (1970), p. 29.
[13] G. B. FIELD and W. C. SASLAW: *Astrophys. J.*, **142**, 568 (1965).
[14] G. B. FIELD and J. HUTCHINS: *Astrophys. J.*, **153**, 737 (1968).
[15] F. H. SHU, V. MILIONE, W. GEBEL, C. YUAN, D. W. GOLDSMITH and W. W. ROBERTS: *Astrophys. J.*, **173**, 557 (1972).
[16] M. JURA: *Astron. J.*, **81**, 178 (1976).
[17] T. CH. MOUSCHOVIAS: *Astrophys. J.*, **206**, 753 (1976).
[18] T. CH. MOUSCHOVIAS: *Astrophys. J.*, **207**, 141 (1976).
[19] B. G. ELMEGREEN and C. J. LADA: *Astron. J.*, **81**, 1089 (1976).
[20] B. G. ELMEGREEN: paper given at *Conference on Giant Molecular Clouds, Gregynog, Wales, August 1977*.
[21] A. H. ROTS and W. W. SHANE: *Astron. Astrophys.*, **45**, 25 (1975).
[22] A. SEGALOWITZ: unpublished thesis, University of Leiden, Netherlands (1976).
[23] C. C. LIN, C. YUAN and F. H. SHU: *Astrophys. J.*, **155**, 721 (1969).
[24] B. VAN LEER: *J. Comp. Phys.*, **32**, 101 (1979).
[25] P. R. WOODWARD: *Astrophys. J.*, **195**, 61 (1975).
[26] P. R. WOODWARD: *Astrophys. J.*, **207**, 484 (1976).
[27] W. F. NOH: in *Methods in Computational Physics*, Vol. **3**, edited by B. ADLER, S. FERNBACH and M. ROTENBERG (New York, N. Y., 1964).
[28] P. R. WOODWARD: invited paper, *IAU Symposium No. 84* (1978).
[29] W. G. SUTCLIFFE: Tech. Rep. UCID-17013, Lawrence Livermore Laboratory, Livermore, Cal. (1973).
[30] R. DeBAR: Tech. Rep. UCIR-760, Lawrence Livermore Laboratory, Livermore, Cal. (1974).
[31] W. F. NOH and P. R. WOODWARD: *Proceedings of the V International Conference on Numerical Methods in Fluid Dynamics, Enschede, Netherlands, 1976*, p. 330.
[32] J. W. MILES: *J. Fluid Mech.*, **4**, 538 (1958).
[33] J. A. FEJER and J. W. MILES: *J. Fluid Mech.*, **15**, 335 (1963).
[34] B. G. ELMEGREEN and C. J. LADA: *Astrophys. J.*, **214**, 725 (1977).
[35] B. J. BOK and E. F. REILLY: *Astrophys. J.*, **105**, 255 (1947).
[36] B. J. BOK, C. S. CORDWELL and R. H. CROMWELL: in *Dark Nebulae, Globules, and Protostars*, edited by B. T. LYNDS (Tucson, Ariz., 1971), p. 33.
[37] J. E. DYSON: *Astrophys. Space Sci.*, **1**, 388 (1968).
[38] D. A. MENDIS: *Mon. Not. R. Astron. Soc.*, **142**, 441 (1969).
[39] F. D. KAHN: *Physica (The Hague)*, **41**, 172 (1969).
[40] J. S. BERRY: *Astron. Astrophys.*, **44**, 401 (1975).
[41] G. TENORIO-TAGLE: *Astron. Astrophys.*, **54**, 517 (1977).
[42] A. BLAAUW: *Annu. Rev. Astron. Astrophys.*, **2**, 213 (1964).
[43] C. J. LADA, B. G. ELMEGREEN, H.-I. CONG and P. THADDEUS: *Astrophys. J.*, **226**, L39 (1978).
[44] W. HERBST and G. E. ASSOUSA: *Astrophys. J.*, **217**, 473 (1977).
[45] R. A. CHEVALIER: *Astrophys. J.*, **188**, 501 (1974).
[46] T. LEE, D. A. PAPANASTASSIOU and G. J. WASSERBURG: *Astrophys. J.*, **211**, L107 (1977).
[47] R. N. CLAYTON, N. ONUMA, L. GROSSMAN and T. K. MAYEDA: *Earth Planet. Sci. Lett.*, **34**, 209 (1977).

[48] A. G. W. CAMERON and J. W. TRURAN: *Icarus*, **30**, 447 (1977).

[49] J. M. LATTIMER, D. N. SCHRAMM and L. GROSSMAN: *Nature (London)*, **269**, 116 (1977).

[50] P. R. WOODWARD: *Annu. Rev. Astron. Astrophys.*, **16**, 555 (1978).

[51] R. B. LARSON: *Fundam. Cosmic Phys.*, **1**, 1 (1973).

[52] R. B. LARSON: in *Star Formation, IAU Symposium No. 75*, edited by T. DE JONG and A. MAEDER (Boston, Mass., 1977), p. 249.

[53] P. BODENHEIMER: *Rep. Prog. Phys.*, **35**, 1 (1972).

[54] D. McNALLY: *Rep. Prog. Phys.*, **34**, 71 (1971).

[55] A. G. W. CAMERON and M. R. PINE: *Icarus*, **18**, 377 (1973).

[56] R. B. LARSON: *Mon. Not. R. Astron. Soc.*, **156**, 437 (1972).

[57] D. C. BLACK and P. BODENHEIMER: *Astrophys. J.*, **206**, 138 (1976).

[58] M. L. NORMAN and J. R. WILSON: *Astrophys. J.*, **224**, 497 (1978).

[59] J. M. BARDEEN: *Astrophys. J.*, **167**, 425 (1971).

[60] J. P. OSTRIKER and P. BODENHEIMER: *Astrophys. J.*, **180**, 171 (1973).

[61] K. J. FRICKE, C. MÖLLENHOFF and W. TSCHARNUTER: *Astron. Astrophys.*, **47**, 407 (1976).

[62] W. TSCHARNUTER: *Astron. Astrophys.*, **39**, 207 (1975).

[63] D. C. BLACK and P. BODENHEIMER: *Astrophys. J.*, **199**, 619 (1975).

[64] R. LYNDS and A. TOOMRE: *Astrophys. J.*, **209**, 382 (1976).

[65] P. GOLDREICH and D. LYNDEN-BELL: *Mon. Not. R. Astron. Soc.*, **130**, 125 (1965).

« The Orion Museum »
or « Six Early Phases of the Solar Life ».

H. REEVES

Section d'Astrophysique, Centre d'Etudes Nucléaires de Saclay
B.P.2 - 91190 Gif-sur-Yvette, France
Institut d'Astrophysique - Paris

Let us for the moment imagine that the Sun and the solar system were born in a stellar *OB* association resembling the Orion association. We could look at the observations as a reconstitution of the very early phases of our solar system, just as when we go to a Museum of Natural History we have, say, a reconstitution of the phases of the development of the human embryo made for us by showing the embryo after one day, one week, one month, etc.

The first phase is, of course, just the very massive cloud shining in CO, which cover a large part of the Orion constellation. Its density is $\simeq 10^3$ cm^{-3} and its mass $\simeq 10^5 \, M_\odot$. We do not really know how it was formed out of general interstellar matter (density 1 cm^{-3}), but we suspect that this may have something to do with spiral nature of our Galaxy. Orion is near an arm (it is sometimes said to form a sort of extension to an arm).

This cloud may then have been initiated by the motion of the spiral pattern, as expected in the density wave theory.

There are actually two CO clouds in Orion, named Orion A and Orion B. They are probably the remnants of an initially much larger cloud, which has been already largely eaten by the process of star formation and the appearance of the subgroups of stars.

Inside Orion A, but rather close to its boundary (less than one pc), there is a subcondensation of $10^3 \, M_\odot$ and 10^5 cm^{-3} density identified as the formaldehyde cloud, because it is by the radiation of the H_2CO molecule that it has been discovered.

Inside this subcloud, the infra-red cluster is the most early phase of stellar evolution accessible to us, thanks mostly to the power of infra-red and millimetric astronomy, which is capable of penetrating the opaque cloud in which it is embedded. This cluster is probably about a few 10^4 years old, not more. It is phase No. 2 in our museum. This cluster contains not only infra-red sources but also sources of radio recombination lines (the so-called compact H II regions) and sources of maser emission of the molecules OH and H_2O. We are not sure whether each source corresponds to a distinct protostar or to different locations in extended bodies, but we are fairly confident that the mean distances between these different bodies is less (and perhaps much less) than a few 10^{16} cm (*i.e.* a few thousand astronomical units: 1 AU = the Earth-Sun distance). The distance between individual maser sources may be only a few AU.

It is probably reasonable to assume that solar systems, such as ours, have their early steps of formation during this phase. We may imagine that the protostars of the IR cluster are surrounded by somethig akin to the standard protosolar nebula of Laplace and Cameron and that the gravitational settling of dust grains in the equatorial plane *à la* Safronov or *à la* Goldreich and Ward has already started there.

Since we have only vague ideas on the time scale required for the full transformation of a protosolar nebula into a full-scale, neat and clean planetary system, we do not know whether this operation is terminated during this phase, or whether it extends to the later phases.

However, in view of the high proximity of the stars during that period, we may well depart from the standard picture of a regular disk, closed upon itself, and consider the possibility of important border interactions between systems especially as far as the comets are concerned. If we ask the question « where were our present-day long-period comets (with orbits extending up to 50 000 AU) when the stars were so close? », we immediately open the door to the possibility that, at that moment, they had perhaps highly nonelliptic orbits bringing them in the vicinity of the solar system, much as the electrons cruising amidst the nuclei of a metallic lattice structure.

Stellar statistics of the Orion *OB* association indicates that about twenty supernovae, issued from stars with $M = 15 M_\odot$, have already exploded in its midst. The remnants of these supernovae, incorporating the nuclear cooking specific to massive stars, are likely to have contaminated the gas from which the infra-red cluster originated. The overall addition of new matter to this gas from these supernovae is expected to be a few percent in fractional mass.

This contamination does not need to be homogeneous within the cloud; supernovae explode here and there and their remnants pervade to a varying degree the different mass element of the original dark cloud.

Therefore, we should not be surprised to find small isotopic differences (up to a few percent) between neighbouring protostars and protosolar nebulae, even if each individual protosolar nebula is largely mixed and homogenized by its own turbulent motions. These considerations give renewed interest to the study of comets as potential carriers of isotopic anomalies. But we do not even have to go that far. The fact that the isotopic composition of oxygen is slightly different for different families of solar-system bodies such as the Earth-Moon system, the achondrites, etc., could be understood by multiple chapters of contaminations taking place before the complete solidification of one family, but after the contamination of another family, etc.

I do not want to be too specific about details and to oversell the story. I just want to show how these various astronomical events can give us much more freedom to understand the solar-system data. In general, it is advisable to keep a story at a simple level, but, when the data appear complicated, one should open the door to complexity in a way which is guided by other observations.

Just on the border of the CO cloud Orion A there is another subcondensation of again $10^3 M_\odot$ with density 10^5 cm^{-3}. This subcloud is a H II region, called the Orion Nebula so beautiful in colour pictures. This H II region is lighted by the blue photons of hot stars found in the small constellation called the Trapezium.

This represents the third phase in which the largest stars of a cluster have already reached the main sequence and are already shining as blue giants.

The Trapezium also contains many smaller stars (a few hundreds) which are confined to a volume of ~ 1 light year in dimension, implying that the mean distance between these objects must be $\simeq 0.1$ light years. The question of whether the cluster contains smaller stars in the same proportion as, say, the general field of stars (or more exactly as the general initial mass function (IMF) of stars) is still debated.

The situation is confused by the fact that it is almost impossible to correctly assign an appropriate mass to a star on the T Tauri phase (still contracting). However, there are arguments suggesting that the OB association contains both large and small stars, probably just, for instance, as in galactic clusters.

The time scale for $O + B$ stars to reach the main sequence is a few 10^5 years and this should also be the age of this cluster.

What happens to the dark matter of the cloud when, upon reaching the MS, the most massive stars start shining vast amount of blue photons, energetic enough to ionize the atoms of hydrogen ($h\nu \simeq 13.6$ eV)? They generate around them an expanding sphere of ionized hydrogen (also called a Strömgren sphere or a H II region) which extends to one light year or more. Within this volume the matter is heated and expands very rapidly. It is probably in this way that the matter left over after star formation is dissipated in space.

What will be the effect of these H II regions on the forming solar system? They would certainly bring large amount of heat to the accreting bodies. Are they reponsible for the heating of the nebula, which later, upon cooling, would have given rise to the condensation sequence of Anders and Lewis? In this case, since we have several stars of different masses which would be responsible for generating H II regions, we would have a sequence of heating and cooling phases giving rise to a complicated thermal history accounting perhaps for events as the early loss of volatiles, dear to ANDERS.

Are they responsible for the loss of H and He in the region of the inner planets too? Are they responsible for the loss of the whole primitive atmospheres of the inner planets too?

The fourth phase is to be seen in the so-called Outer Sword region.
The nebulosity has been entirely dissipated. The subgroup of stars in this
phase is $3 \cdot 10^6$ y old and the motion of the stars suggests a linear expansion.
At this time one or two SN of stars with masses $(30 \div 50)\, M_\odot$ may already
have exploded and their remnants may already have contaminated not only
their immediate neighbourhood, but even the matter which is in the other region
(younger or older). The possible nature of this contamination I have discussed at great length in another paper. The effects of these SN explosions on
the forming solar systems are, of course, multiple. The advancing shock waves
must have deeply perturbed any protosolar nebula by bringing it mechanical
motions, heat, pressure and new matter.

There is presently a debate on the nature of the agent which seems to propagate the onset of star formation from one region to another. Some authors
like to see, there, the effect of the expanding Strömgren spheres of new born O
and B stars. Others would rather believe that star formation is propagated
by SN explosions and subsequent shock waves. Personally I prefer the second
version, since a) it explains the time delay of $\simeq 3 \cdot 10^6$ years between the birth
of each subgroup, b) consequently it explains why we have discrete subgroups
and not a continuum of subgroups, c) it also explains why the average distance
between each subgroup is $\simeq 10$ pc.

One other interesting question is: why do individual stellar subgroups
expand? As mentioned before, the binding energy of the Trapezium cluster is
$\simeq 10^{48}$ erg. In a million years the OB stars of the cluster will have emitted
$10^4 \div 10^5$ more energy in the form of ionizing radiation. One exploding SN
will also give $\simeq 10^4$ more energy, kinetic and magnetic, to the remnant. Again
we have to decide between both possibilities. The problem is more complicated,
however, because we do not really know how to link these energies to a process
of stellar motions. We need a « clutch » (which may have only a very low efficiency). We could imagine that these stars have extended protostar nebulae,
with magnetic lines of forces still attached to the general gas of the cloud. The
shock wave of the SN remnant may then carry the whole structure with it
when it comes along ...

In phase 5 (the Belt region) and phase 6 (the Northwest region) we see a
gradual depletion of massive stars and we infer a large amount of supernova
explosions. Because of the continuing expansion motion of the stars, the subgroups occupy a larger and larger volume of space. After this phase it is probably
difficult to identify older subgroups. The stars venture alone in the Galaxy.
The sequence of exhibits of stellar embryos ends here in our Orion museum.

The last feature to mention is of course the Barnard's loop which still shines
quietly, possibly sporadically reactivated by the sequence of SN remnants
which at the end of their expansion brings there their last drops of energy...

Accumulation of the Protoplanetary Bodies.

V. S. Safronov

O. J. Schmidt Institute of Physics of the Earth
Academy of Sciences of the USSR - Moscow, USSR

1. – Introduction.

The concept of formation of planets by accumulation of solid bodies, widely recognized at present, was most clearly and definitely formulated in 1944 by Schmidt [1-3]. A similar picture was suggested also by Edgeworth [4] in 1949. Schmidt has made an important methodological step, considering the cosmogony as a complex problem of astronomy, geophysics, geology, etc., supplied by the many data of all Earth's sciences. In 1946 he headed a small department in the Geophysical Institute working on the origin of the Earth, at the same time remaining Director of the Institute. In 1950 Schmidt has organized an all-union conference in Moscow, where his theory was critically discussed. He encouraged some scientists to work on the theory: Gurevich and Lebedinsky [5], Safronov [6], etc.

The next important step in planetary cosmogony was the physico-chemical study of the structure and composition of meteorites by Urey in the USA [7, 8]. This rich source of new data allowed the determination of the physical conditions during planetary accumulation. Independently of Schmidt, Urey has come to the conclusion that the planets have formed from solid bodies.

Schmidt's theory, as it was at the end of his life (1956), has been published in the posthumous edition of his *Four Lectures on the Theory of Origin of the Earth* [3]. The theory turned out to be perspective and it was worked out by Schmidt's followers in the Institute of Physics of the Earth, now named after him. The main stages of the evolution of the circumsolar protoplanetary cloud and of the process of accumulation of planets in the cloud were quantitatively investigated in detail. A qualitative description of the theory was given in a popular book by Levin [9]. The most complete publication of the results was given in our book [10] and then a brief description in [11]. The theory of the formation of circumplanetary satellite swarms was also developed [12, 13]. In the 1970's an active work on the study of the accumulation of planets and

satellites began also in the USA (W. W. KAULA, A. W. HARRIS, G. W. WE-
THERILL, S. J. WEIDENSCHILLING, A. G. W. CAMERON, P. GOLDREICH, W.
R. WARD, W. K. HARTMANN and others) and in Japan (C. HAYASHI with col-
leagues, SCHIMAZU and others).

The study of the earliest stages of the origin of the solar system (formation
and evolution of the solar nebula, formation of the protoplanetary cloud) has
begun later than the study of the accumulation of the planets. Hypotheses
on the capture of the cloud by the Sun (including also that by SCHMIDT) appeared long ago. But they did not receive any recognition. Present-day hypotheses of a common origin of the Sun and the preplanetary cloud date from
models by HOYLE [14], CAMERON [15], SCHATZMAN [16].

In the present lecture we do not touch problems on the early evolution
beginning from the formation and collapse of the solar nebula and up to the
formation and evolution of the dust layer. They are discussed in other lectures.
Our consideration begins with the stage when a swarm of numerous protoplanetary bodies formed with sizes large enough for their effective gravitational
interactions. Hence the evolution of the swarm was mainly determined by
dynamical and not by physical or chemical processes.

2. – Relative velocities of preplanetary bodies.

Important characteristics of preplanetary bodies are their mass distribution
and their relative velocities. The latter determined the rate of growth of planets
and the efficiency of fragmentation of colliding bodies. The distribution of
masses of bodies substantially influenced the accumulation process and determined the initial temperature of the planets. Strictly speaking, these characteristics should be estimated in common, because the mass distribution depends
on the relative velocities of bodies and the velocities in turn depend on the
mass distribution. This estimate is hardly possible in analytical form. Now
there are attempts to calculate numerically this coupled evolution of the size
and velocity distribution in the first stage of the planetary accumulation [17].
Analytically these characteristics were estimated separately: the mass distribution function was determined for given values of the velocities, and the
velocities were determined for assumed mass distributions (usually an inverse
power law) of bodies [10].

The bodies had formed in a flattened dust layer and moved originally on
nearly circular orbits around the Sun, thus their relative velocities were very
small. As the bodies grew due to collisions, their gravitational interactions
increased and accordingly increased their relative velocities and the eccentricities
of their orbits. In a differentially rotating system with inelastic collisions
(coalescence of bodies) the relative velocities of bodies are determined by balancing the energy of random motions gained in encounters and the energy

lost in collisions. An approximate expression for the system of bodies of equal masses m and radii r was found by GUREVICH and LEBEDINSKY [5] assuming that the change of eccentricity of a body during its closest encounters with other bodies is of the order of the eccentricity itself ($\Delta e \approx e$). Then the average velocities v of bodies relative to a circular Keplerian motion are

$$(1) \qquad v \approx \sqrt{\frac{Gm}{2r}} \, .$$

The increase of the energy of random («thermal») motion of bodies can be estimated from the usual formulae of hydrodynamics for the dissipation of the energy of mechanical motion in a viscous fluid with a velocity gradient. When the mean free path of particles is small, the energy E liberated in 1 cm³ per second equals

$$(2) \qquad E = \eta R^2 \left(\frac{\mathrm{d}\Omega}{\mathrm{d}R}\right)^2 ,$$

where η is the viscosity of the medium, Ω is the angular velocity at a distance R from the axis of rotation. Application of this expression to the swarm of protoplanetary bodies shows that, in the case of small bodies $r < r_c$ (in the Earth's zone r_c is about several centimetres), the relative velocities of bodies decrease with time, the swarm flattens and becomes gravitationally unstable. But, if $r > r_c$, the velocities of bodies increase, the thickness of the swarm increases, the mean free path becomes comparable with R and eq. (2) becomes invalid. We should have then a much more complicated expression for E, but for randomizing the Keplerian shear of the swarm it can be reduced to the form (2) by introducing a correction factor β [10]:

$$(3) \qquad E = \beta\eta R^2 \left(\frac{\mathrm{d}\Omega}{\mathrm{d}R}\right)^2 .$$

For a system of bodies of equal size β was estimated to be 0.2. We should note that in the semi-empirical theory by PRANDTL the dissipation of energy in turbulent rotating fluids (also the case of large mean free path) is determined not by the gradient of the angular velocity but by the gradient of the angular momentum ΩR^2. At the Keplerian rotation this corresponds to $\beta = 1/9$.

It was shown that, when the mean free path λ is large, in the expression for the viscosity $\eta = \varrho\lambda^2/3\tau$ it should be taken a squared average radial displacement $\overline{\Delta R^2} = 2e^2 R^2$ of bodies (e is the eccentricity of the orbit) instead of λ^2. The increase of energy of relative motions of bodies per unit mass $v^2/2$ equals $2\beta v^2/3$ for a relaxation time τ_g—the time interval during which the vector $\boldsymbol{v}$ changes its direction by the angle $\pi/2$. In inelastic collisions the energy of bodies diminishes by the value $\zeta v^2/2$. The loss of energy is maximum when the bodies coalesce. Then $\zeta \approx 0.4$. The velocities of bodies tend to an equilibrium value when the gain and the loss of energy are equal. This value can be written in the same form as (1):

$$(4) \qquad\qquad v = \sqrt{Gm/\theta r} \, .$$

The nondimensional parameter θ depends on the properties of the system. For a system of bodies of equal size it was found $\theta \sim 1$. For a system of bodies of different sizes expression (4) also can be used, where m and r now are the mass and the radius of the largest body in the zone under consideration. The equation for θ is much more complicated in this case. For an inverse power law of distribution of masses of bodies

$$(5) \qquad\qquad n(m) = cm^{-q} \, ,$$

it leads to values $\theta \approx 3 \div 5$ at $q < 2$. When $q > \frac{5}{3}$ the velocities of the smaller bodies are a little less than those of the larger ones. If the bodies are in a gas which slows down their motion, then θ can arrive to a few tens. It was shown also that, when the mass distribution is not continuous and the largest body m has considerably overrun in mass all other bodies in the zone and moves in a nearly circular orbit, its gravitational perturbations increase the velocities of the rest of bodies less effectively. In this case in formula (4) the values of m_1 and r_1 for the second largest body should be taken instead of m and r.

From this consideration it was concluded that the collisions of bodies cannot lead to the formation of « jet streams » in which $v \to 0$. They only limit the relative velocities by the value (4). Smaller values are prevented by perturbations.

Repeated encounters of a body with a planet moving in a circular orbit do not lead to a systematic increase of the random velocity of the body, because encounters occur at the same distance from the Sun. But ÖPIK [18] has shown that, when the eccentricity e_0 of the planet's orbit is not zero, such encounters increase the velocity of the body. During the time interval τ_{g0}, when, due to encounters, the vector v turns on the average by $\pi/2$, the energy of random motion of the body increases by $\beta_0 v_0^2$, where v_0 is the velocity of the planet relative to the Keplerian circular velocity and $\beta_0 \sim 1$.

The ratio of the energy gained due to encounters only with the planet to the energy gained due to encounters with all other bodies is equal to

$$\frac{\beta_0 v_0^2 \tau_g}{\beta v^2 \tau_{g0}} = \frac{\beta_0}{\beta} v^2 \frac{\tau_g}{\tau_{g0}} \, .$$

Because $\beta \approx 0.2$, we obtain $\beta_0/\beta \approx 5$. However, these quantities were estimated in different ways and it may be really $\beta_0/\beta \approx 1$. At the last stage of accumulation, when the mass of a growing planet (planet's embryo) becomes comparable with the whole mass of other bodies in its zone, we have $\tau_{g0} \ll \tau_g$ and the mechanism of Öpik is more effective if v is not too small. But encounters of a body with two neighbouring planets moving in nearly circular orbits increase its velocity more effectively than Öpik's mechanism. At the earlier stages of accumulation $\tau_{g0} > \tau_g$, $v^2 \sim 10^{-1}$ and the mechanism of Öpik is not effective.

When the swarm is not very thin (*i.e.* the largest bodies exceed one kilometre in size), the gravity acceleration along z-axis parallel to the axis of rotation is determined mainly by the Sun and for $z \ll R$ is equal to

$$(6) \qquad Z \approx -\Omega^2 z \,.$$

When the velocities v do not depend on z, the density distribution in the swarm along z is

$$(7) \qquad \varrho(z) = \varrho_0 \exp\left[-\frac{3\Omega^2}{2\bar{v}^2} z^2\right]$$

and the thickness of a homogeneous layer is

$$(8) \qquad H = \frac{\sigma}{\varrho_0} = \frac{1}{\varrho_0}\int_{-\infty}^{+\infty} \varrho \, \mathrm{d}z = \frac{\pi}{2\Omega}\, \bar{v} \,,$$

where σ is the surface density of the swarm. Therefore, the thickness of the swarm is proportional to the relative velocities of bodies and grows with time proportionally to the radius of the largest body m.

3. – Mass distribution of preplanetary bodies.

The accumulation process of the protoplanetary bodies is in some sense similar to the coagulation process of colloids in chemistry or of raindrops in meteorology, thus the methods of coagulation theory were used for its study. Unfortunately, the specifity of the problem does not allow the use of the solutions already found in other fields of science. There is substantial additional complication connected with the fragmentation of bodies at the second stage of accumulation when the velocities of bodies are high.

The generalized integro-differential equation of coagulation which takes into account the fragmentation of bodies in collisions has the form

$$(9) \qquad \frac{\partial n(m, t)}{\partial t} = \int_{0}^{m/2} w(m', m-m')\, A(m', m-m')\, n(m',t)\, n(m-m', t)\, \mathrm{d}m' -$$

$$- n(m, t) \int\limits_0^\infty A(m, m')\, n(m', t)\, \mathrm{d}m' + \int\limits_m^\infty n_1(m, m'') \cdot$$

$$\cdot \int\limits_0^{m''} [1 - w(m', m'' - m')]\, A(m', m'' - m')\, n(m', t)\, n(m'' - m', t)\, \mathrm{d}m'\, \mathrm{d}m'',$$

where $A(m, m')$ is a coagulation coefficient which characterizes the probability of collisions of two bodies with masses m and m', $w(m, m')$ is the probability of coalescence of these bodies in the collision and, accordingly, $1 - w(m, m')$ is the probability of their fragmentation, $n_1(m, m'')$ is the mass distribution of fragments which form when two bodies with the sum of their masses m'' disintegrate after collision. In the first stage of accumulation, the relative velocities of bodies are low and the bodies only coalesce in collisions without fragmentation. Then one assumes $w = 1$. At the last stage only the largest bodies do not undergo fragmentation.

A well-known solution of the coagulation equation without fragmentation $(w \equiv 1)$ for the case $A = \mathrm{const}$ had been found by SMOLUCHOWSKY. In 1962 we have obtained the analytical solution for the other important case when the coagulation coefficient is proportional to the sum of masses of colliding bodies: $A(m, m') = c(m + m')$. This expression for A is intermediate between that for small bodies (geometrical cross-section) and that for large bodies which have much larger effective cross-sections due to gravitation. The mass distribution function, found for the initial distribution $n(m, 0) = a \exp\left[-bm\right]$, can be written almost for the whole region of variation of m and t as the product of a power function of m and an exponential one

$$(10) \qquad n(m, t) \approx \frac{N_0(1 - \tau)}{2\sqrt{\pi b}\tau^{\frac{3}{4}}}\, m^{-\frac{3}{2}} \exp\left[-\left(1 - \sqrt{\tau}\right)^2 bm\right],$$

where $\tau = 1 - \exp\left[-c\varrho t\right]$. The exponential function cuts down the distribution at high values of m and practically does not affect it at small m. With the lapse of time the region where the power law approximation for the distribution function is satisfactory extends to larger and larger values of m. From this it was concluded that the inverse power law is an asymptotic solution of the coagulation equation which does not depend on the initial mass distribution. A qualitative investigation of the equation confirmed this conclusion. Then a qualitative investigation of the equation with more general assumptions about the coagulation coefficient was fulfilled [19]. It was found that, except for a rather narrow interval of the most massive bodies, the mass distribution tends to the inverse power law (5) with the index $q \approx 1.55 \pm 0.15$.

A more general equation including the fragmentation of bodies was also studied and its asymptotic solutions were found by taking smooth functions of m for the parameters characterizing the fragmentation. The solution may be approximated in the form (5) with index $q \approx 1.8$ everywhere except the interval of the largest masses [20]. From this solution as a special case when there is no accumulation and bodies fragment in collisions ($w \equiv 0$) we obtain the asymptotic solutions for the mass distribution of asteroids found by DOHNANYI [21], HELLYER [22] and BANDERMANN [23]. However, the solution with $q = 11/6$ is obtained only if it is assumed that the criterion of fragmentation does not depend on the masses of bodies. Numerical computations of the evolution of the mass distribution have shown [24] that the deviation of the distribution function from the power law in the region of the largest bodies is similar to the appearance of an additional exponential factor in the analytical solution (10) of the equation without fragmentation of bodies.

4. – Accelerated growth of the largest bodies. Formation of planet embryos.

As was pointed out above, the inverse power law (5) is not valid for the description of mass distribution in the range of the largest bodies. The concept of distribution function itself, being essentially a statistical one, cannot be applied to the largest bodies. Hence special consideration is needed to find out the main features of growth of these bodies.

The effective cross-section of collision πl^2 for the largest body (m, r) due to gravitational focusing of orbits in its vicinity is much larger than the geometric cross-section

$$(11) \qquad \pi l^2 = \pi (r + r')^2 \left[1 + \frac{2G(m + m')}{V^2(r + r')} \right],$$

where V is the velocity of the body m' relative to m before the encounter. Because the largest body moves in a much less eccentric orbit, V is approximately equal to the velocity v of a body m' relative to the circular Keplerian velocity and is determined according to (4). Assuming $m + m' \approx m$ and $r + r' \approx r$, we obtain

$$(12) \qquad \pi l^2 \approx \pi r^2 (1 + 2\theta) .$$

The cross-section of the second largest body (m_1, r_1) is written in a similar way

$$(13) \qquad \pi l_1^2 \approx \pi r_1^2 (1 + 2\theta r_1^2/r^2) .$$

As long as r_1 does not differ considerably from r, we have $l^2/l_1^2 \approx r^4/r_1^4$. Bodies colliding with the largest bodies coalesce with them. Therefore,

$$(14) \qquad \frac{\mathrm{d}m}{\mathrm{d}m_1} \approx \frac{r^2(1 + 2\theta)}{r_1^2(1 + 2\theta r_1^2/r^2)}$$

and

$$(15) \qquad \frac{\mathrm{d}(m/m_1)}{\mathrm{d}m} = \frac{1}{m}\left(\frac{m_1}{m} - \frac{\mathrm{d}m_1}{\mathrm{d}m}\right) = \frac{r_1^2}{mr^2}\left(\frac{r_1}{r} - \frac{1 + 2\theta r_1^2/r^2}{1 + 2\theta}\right).$$

The ratio of masses m/m_1 remains constant if the expression in brackets in the right-hand side of (15) is zero. This takes place when $r = r_1$ (the trivial case) and when $r = 2\theta r_1$, if $\theta > \frac{1}{2}$. As soon as r_1 is slightly less than r, the second term in brackets becomes smaller than the first one and m/m_1 begins to increase. This more rapid growth (runaway) of the largest body m stops only when $r = 2\theta r_1$, i.e. when the limiting mass ratio $(m/m_1)_{\mathrm{lim}} \approx (2\theta)^3$ is reached.

If $\theta < \frac{1}{2}$, the limiting ratio is $(m/m_1)_{\mathrm{lim}} = 1$ as r_1 cannot be larger than r. In this case there is a tendency to the equalization of the masses during their accumulation. At small θ, the velocities are large and the cross-sections of collisions are nearly geometrical. Then the radii of bodies increase with an equal rate $(\mathrm{d}r_1 = \mathrm{d}r)$ and $m/m_1 \to 1$.

As was shown above, the values of θ during the accumulation process were of the order of several units. So there was a considerable runaway in mass of the largest bodies in their feeding zones. In this way many potential « embryos » of the planets have been formed. They had much larger masses and much smaller eccentricities e of orbits than other bodies. The half-width ΔR_f of the ring-shaped zone of feeding of the embryo is determined mainly by the average eccentricity $\bar{e}$ of the bodies

$$(16) \qquad \Delta R_t = (\bar{e} + e)R \approx \frac{\sqrt{2}(1 + \nu)v}{\Omega} = \frac{1 + \nu}{\Omega}\left(\frac{8\pi G\delta}{3\theta}\right)^{\frac{1}{2}} r\,,$$

where $\nu = e/\bar{e}$, δ is the density of the embryo and v is taken from (4). All bodies with the major semi-axes in the range $2\Delta R_t$ can fall onto the embryo.

The embryo has also a zone of gravitational influence (with a half-width ΔR_g). A body moving in this zone originally on a nearly circular orbit when approaching the embryo undergoes considerable gravitational perturbations and begins to move over the whole feeding zone $2\Delta R_t$. We can express ΔR_g, for example, in terms of the radius of the gravitational Hill's sphere. For the case of a circular orbit of the embryo

$$(17) \qquad \Delta R_{g0} = kr_{L_1} = k\left(\frac{4\pi G\delta}{9\Omega^2}\right)^{\frac{1}{3}} r\,,$$

where $r_{L_1} = (m/3M_\odot)^{\frac{1}{3}}R$ is the distance to the first Lagrangian point. From an analytical study on the stability of quasi-circular orbits in the planar restricted three-body problem by DOLE [25] it can be found $k \approx 3$. Due to many repeated encounters the zone of instability should be somewhat wider. Two neighbouring embryos m and m_1 with orbital eccentricities e and e_1 begin to interact gravitationally when the difference ΔR of their distances from the

Sun becomes less than ΔR_g:

(18) $$\Delta R_g = \Delta R_{g0} + (e + e_1) R = k_{ef} r_{L_1}$$

and

$$k_{ef} \approx k + 20(\nu + \nu_1) \left(\frac{3}{\theta} R_{ae} \right)^{\frac{1}{2}} ,$$

where R_{ae} is expressed in astronomical units. It can be shown that ΔR_f is about twice ΔR_g. The embryos grew, their zones widened proportionally to their radii and neighbouring zones overlapped more and more. Hence in the feeding zone of the embryo there were always 2 or 3 other, neighbouring embryos which were not yet inside its influence zone. This provided relatively small values of θ in eq. (4), because bodies underwent gravitational interactions with several largest bodies (embryos).

The regular system of embryos with ring-shaped feeding zones formed only when the embryos became large enough. In the earlier stage, when the bodies were small, their feeding and gravitational zones comprised only a small part of the ring. Inside the same ring a lot of other « largest » bodies of comparable mass existed. Due to the small width of the ring ($\Delta R_f \propto r$) their synodic periods with respect to m were very long—more than the characteristic time of growth of the bodies themselves. Until the close encounters with m, they grew on the average with the same rate. During the rare encounters with m, their lag behind in growth from m was not considerable. Due to differential rotation the bodies in the zone of m continuously renewed and there were no conditions for a considerable runaway of m from other « largest » bodies of the ring at this stage. Accordingly, the eccentricity of m orbit due to encounters with these bodies did not differ significantly from that of the rest of bodies.

It can be shown that, only when the radius of m increases up to $r \sim 100$ km, the zone $2 \Delta R_g$ around its orbit becomes free from other bodies comparable in mass with m. Then the further growth of m takes place without strong perturbations of its orbit, the eccentricity begins to decrease and the body becomes a « potential » embryo of a planet.

Large bodies which have occurred inside the feeding zone of the embryo due to its widening begin to be backward considerably from m. However, as was shown by WETHERILL [26], even during the very long period of growth of two bodies in the same zone (an increase of m by a factor of more than 10^6) the ratio m/m_1 does not reach the limiting value $(20)^3$ and remains $2 \div 3$ times smaller. Therefore, this formal limit is not so informative as we interpreted it earlier.

5. – Growth of embryos and formation of planets.

We already mentioned that the growth of embryos caused the increase of relative velocities of bodies and their fragmentation in high-velocity collisions. But for the embryos themselves collisions were not dangerous. Let us consider the most unfavourable case of collision of two embryos of equal mass which move before the encounter with the velocities determined by eq. (4). The energy of impact is equal to

$$(19) \qquad E_1 = \frac{3}{5} G \left[\frac{(2m)^2}{2^{\frac{1}{3}} r} - \frac{2m^2}{r} \right] + \frac{2mv^2}{2} = \left(0.71 + \frac{1}{\theta} \right) \frac{Gm^2}{r} \, .$$

Only a part η_M of this energy remains as mechanical energy of ejection of material. For the complete scattering of the whole mass $2m$ into infinity this energy should be greater than the potential energy of the globe with mass $2m$ and radius $2^{\frac{1}{3}} r$:

$$(20) \qquad \eta_M (0.71 + 1/\theta) > \frac{3}{5} \cdot \frac{4}{2^{\frac{1}{3}}} = 1.9 \qquad \text{and} \qquad \eta_M > 1 \, .$$

This is impossible because $\eta_M < 1$. Therefore, the collision of the embryo m with any other body cannot lead to its fragmentation into separate pieces because of the gravitation, though it is accompanied by the crushing-up of material. All collisions of the embryos lead to their coalescence with colliding bodies.

Initially there were many potential embryos in the zone of the planet m_p. They grew on the average with an equal rate and densely filled the whole zone of the planet. Their total number N in the zone was of the order

$$(21) \qquad N \sim \frac{R}{\Delta R_f} \sim \left(\frac{m_p}{m} \right)^{\frac{1}{3}} \, .$$

The mass of other bodies per one embryo (except the embryo m itself) was equal to

$$(22) \qquad m_b \approx m_p / N - m \approx m \left[\left(\frac{m_p}{m} \right)^{\frac{2}{3}} - 1 \right] \, .$$

While $m \ll m_p$, the masses of all embryos comprise only a small part of the mass of other bodies. The embryos grew sweeping out all bodies colliding with them and their feeding zones ΔR_f widened proportionally to their radii. The zones of neighbouring embryos overlapped and, when the smaller embryo m_1 became covered by the zone of the larger embryo m, it began to lag behind in mass from m in accordance with (14). After the increase of the mass of m by the factor $(\Delta R_f / \Delta R_g)^3$ (about ten), the embryo m_1 turned out to be already inside the zone ΔR_g of m and, under the strong perturbations of m, it left its nearly circular orbit and loose its privileged position of embryo. The ratio m/m_1, which had increased to that moment $2 \div 3$ times, went on to increase further. Because the close encounters are more probable than the collisions,

the embryo m_1 before the collision with m passed near it inside the Roche limit
and disintegrated into smaller pieces. In such a way a rarefaction of the embryos proceeded. On the average the number of embryos decreased two times
after the doubling of ΔR_f, *i.e.* after the increase of the masses of the larger
embryos by a factor 8. Therefore, the embryos grew mainly due to sweeping
out smaller bodies in their zones and not the neighbouring embryos. The
number of embryos decreased until they consumed all solid material between
them and until the distances between their orbits had become so large that the
system of the remaining embryos (now the planets) could be stable in spite
of their gravitational interactions during billions of years. This requirement
was the main condition for a law of planetary distances after completing the
accumulation process.

In the course of growth of the embryo's mass m, the width of its feeding
zone ΔR_f enlarged proportionally to its radius, and thus the total mass of the
matter in the zone increased proportionally to $m^{\frac{1}{3}}$, that is considerably slower
than m itself. Therefore, ΔR_f and m do not grow without limits when the
embryo consumes the matter of the zone, but they tend to a certain limit value
which is determined by the surface density σ and the value of θ [27]. At
$\sigma_0(R) = \sigma_1 R^{-\nu'}$

$$
(23) \quad
\begin{cases}
\Delta R_{f\,\max} = \left(\dfrac{2\delta}{3}\right)^{\frac{1}{4}} \left(\dfrac{2\pi}{\theta M_\odot}\right)^{\frac{3}{4}} (2\sigma_0)^{\frac{1}{2}} R^{(11-2\nu')/4}\,, \\[3mm]
m_{\max} = \dfrac{4}{3}\,\pi\delta \left(\dfrac{6\pi}{\theta\,\delta M_\odot}\right)^{\frac{3}{4}} (2\sigma_0)^{\frac{3}{2}} R^{(15-6\nu')/4}\,.
\end{cases}
$$

Taking $\theta \approx 2$ for the region of terrestrial planets, one can get $\Delta R_{f\,\max}$ in good
agreement with the present distances between the planets. But the process
of interaction of embryos has been not yet investigated in detail; it should have
provided sufficiently large ΔR_g to exclude, for example, the existence of one
more planet between the Earth and Mars.

Recently LEVIN [28] has suggested a different model of the process, assuming
a single embryo in the whole zone of the planet, which grew exceptionally
rapidly as compared to the rest of bodies. He has supposed that, due to the
runaway of the embryo in mass from other bodies, the velocities of the latter
would be extremely small (which corresponded to very large $\theta \sim 10^2 \div 10^3$), and
this would accelerate the growth of the embryo. It would sweep out rapidly
all the matter of its narrow zone and then continue to grow further, at the
expense of small planetesimals penetrated into its zone from other parts of
the zone of the planet.

But this model is not in accord with the main features of the accumulation
process. The preponderant growth of the largest body takes place only relative
to bodies moving inside its feeding zone $2\Delta R_f$. The bodies moving outside
this zone grow quite independently and essentially under the same conditions,
because the average rate of growth is determined by the surface density of

the matter σ and by the angular velocity Ω, which do not vary appreciably across the whole zone of the planet. Moreover, as we have already mentioned, there is an opposite trend toward the equalization of the masses of the two largest bodies, growing independently in different zones, because the radii of bodies increase with the same rate. Therefore, the embryos grew in the whole planet's zone in the same way and their number was as great as possible. The permanent overlapping of feeding zones of neighbouring embryos provided the encounters of bodies at least with the two of embryos and made implausible the low relative velocities of bodies with high θ in (4).

Additional evidence about the process of accumulation may be drawn from the consideration of the rotation of planets. There are two different components of the axial rotation—a regular component (direct rotation), connected with the rotation of the swarm of bodies from which the planets formed, and a nonregular, random component, which is manifested in planetary obliquities. The latter is related to the random directions of impacts of discrete bodies. The larger the falling bodies, the greater should be the obliquities, because the sum of randomly oriented angular momenta contributed by a few individual large bodies considerably differs from the average value. It was found [10] that, depending on the distribution of masses (especially in the range of big bodies), the masses of the largest bodies, impacted on the Earth in the final stage of its growth, reached $10^{-3} \div 10^{-2}$ of the Earth's mass: 10^{-3} for the continuous power law for the mass distribution and 10^{-2} in the discrete impacts of a few big bodies. The above-described model of the planet's accumulation with several embryos corresponds rather to the second case. From (21) it is seen that, when in the zone of the planet only two embryos remained, their masses were $m \sim$ $\sim 10^{-1} m_\mathrm{p}$. The limitation $m_1/m_\mathrm{p} \leqslant 10^{-2}$ relates to the case when the body m_1 falls onto the planet of mass m_p, i.e. at the end of its growth. If the planet has not yet grown and its mass is m, the angular momentum imparted by the body m_1 is smaller by the factor $(m_\mathrm{p}/m)^{\frac{3}{}}$. The previous limitation then reduces to

$$\frac{m_1}{m_\mathrm{p}}\left(\frac{m}{m_\mathrm{p}}\right)^{2/3} = \frac{m_1}{m}\left(\frac{m}{m_\mathrm{p}}\right)^{5/3} \leqslant 10^{-2}.$$

For $m/m_1 \approx 4$ we find $m/m_\mathrm{p} \leqslant 0.15$ or $m_1/m_\mathrm{p} \leqslant 1/28$. Therefore, the limitation is easily fulfilled if the last fallen embryo m_1 fell onto the embryo m relatively early. In the case of later collision we should suppose that the limitation on m_1 is fulfilled due to its disintegration into smaller pieces in close approaches with m inside its Roche limit.

There are numerous attempts to solve the problem of regular rotation. Recent analytical estimates [29] and numerical calculations [30] have shown that the angular momentum imparted due to the direct fall onto the planet is less than the observed angular momentum of planets. Presumably a significant part of the angular momentum was acquired by planets during the fall out of matter from inner parts of satellite swarms surrounding them. Such a source of angular momentum for the planets was envisaged by ARTEMJEV and RADZIEVKY [31], and is supported by the study of the formation of satellite swarms around the growing planets fulfilled by RUSKOL [12, 13] and HARRIS and KAULA [32].

6. – The duration of the process of accumulation of planets.

We have already mentioned that during the whole accumulation process the planet's embryo grows up most effectively, consuming all bodies colliding with it. The velocities of other bodies in its zone are determined by expression (4), *i.e.* by the mass and radius of the embryo. In this case the growth of an embryo (*i.e.* of a planet) is described by the simple formula

$$(24) \qquad \frac{\mathrm{d}m}{\mathrm{d}t} = 4\pi r^2 (1 + v_e^2/v^2)\,\sigma/P = 4\pi r^2 (1 + 2\theta)\,\sigma/P\ ,$$

where $P = 2\pi/\Omega$ is the period of revolution of the planet around the Sun, and $\sigma = \sigma_0[1 - (m/m_\mathrm{p})^{\frac{2}{3}}]$. At $\theta = 3$ and $\sigma_0 = 10$ g/cm^2 the time scale of sweeping-out of bodies (the decrease of their spatial density by a factor of e) in the Earth's zone at the final stage is $\tau \sim 20$ million years, and the accumulation of 98 % of the total Earth's mass takes 100 million years. This time scale, estimated by us as early as in 1958, was thereafter questioned many times. Some geophysicists and geochemists adopted a much shorter time scale $\sim 10^5$ y in order to obtain an initially hot Earth. The short time scale was also deduced from certain cosmogonic models assuming low relative velocities of preplanetary bodies, *i.e.* high values of θ in (24). But a detailed study of the problem by WEIDENSCHILLING [33] has confirmed the time scale of 10^8 y. Low velocities of bodies at the final stage of accumulation are shown to be implausible (WETHERILL [26] and this volume).

For the outer planets Uranus and Neptune eq. (24) gives an unacceptably large time span for the growth: $\sim 10^{11}$ y, due to large P and small σ_0, calculated from the present masses of these planets. The accumulation process of the giant planets was more complicated than that of terrestrial planets, and there exists no complete theory for it. When the masses of growing giant planets reach about one tenth of their present values, the velocities of bodies in their zones approach the velocity of escape and ejection of bodies out of the solar system begins. About 1 % of the ejected bodies has remained on the periphery of the solar system in the cometary cloud.

The effective ejection of bodies implies that the initial surface density of matter σ_0 in the region of giant planets was considerably higher than that found from the present masses of the planets. When the velocities of bodies v reach the maximum possible value, the escape velocity, a further increase of m in eq. (4) at constant v means that the parameter θ also should be higher. It is seen from (24) that the increase of the product $\sigma\theta$ by a factor of 10^2 is sufficient to obtain a plausible time scale of growth of Uranus and Neptune. There exists one more possibility of acceleration of the growth of these planets—namely a more lengthy initial stage of low-density dust condensations with large cross-sections for collision (see [34]). We conclude that further work on the theory of formation of giant planets is needed.

REFERENCES

[1] O. J. SCHMIDT: *Dokl. Akad. Nauk SSSR*, **45**, 245 (1944).
[2] O. J. SCHMIDT: *Mem. Soc. R. Sci. Liege*, Collect. 4°, **15**, 638 (1955).
[3] O. J. SCHMIDT: *Four Lectures on the Theory of Origin of the Earth* (Moscow, 1957), translation from the Russian: *A Theory of the Origin of the Earth, Four Lectures* (London, 1959).
[4] K. E. EDGEWORTH: *Mon. Not. R. Astron. Soc.*, **109**, 600 (1949).
[5] L. E. GUREVICH and A. I. LEBEDINSKY: *Izv. Akad. Nauk SSSR Ser. Fiz.*, **14**, 765 (1950).
[6] V. S. SAFRONOV: *Astron. Ž.*, **31**, 499 (1954).
[7] H. C. UREY: *The Planets. Their Origin and Development* (New Haven, Conn., 1952).
[8] H. C. UREY: *Proc. Chem. Soc. London* 67 (1958).
[9] B. J. LEVIN: *Origin of the Earth and the Planets* (Moscow, 1964).
[10] V. S. SAFRONOV: *Evolution of the Protoplanetary Cloud and Formation of the Earth and the Planets* (Moscow, 1969); NASA TT F-677, Va. (1972).
[11] V. S. SAFRONOV: in *On the Origin of the Solar System*, edited by H. REEVES (Paris, 1972), p. 89.
[12] E. L. RUSKOL: *Origin of the Moon* (Moscow, 1975).
[13] V. S. SAFRONOV and E. L. RUSKOL: in *Planetary Satellites*, edited by J. BURNS (Tucson, Ariz., 1977), p. 501.
[14] F. HOYLE: *Q. J. R. Astron. Soc.*, **1**, 28 (1960).
[15] A. G. W. CAMERON: *Icarus*, **1**, 13 (1962).
[16] E. SCHATZMAN: *Ann. Astrophys.*, **30**, 963 (1967).
[17] R. GREENBERG, J. F. WACKER, W. K. HARTMANN and C. R. CHAPMAN: *Icarus*, **35**, 1 (1978).
[18] E. J. ÖPIK: *Contrib. Armagh Observatory*, No. 34, 185 (1961).
[19] E. V. ZVJAGINA and V. S. SAFRONOV: *Astron. Ž.*, **48**, 1023 (1971).
[20] E. V. ZVJAGINA, G. V. PECHERNIKOVA and V. S. SAFRONOV: *Astron. Ž.*, **50**, 1261 (1973).
[21] J. S. DOHNANYI: *J. Geophys. Res.*, **74**, 2531 (1969).
[22] B. HELLYER: *Mon. Not. R. Astron. Soc.*, **148**, 383 (1970).
[23] L. W. BANDERMANN: *Mon. Not. R. Astron. Soc.*, **160**, 321 (1972).
[24] G. V. PECHERNIKOVA, V. S. SAFRONOV and E. V. ZVJAGINA: *Astron. Ž.*, **53**, 612 (1976).
[25] S. H. DOLE: *Am. Rocket Soc. J.*, **31**, 214 (1961).

[26] G. W. WETHERILL: *Proceedings of the VII Lunar Science Conference* (Houston, Tex., 1976), p. 3245.

[27] A. V. VITJAZEV, G. V. PECHERNIKOVA and V. S. SAFRONOV: *Astron. Ž.*, **55**, 107 (1978).

[28] B. J. LEVIN: *Astron. Ž. Pis'ma*, **4**, 102 (1978).

[29] A. W. HARRIS: *Icarus*, **31**, 168 (1977).

[30] R. I. KILADZE: *Bull. Abastumanskaya Astrofiz. Obs. Akad. Nauk Gruz. SSR*, **48**, 191 (1977).

[31] A. V. ARTEMJEV and V. V. RADZIEVSKY: *Astron. Ž.*, **42**, 124 (1965).

[32] A. W. HARRIS and W. M. KAULA: *Icarus*, **24**, 516 (1975).

[33] S. J. WEIDENSCHILLING: *Icarus*, **27**, 161 (1975).

[34] V. S. SAFRONOV: *Proceedings of the Soviet-American Conference on Cosmochemistry of the Moon and Planets*, edited by A. P. VINOGRADOV (Moscow, 1975), p. 624; edited by J. H. POMEROY and N. J. HUBBARD, NASA (Washington, D. C., 1977), p. 797.

Some Problems of Evolution of the Solar Nebula and of the Protoplanetary Cloud.

V. S. SAFRONOV

O. J. Schmidt Institute of Physics of the Earth
Academy of Sciences of the USSR - Moscow, USSR

1. – Initial mass and angular momentum of the solar nebula.

Already in the first models of the 1960's of the origin of the Sun and planets from a single solar nebula very different values of the initial mass and angular momentum of the nebula were assumed. In the model by HOYLE [1] the smallest possible values were taken: the mass of the nebula $M_n = 1.01 M_\odot$, which is mainly the sum of the masses of the Sun, Jupiter and Saturn and of the 10^2 times increased masses of Uranus and Neptune (to get the solar composition); the angular momentum corresponding to this mass distribution is $K_n = 4 \cdot 10^{51} \, \mathrm{gcm^2 \, s^{-1}}$. HOYLE pointed out that the nebula with initial dimensions of the order of the distance to the nearest stars rotating with the angular velocity of the Galaxy would possess an angular momentum two orders of magnitude higher. But he suggested that, at the initial stage of the contraction, the solar nebula would be coupled by magnetic lines of force with the surrounding interstellar clouds and due to that would have lost about 99 % of the initial angular momentum. Opposite to HOYLE, taking a much lower value of the galactic magnetic field $(3 \cdot 10^{-6} \, \mathrm{G})$, CAMERON [2] concluded that the angular deceleration of the nebula at this stage would be not significant. In his model the nebula has get rid of the enormous initial angular momentum only after the collapse. The loss of angular momentum proceeded due to the mass loss. This is why the initial mass of the nebula in Cameron's model [3] was taken $M_n = 2 M_\odot$ and the initial angular momentum $K_n = 10^{54} \, \mathrm{gcm^2/s}$. SCHATZMAN [4] has suggested an effective mechanism of angular-momentum loss by the newly forming electromagnetically active Sun with the aid of the ejection of charged particles driven by magnetic-force lines. In his nebular model [5] a « moderate » initial mass of the nebula was adopted, $M_n \approx 1.1 M_\odot$. Our estimate of the mass of the protoplanetary cloud, based on the consideration of its evolution, is similar: $M_c \approx (0.05 \div 0.1) M_\odot$ [6].

Cameron's model was used widely by specialists in adjacent fields. Many numerical calculations of the collapse had been fulfilled for the nebulae with high angular momentum. Now Cameron's model is subject to considerable modifications, connected in part with the ideas on the initiation of collapse by a shock wave from a new-born supernova star, and the nebula is supposed already less massive. It should be mentioned that until now in the models of the massive nebulae one has not succeeded in explaining the mass and angular-momentum transfer. It was shown [7] that neither meridional circulation nor the Goldreich-Schubert instability do provide a necessary mass and momentum transfer in the solar nebula. Thus there are serious reasons for the transition from the concept of a massive nebula to that of a moderate-mass nebula.

A similar conclusion must be drawn concerning the initial angular momentum of the nebula. There are arguments against its estimation from the value of the momentum of the rarefied interstellar cloud, in which internal motions are reduced only to the regular rotation of the Galaxy. We should recall the following arguments:

1) It was emphasized by LARSON, REEVES [8] and other authors that the stars in clusters have no preponderant orientation of rotation axes parallel to the axis of the Galaxy. Therefore, the rotation of stars must have been governed by some factors more efficient than the galactic rotation.

2) The interstellar clouds with mass of the order of a solar mass cannot condense into more dense objects in ordinary conditions in the Galaxy. As was mentioned by REEVES [8] in his excellent review, this condensation must surpass three difficult barriers: a thermal one, a magnetic field and a barrier of angular momentum. Therefore, the original condensation takes place only in large cold aggregates of the interstellar medium with masses of $\sim (10^3 \div \div 10^4) M_\odot$. The way towards such an aggregate and through it to the last fragment, the solar nebula, is so long and complicated that any relation between the fragment's properties and the average conditions in the Galaxy vanishes. The existence at the same time of chaotic turbulent motions, of shock fronts and of twisted magnetic fields imparts a statistical character to the evolution of those aggregates: regions with very different values of density, angular momentum, intensity of magnetic field will arise and thus will have very different capability for contraction.

3) The variety of conditions results in the diversity of the forming objects—multiple systems, double stars, stars with planetary systems, single stars. Being interested first of all in the origin of our solar system, we must choose appropriate boundary conditions, which can lead to its formation. But in such a case we must reject the large angular momentum of the nebula, because in the process of collapse of this nebula a ring originates [9, 10], and it disintegrates into a double-star system.

For this reason we believe that the problem of the initial state of the solar nebula must be solved by starting not from the average data on the galactic rotation, but by taking into account more definite modelling of the latest stages of the origin of the Sun and planets, of the evolution of the preplanetary cloud. In this case it is suitable to use the principle of actualism suggested by ALFVÉN and ARRHENIUS as a main principle in cosmogony [11].

Sometimes a warning is expressed that the suggestion to consider the origin of the Sun from a slowly rotating nebula makes the theory implausible. Actually this suggestion transfers the problem of getting rid of the excess of angular momentum to a more early stage, preceding the formation of the solar nebula, and the judgement on its plausibility must be made on the basis of the study (mostly statistical) of the main processes at that stage.

Going back from the present planets and the Sun into the past, we get the following statement. The present angular momentum of the planets is $K_p = 3.1 \cdot 10^{50}$ gcm^2/s and that of the Sun is $K_s \sim 6 \cdot 10^{48}$ gcm^2/s. The angular momentum of the preplanetary cloud, having the present dimensions and the mass $0.1 M_\odot$, equals $K_c \approx 2.4 \cdot 10^{52}$ gcm^2/s. One can suppose that the effective braking of the rotation of the protosun (after SCHATZMAN) would begin at the moment of its entering the phase of complete mixing (the Hayashi stage) which evolved from $R \approx 50 R_\odot$ [12]. A protosun of such dimensions rotating with the Keplerian velocity would have an angular momentum $K_{ps} \approx 2 \cdot 10^{52}$ gcm^2/s. This momentum is partially removed by the solar wind and partially retained in the preplanetary cloud. Therefore, the initial total angular momentum of the solar nebula was less than the sum $K_c + K_{ps} \approx 4.4 \cdot 10^{52}$ gcm^2/s. As on the whole preceding stage of collapse of the nebula the angular-momentum transfer was not efficient, the above value may be considered as an upper limit for it, i.e. we can take

$$K_n \approx (2.4 \div 4) \cdot 10^{52} \text{ gcm}^2/\text{s} .$$

Now it is interesting to see what is the result of the collapse of such a nebula. The basic features of the collapse are discussed in other lectures. We are interested mainly in the conditions of formation of a ring, from which a double star originates.

In the early, isothermal stage of the collapse the nebula consists of a homogeneous core with dimensions decreasing with time and of an envelope with the density diminishing outwards. At a certain moment of time at the edge of the core a ring-shaped condensation arises. The greater is the initial amount of the angular momentum of the nebula, the earlier arises the ring. If the nebula rotates slowly, then the condensing nucleus becomes rather opaque earlier than a ring begins to appear. The extrapolation of the calculations of collapse by BLACK and BODENHEIMER [10], made for large initial ratios β_0 of the rotational to the potential energy, to the interval of small β_0 permits to evaluate approximately the values β_0 at which the ring does not form during the isothermal stage of collapse. For the radius of the nucleus at the moment of the ring formation taken to be proportional to β_0 we obtain[13]

$$
(1) \qquad \beta_0 \leqslant 4 \cdot 10^{-6} \left(\frac{M}{M_\odot} \right)^{-2/3} \varrho_0^{1/6} \varrho_{\mathrm{op}}^{-1/2} \; .
$$

For an initial density of the nebula $\varrho_0 \approx 10^{-18}$ and a density at the beginning of opacity $\varrho_{\mathrm{op}} \approx 10^{-12}$ the nucleus becomes opaque earlier than the ring originates, if $\beta_0 < 4 \cdot 10^{-3}$, *i.e.* for an angular momentum $K_{\mathrm{n}} < 10^{53}$ gcm^2/s. If we take the radius of the nucleus at the ring appearance proportional to $\beta_0^{\frac{3}{2}}$ (after LARSON and BLACK and BODENHEIMER), then we get correspondingly

$$
(2) \qquad \beta_0 \leqslant 7 \cdot 10^{-5} \left(\frac{M_\odot}{M} \right)^{4/9} \varrho_0^{1/9} \varrho_{\mathrm{op}}^{-1/3}
$$

and for the same conditions the ring does not appear in the isothermal stage of collapse, if $\beta_0 \leqslant 7 \cdot 10^{-3}$, that is at an angular momentum $K_{\mathrm{n}} < 1.3 \cdot 10^{53}$ gcm^2/s. The angular momentum of the solar nebula estimated by us is $3 \div 4$ times lower than the value necessary for the formation of a ring in the isothermal stage of the collapse.

The opaque core collapses adiabatically and the conditions for the formation of a ring in it are considerably more severe than in the isothermal nucleus. According to TAKAHARA *et al.* [14] in this case a rapid rotation and a low temperature are necessary. The local value of the parameter β should be greater than 0.2. Thus, after OSTRIKER [15], a rotating spheroid with a polytropic structure becomes dynamically unstable with respect to nonaxisymmetrical perturbations at $\beta > 0.26$ and with respect to axisymmetrical (ringlike) ones at $\beta > 0.46$. Because $\varrho \propto \beta^3$, in the central part of the nebula this instability would arise at $\varrho \gtrsim 10^{-6}$ g/cm^3. If there is a sufficiently effective transfer of angular momentum at that stage, this instability cannot arise. Such a transfer seems to be quite plausible due to the high temperature of the nebula at this stage, considerable ionization of gas and creation of strong magnetic fields.

The above considerations seem to support our favouring the model of the solar nebula with a moderate mass $((1.1 \div 1.2)\, M_\odot)$ and a comparatively low angular momentum $((2.4 \div 4) \cdot 10^{53}\,\mathrm{gcm^2/s})$.

2. – The gravitational instability in the preplanetary cloud.

The gravitational instability is one of the most frequent and most important instabilities in cosmogony. But the same idea is applied in different ways in the stellar and in the planetary cosmogony. In the stellar cosmogony one deals with the gravitational instability in a gas—that is with the formation of self-gravitating gaseous condensations. In the planetary cosmogony one had to reject such a picture, and now one considers the gravitational instability of a dust layer and its fragmentation into a great number of dust condensations. In general, the evolution may be described as follows. After the end of the collapse of the solar nebula and of its division into the protosun and the protoplanetary cloud, the latter shortly transforms into a laminar rotating gaseous cloud, because the chaotic turbulent motions in it have to decay rather rapidly, *i.e.* in a time interval of several revolutions around the Sun. The cloud after heating in the process of collapse cools and the condensation of solid particles begins in it. The particles which were present in the cloud before the collapse could be preserved only in the outer, more cold regions. In a quiescent gas the particles begin to precipitate toward the central plane of the cloud, at the same time continuing to grow up. The growth of small particles presents an additional problem, which is discussed in other lectures. The result of the process of « sedimentation » depends on the established size distribution of particles. If a comparatively small number of large particles were formed, then such particles, precipitating toward the plane $z = 0$ faster than the smaller ones, sweep out all these particles and for a time interval $(10^3 \div 10^4)$ years grow up to a size ~ 1 cm at $z \to 0$, forming a thin disk of solid material.

If, however, a large number of particles of average size forms, their settling velocities are almost the same, the particles grow more slowly and the dust precipitates also more slowly. In either case the settling of particles through the gas towards the equatorial plane results in the formation of a dust layer (disk) of higher density.

The further evolution of the layer depends on the velocities of the particles. In the gas with laminar rotation without any random (turbulent) motions the velocities of the particles are very small, the dust disk reaches the critical density and becomes gravitationally unstable. However, even small random velocities in the gas can prevent such a high flattening. At first we describe briefly the main results of the theory of gravitational instability in a gas. With some corrections they can be applied to the disk of solid particles.

The linearized theory of gravitational instability in an infinite homogeneous nonrotating medium was developed by JEANS in 1902. The dispersion relation for a sound wave propagating in such a medium is

(3) $$\omega^2 = k^2 c^2 - 4\pi G \varrho \, ,$$

where $k = 2\pi/\lambda$ is the wave number, c is the velocity of sound. The perturbation is assumed in the form exp $[i\omega t]$ and an instability takes place at $\omega^2 < 0$. Though an infinite medium with uniform density cannot be in equilibrium, Jeans' criterion is widely applied for various cases. For an infinite homogeneous medium uniformly rotating with an angular velocity Ω the instability criterion was obtained by CHANDRASEKHAR [16]. The equation for the waves propagating in a plane perpendicular to the axis of rotation is

$$(4) \qquad\qquad \omega^2 = 4\Omega^2 + k^2 c^2 - 4\pi G \varrho \,.$$

A more general solution when the rotation is nonuniform $\left(\Omega = \Omega(R)\right)$ was found by BEL and SCHATZMAN [17]. For axisymmetric perturbations

$$(5) \qquad\qquad \omega^2 = \varkappa^2 + q^2 c^2 - 4\pi G \varrho \,,$$

where $\varkappa$ is the epicyclic frequency, $\varkappa^2 = \mathrm{d}(\Omega R^2)^2/R^3 \, \mathrm{d}R$, $q \approx k$. The systems considered here have an infinite extension along the axis of rotation and the results are mainly of mathematical interest.

The protoplanetary cloud is a very « flat » system $(z \ll R)$. In 1960 we have shown [18] that the instability criterion for flat systems can be written in the form (5) with an additional factor $f(kH)$ in the last term of the right-hand side of the equation. Because the dispersion relation represents the balance of forces acting upon an element radially displaced due to the perturbation, the function $f(kH)$ was calculated directly from the gravitational field of a ring-shaped perturbation. In 1975 GENKIN and SAFRONOV [19] have found a satisfactory approximation for this function, $f(kH) \approx (1 + 2/kH)^{-1}$, and have written the dispersion relation in the form

$$(6) \qquad\qquad \omega^2 = \varkappa^2 + q^2 c^2 - 4\pi G \varrho_0 (1 + 2/kH)^{-1} \,,$$

where $H = \sigma/\varrho_0$ is the thickness of the homogeneous disk, ϱ_0 is the density in the central plane $z = 0$. When $H \to \infty$ the equation reduces to that of Bel and Schatzman (5).

The dispersion equation for an infinitely thin gravitating disk was obtained in the course of an investigation of the stability of the Galaxy [20, 21]:

$$(7) \qquad\qquad \omega^2 = \varkappa^2 + q^2 c_\mathrm{R}^2 - 2\pi G \sigma q \,,$$

where σ is the surface density of the disk, c_R is the thermal velocity of particles in the radial direction. This relation is obtained from our eq. (6) when taking $H = 0$ but conserving $c \neq 0$. There is some inconsistency in eq. (7). An infinitely thin disk is a cold disk with $c_z = 0$. At the same time it is assumed $c_\mathrm{R} \neq 0$. Physically it seems more preferable to take $c_\mathrm{R} \approx c_z$. This can be reached in eq. (6) by taking H depending on $c_z \approx c = c_\mathrm{R}$. On the other hand, eq. (7) is simpler and the difference between these equations is not large. In a thin disk of finite thickness the instability begins at a value of σ 40 % higher than in an infinitely thin disk.

In a protoplanetary cloud of small mass as compared to that of the Sun the rotation is almost the Keplerian one and

$$(8) \qquad \varkappa^2 \approx \Omega^2 = GM_\odot/R^3 = \tfrac{4}{3}\pi G\varrho^* \,,$$

where $\varrho^* = 3M_\odot/4\pi R^3$. The quantities H and c are connected by

$$(9) \qquad H = \left(\frac{2}{\pi G\varrho_0}\right)^{\frac{1}{2}} cJ \,,$$

where at the state approaching the instability $J \approx 0.9$ [6]. From eq. (6) at $\omega^2 = 0$ we find the critical density when the instability sets in. It has a minimum at a critical value of λ

$$(10) \qquad \varrho_{\mathrm{cr}} \approx 2\varrho^* \qquad\qquad \text{for } \lambda_{\mathrm{cr}} \approx 8H \,.$$

When $\varrho > \varrho_{\mathrm{cr}}$ the instability takes place for all values of λ which are in some interval containing the value λ_{cr}. This value of the critical density is substantially lower than the so-called Roche density $\varrho_{\mathrm{R}} \approx (10 \div 15)\varrho^*$, which is needed for the stability of local self-gravitating condensations.

Application of these results to the gaseous component of the protoplanetary cloud shows that for the gravitational instability a large mass of gas at low temperature is needed. If we assume that Jupiter or Saturn could form due to this instability, it would be inexplicable why only 1% of the gas in these zones entered the planets. There are also other objections which force to reject the idea of instability in the gas.

In the dust disk gravitational instability due to axisymmetric perturbations is possible. However, the disk in this case should be highly flattened. For the Earth zone it is needed $H_{\mathrm{p}}/R \sim 2 \cdot 10^{-6}$, i.e. thickness of the disk $H_{\mathrm{p}} \sim 0.05 R_\odot$ and velocities of particles $v_{\mathrm{cr}} \sim 10$ cm/s. In Jupiter's zone one needs $H_{\mathrm{p}}/R \sim$ $\sim 10^{-4}$ and $v_{\mathrm{cr}} \sim 270$ cm/s, while in the zone of Mercury it should be $H_{\mathrm{p}}/R \sim$ $\sim 4 \cdot 10^{-8}$, $H_{\mathrm{p}} \sim 7$ km and $v_{\mathrm{cr}} \sim 0.4$ cm/s. The closer to the Sun, the more unfavourable were the conditions for the instability. However, just the Sun is the source of various perturbations in the cloud. Therefore, we can conclude that gravitational instability in the dust disk did not occur in the region of the inner planets—Mercury, possibly Venus, etc.—and the particles instead grew by coalescing in collisions.

In the region of giant planets the conditions for instability were more suitable. Dust condensations which had been formed due to the instability had masses and equatorial radii

$$(11) \qquad m_0 \approx \sigma^3/\varrho^{*2} \,, \qquad a_0 \approx \sigma/2\varrho^* \,.$$

Due to rotation they could not contract directly to reach the density of solid bodies. Only after the coalescence and the increase in mass by a factor of 10^2 or 10^3 they transformed into normal bodies. The estimates give for the Earth zone $m_0 \approx 5 \cdot 10^{16}$ g, $a_0 \approx 4 \cdot 10^7$ cm and for the zone of Jupiter 10^{22} g and 10^{10} cm, respectively.

In this consideration it is assumed that the dust behaves similar to the gas. For mixture of gas and dust the situation is more complicate, though, as was shown by SPIEGEL [22], the results are essentially similar. The quiescent nonturbulent gas transfers perturbations to the dust. Then it slows down their development, but does not prevent their further increase. At the beginning of the instability the spatial density of the dust (ϱ_{cr}) is much higher than the gas density. Thus the influence of the gas on the instability criterion is not considerable. The most important thing is that the gas determines the relative velocities of the particles. In the quiescent gas they can be very small. In this case the criterion should be changed towards $c \approx 0$. This can lead to a decrease of the critical wavelength λ_{cr} [23] and can break the relation between c and H permitting the instability for larger H and lower ϱ. Then, for example, unstable modes with $\lambda \sim H$ and for $\varrho \sim 0.5 \varrho^*$ may be possible. However, one can seriously doubt that the gas could be so quiescent as to secure velocities of particles $c \approx 0$.

One more interesting question concerns the possibility of different rates of growth of iron and silicate particles, their different rates of settling to the central plane $z = 0$ and, accordingly, different degrees of their participation in the process of gravitational instability. One cannot exclude then the possibility of a fractionation of metals and silicates at this stage, which may be revealed in some form in the internal constitution of the planets.

REFERENCES

[1]   F. HOYLE: *Q. J. R. Astron. Soc.*, **1**, 28 (1960).
[2]   A. G. W. CAMERON: *Icarus*, **1**, 13 (1962).
[3]   A. G. W. CAMERON and M. R. PINE: *Icarus*, **18**, 377 (1973).
[4]   E. SCHATZMAN: *Ann. Astrophys. Ser. B*, **25**, 18 (1962).
[5]   E. SCHATZMAN: *Ann. Astrophys.*, **30**, 963 (1967).
[6]   V. S. SAFRONOV: *Evolution of the Protoplanetary Cloud and Formation of the Earth and the Planets* (Moscow, 1969); NASA TT F-677, Va. (1972).
[7]   T. V. RUZMAIKINA and V. S. SAFRONOV: *Astron. Ž.*, **53**, 860 (1976).
[8]   H. REEVES: in *On the Origin of the Solar System*, edited by H. REEVES (Paris, 1972), p. 28.
[9]   R. B. LARSON: in *On the Origin of the Solar System*, edited by H. REEVES (Paris, 1972), p. 142.
[10] D. C. BLACK and P. BODENHEIMER: *Astrophys. J.*, **206**, 138 (1976).
[11] G. ARRHENIUS: in *On the Origin of the Solar System*, edited by H. REEVES (Paris, 1972), p. 80.

[12] C. HAYASHI, R. HOSHI and D. SUGIMOTO: *Prog. Theor. Phys. Suppl.*, **22**, 1 (1962).
[13] V. S. SAFRONOV and T. V. RUZMAIKINA: in *Protostars and Planets*, edited by T. GEHRELS (Tucson, Ariz., 1978).
[14] M. TAKAHARA, K. NAKAZAWA, S. NARITA and CH. HAYASHI: *Prog. Theor. Phys.*, **58**, 536 (1977).
[15] J. P. OSTRIKER: in *On the Origin of the Solar System*, edited by H. REEVES (Paris, 1972), p. 154.
[16] S. CHANDRASEKHAR: *Vistas Astron.*, **1**, 344 (1955).
[17] N. BEL and E. SCHATZMAN: *Rev. Mod. Phys.*, **30**, 1015 (1958).
[18] V. S. SAFRONOV: *Dokl. Akad. Nauk SSSR*, **130**, 53 (1960); *Ann. Astrophys.*, **23**, 901 (1960).
[19] I. L. GENKIN and V. S. SAFRONOV: *Astron. Ž.*, **52**, 306 (1975).
[20] A. TOOMRE: *Astrophys. J.*, **139**, 1217 (1964).
[21] P. GOLDREICH and D. LYNDEN-BELL: *Mon. Not. R. Astron. Soc.*, **130**, 125 (1965).
[22] E. A. SPIEGEL: in *On the Origin of the Solar System*, edited by H. REEVES (Paris, 1972), p. 165.
[23] P. GOLDREICH and W. R. WARD: *Astrophys. J.*, **183**, 1051 (1973).

The Smaller Bodies of the Solar System and Their Role as Meteorite Sources.

G. W. WETHERILL

Carnegie Institution of Washington, Department of Terrestrial Magnetism
5241 Broad Branch Road, N. W. Washington, D. C. 20015

1. – Introduction.

Most descriptions of the solar system quite properly emphasize the larger bodies: the Sun, planets and to some extent the satellite systems. However, from some points of view, including that of this course, the smaller bodies which traverse the regions between the planets are of comparable importance. The comets, asteroids and Earth-approaching Apollo-Amor objects are our only surviving relics of the planetesimals which accreted to form the planets. The record of the impacts of these smaller bodies with the planets is provided by the cratered planetary surfaces, and is a principal tool in correlating the geological evolution of these planets. Most of our detailed knowledge concerning conditions in the formative solar system is obtained from petrographic, chemical and isotopic studies of meteorites, which are fragments of some, and possibly all, classes of these smaller bodies.

In order to place these smaller bodies in context, their masses and general chemical composition are compared with those of the Sun, planets and satellites in table I. Most of these mass estimates can be considered to be reliable, with the exception of the comets. The value given, $\sim 10^{27}$ g, is based on the estimate that there are $\sim 10^{11}$ comets in the Oort cloud of comets, and that their average mass is $\sim 10^{16}$ g. These estimates are based on the number believed necessary to provide the observed flux of comets, the inferred capability of the primordial outer solar system as a source of the cometary cloud, the quantity of volatile material and dust emitted by comets near perihelion and the brightness of comet nuclei at great heliocentric distances. However, all of these quantities are uncertain, and the total estimate may well err in either direction by several orders of magnitude.

TABLE I. – *Large and small bodies of the solar system.*

Type of object	Total mass in class (g)	Chemical-composition class
Sun (*G* star)	$2 \cdot 10^{33}$	H
Jupiter, Saturn	$2 \cdot 10^{30}$	H
Uranus, Neptune	$2 \cdot 20^{29}$	C
Terrestrial planets	$1 \cdot 10^{28}$	S
Satellites	$6 \cdot 10^{26}$	C, S
Interplanetary, «small» bodies:		
Pluto, Chiron	$\sim 1 \cdot 10^{25}$	?
Comets	$\sim 10^{27}$	C
Asteroids	$3 \cdot 10^{24}$	C, S
Apollo-Amor objects	$\sim 10^{19}$	S, C
Earth-crossing meteoroids	$\sim 10^{18}$	C, S
Chemical composition classes:		
Abundance ($Si \equiv 10^6$) $4 \cdot 10^{10}$	$4 \cdot 10^7$	$4 \cdot 10^6$
H H, He	CNO	Mg, Si, Fe, S
C	CNO	Mg, Si, Fe, S
S		Mg, Si, Fe, (S)

2. – Comets.

Comets are small bodies (~ 1 to 10 km diameter) consisting of a mixture of volatile compounds of H, C, N and O together with dust and larger fragments probably consisting of Mg and Fe silicates. Almost all comets are confined to the outermost regions of the solar system, their typical semi-major axis being $(10^4 \div 10^5)$ AU and their perihelion normally well beyond the orbit of Neptune. This relatively stable storage region is called the «Oort cloud», following Oort's inference [1] of its existence. At these great heliocentric distances comets are only loosely bound to the Sun and perturbations by passing stars can produce large perturbations in their orbits. As a consequence, their perihelia are occasionally perturbed into the inner solar system (about 2 per year). Within ~ 3 AU solar radiation heats the cometary surfaces producing the coma $((10^4 \div 10^5)$ km diameter) and the tail $((10^6 \div 10^8)$ km length), which give observed comets their often spectacular appearance. It should be remembered, however, that comets «look like comets» only for $\sim 10^{-9}$ of their

history. Most of the time they are more inconspicuous than the smallest observed asteroids. The physical characteristics of comets have been reviewed
by WHIPPLE [2] and DELSEMME [3] and their orbits and dynamics by
MARSDEN [4].

A comet which enters the inner solar system for the first time from the Oort
cloud is termed a « new comet », even though all comets are probably actually
as old as the solar system, $4.5 \cdot 10^9$ years. Most new comets survive for only
about 5 orbital periods before they are perturbed into a hyperbolic escape
orbit or undergo disruption [5]. A small fraction, a few per century, are perturbed, principally by Jupiter, into short-period orbits, usually of about 7 years.
About 100 such comets are known at the present time. These orbits are much
more dynamically stable, and the comet will often remain in the inner solar
system long enough ($\sim 10^3$ years) to become severely depleted in its volatile
content, and hence relatively inactive.

A much smaller number of comets, perhaps only one in 10^4 years, evolves
into an orbit with aphelion less than 4.3 AU, relatively secure from the effects
of strong perturbations by Jupiter. Only one such comet is known at present
(Encke). However, there are a number of meteor streams with small aphelia
containing meteors with physical characteristics similar to the Taurids, associated with comet Encke. Therefore, it is unlikely that comets like Encke
are extremely unusual. Encke's orbit is much more stable than those of the
short-period comets which can encounter Jupiter, and its aphelion can be
expected to remain within the orbit of Jupiter as long as $(10^7 \div 10^8)$ years.
Solar radiation will exhaust the comet's volatile inventory on a time scale of
$\sim (10^3 \div 10^4)$ years, as a gradual decline in Encke's brightness by about 2
magnitudes during the past two centuries has been observed. SEKANINA [6]
has modelled the orbital evolution and activity of Encke and concludes that
it has nearly reached the stage of being an extinct comet nucleus of ~ 1 km
radius, probably consisting principally of nonvolatile silicates. As will be
discussed later, from an observational point of view, Encke will then be an
Apollo object. If the estimate of one « Encke » per 10^4 years is correct, this
will suffice to produce a steady-state assemblage of $(10^3 \div 10^4)$ Apollos, somewhat more than the number actually believed to exist at present [7].

Active comets are known to be sources of meteoroids ranging from microscopic (10^{-6} g) radar meteors up to magnitude -10 fireballs with masses in
the kilogram range. Shower meteoroids with similar physical characteristics
but lacking a definite cometary association have estimated masses as large
as 100 tons. It is not known if any of these larger fragments have sufficient
mechanical strength to survive passage through the atmosphere and be re-
covered as meteorites. Most of the ~ 10 µm micrometeoric material collected
in the stratosphere by BROWNLEE *et al.* [8] is probably of cometary origin.
Electron microprobe measurements of this material show it to be very similar
in composition to carbonaceous chondrites and to the nonvolatile component

of average solar matter, in agreement with customary thinking regarding the composition of comets. If any meteorites are actually derived from comets, it is likely that the comet has evolved into an extinct cometary Apollo object, because cosmic-ray exposure ages of meteorites are considerably larger than the $10^3 \div 10^4$ time scale for the duration of cometary activity.

It has been speculated [9] that comets were an important source of volatile matter (inert gases, H_2O, CO_2) on the Earth, during the ejection of the cometary cloud from the Uranus-Neptune region early in solar-system history [10]. This would be expected if the cometary Oort cloud were formed by the ejection of material from the outer solar system in the manner described by OORT [1] and ÖPIK [11]. A corollary of this is that a similar quantity of matter would traverse the inner solar system, some of which would impact the terrestrial planets. A quantitative estimate of the magnitude of this effect is difficult, because of our lack of detailed knowledge concerning the formation of the outer solar system and the comets.

3. – Asteroids.

Asteroids are small (up to 1000 km diameter) bodies confined to orbits with semi-major axes between those of Mars and Jupiter, and usually with $2.2 < a < 3.3$ AU. Unlike the comets, they emit no volatile coma or tail. Spectro-

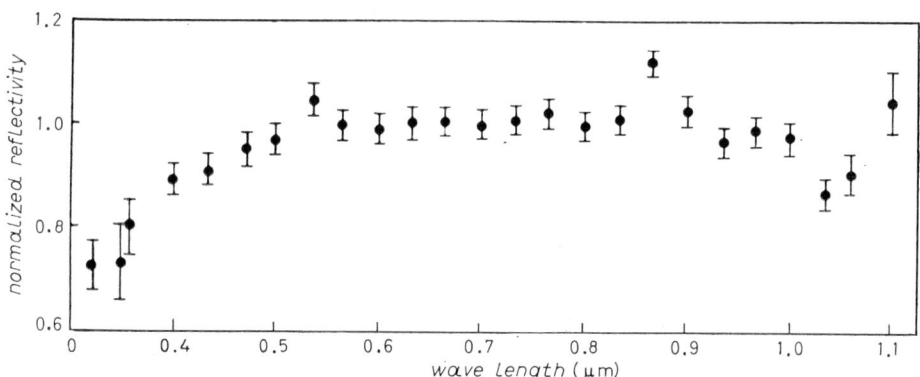

Fig. 1. – Spectral reflectivity of C asteroid 10 Hygiea. Data from ref. [12].

photometric studies of their surfaces shows that their composition usually falls into one of two classes:

1) The C asteroids with low geometric albedos (~ 0.04) and with flat and relatively featureless reflectance spectra in the region between 0.3 and 1.1 μm (fig. 1).

2) The S asteroids with higher (~ 0.15) geometric albedos, higher reflectivity in the red and near IR than in the blue, and a pronounced absorption band at 0.95 μm (fig. 2).

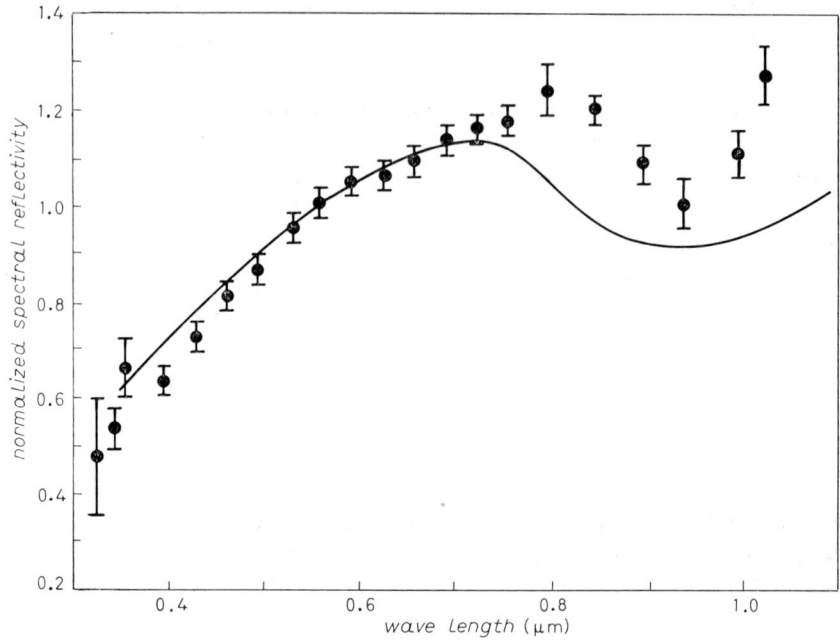

Fig. 2. – Spectral reflectivity of S asteroid 8 Flora. Data from ref. [12].

It is generally thought that the C asteroids are similar in composition to carbonaceous chondrites, whereas the S-asteroid spectra suggest a mineral assemblage containing pyroxene (responsible for the 0.95 μm absorption) and metallic iron. Many asteroid observers, *e.g.* [13], feel the S-asteroid spectra are incompatible with those of ordinary chondrites, and indicate a more iron-rich composition deficient in olivine. This interpretation has been questioned [14] on the grounds that a chondritic spectrum may be altered by asteroidal surface processes (*e.g.* solar-wind and micrometeorite bombardment) to resemble those observed. It is probably best to regard this question unresolved at present. It is obviously central to the question of whether or not ordinary chondrites are derived from the asteroid belt. If the S asteroids are not ordinary chondritic, it appears unlikely that an adequate asteroidal source for this most abundant class of meteorite exists.

The size distribution of asteroids is such that most of the $\sim 3 \cdot 10^{24}$ g mass of the asteroid belt is in the larger bodies. Over the observed mass range of $\sim 10^{16}$ g to 10^{24} g, asteroid masses follow fairly well a power law $\mathrm{d}N \propto m^{-1.8}\,\mathrm{d}m$, although the size distribution cannot be exactly fitted to the same power law over its entire range. Studies of asteroid collision rates [15, 16] show that, at least below 100 km diameter, asteroid sizes probably represent a fragmentation steady state, and that the smaller asteroids are fragments of the larger ones.

Asteroid orbits are usually more eccentric (typically ~ 0.1) and inclined (typically $\sim 10°$) than those of most planets. This results in a mean relative velocity between asteroids of ~ 5 km/s. The origin of this high relative velocity is poorly understood, as both their mutual perturbations and perturbations by the planets are at present inadequate to explain these high velocities.

The orbits of asteroids are not uniformly distributed in semi-major axis, eccentricity, or inclination. This nonuniformity is a consequence of resonant

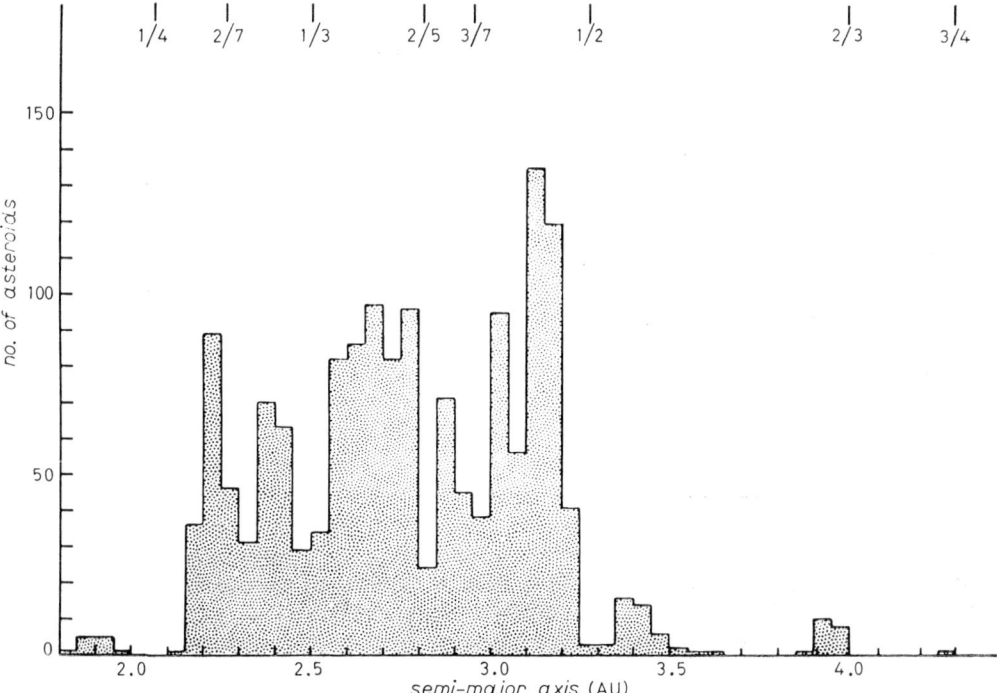

Fig. 3. – Number of asteroids as a function of semi-major axis. The values of a corresponding to bodies having orbital periods with simple integer ratios to the period of Jupiter are depleted in number (Kirkwood gaps).

gravitational perturbations, principally due to Jupiter. The Kirkwood gaps (fig. 3) represent a depletion of asteroids with periods in ratios of $\frac{1}{2}$, $\frac{3}{7}$, $\frac{2}{5}$, $\frac{1}{3}$ and $\frac{2}{7}$ with that of Jupiter. Theoretical investigations of these orbits show them to be stable, and this conclusion is supported by the fact that the only asteroids with $a \gtrsim 3.8$ AU are those corresponding to the $\frac{2}{3}$ (Hilda), $\frac{3}{4}$ (Thule) and $\frac{1}{1}$ (Trojan) resonances. The coincidence of the Kirkwood gaps with the commensurable orbits cannot be an accident, but, to date, attempts to explain the depletion at these positions are either inadequate or require improbable assumptions regarding conditions in the early solar system.

Asteroids are also absent in the vicinity of the secular resonant surfaces discovered by WILLIAMS [17] (fig. 4). These represent resonances between the rates of precession of the longitudes of perihelion or node of the asteroid and the characteristic frequencies of the planetary system [18].

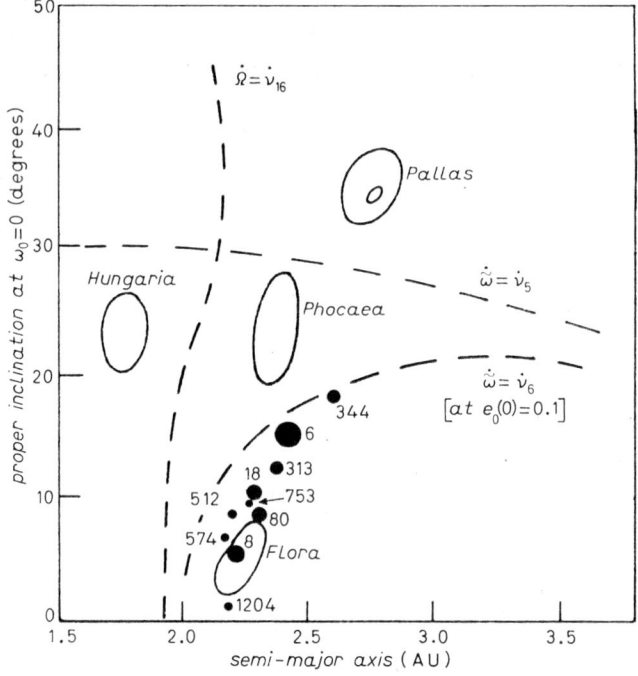

Fig. 4. – Distribution of asteroids in the vicinity of the ν_6 resonance, as calculated by WILLIAMS (ref. [17]).

The asteroid belt is commonly regarded as the source region of meteorites of all classes. If this is true, there must exist gravitational mechanisms adequate to gently transfer asteroidal fragments from the asteroid belt into Earth-crossing orbits. The velocity change of ~ 5 km/s required for direct transfer of collision fragments into Earth-crossing is almost certainly incompatible with the observed relatively unshocked condition of most meteorites and with a reasonable partition between kinetic and internal energy of the collision fragments. Several such mechanisms have been studied and shown to be at least semi-quantitatively of importance. These involve a combination of statistically probable collisions, and $\frac{1}{2}$ Kirkwood gap and Jupiter encounter perturbations [19], direct injection into the ν_6 secular resonance [20], and non-linear combination of ν_6 and Mars perturbations [21, 22]. Work of Scholl and

Froeschlé [23] suggests that a mechanism analogous to that of ref. [17] involving the $\frac{2}{5}$ Kirkwood gap many be more effective than that associated with the $\frac{1}{2}$ commensurability.

The process whereby the interaction of ν_6 and Mars perturbations [21, 22] brings material from S asteroids near the inner edge of the asteroid belt into Earth-crossing on a time scale of $\sim 5 \cdot 10^8$ years appears to provide an adequate supply of iron meteorites. Also, the calculated time scale corresponds to their observed cosmic-ray exposure ages. Of course, it is not known if these asteroids actually contain iron fragments of the size of iron meteorites or in the variety required by the number of classes of these meteorites found [24]. This same mechanism appears capable of providing $\sim 10^7$ g/y of stone meteorites and can, therefore, be considered a potential source of differentiated achondritic meteorites. This yield appears to be too low to represent a major source of chondrites, but this possibility cannot be entirely excluded.

The other resonant mechanisms can place asteroidal fragments into Earth-crossing orbit on the necessary short ($\sim 10^6$ year) time scale required for sources of chondrites. It has proven difficult to calculate very accurately the yield expected for these mechanisms and they may well be insufficient. As noted earlier, there still exists the problem of finding an adequate number of asteroids with ordinary chondritic composition, as indicated by spectrophotometric investigations.

These same mechanisms which can transfer $(100 \div 10^6)$ g meteoritic fragments into Earth-crossing orbits are also qualitatively capable of providing $\sim 10^{16}$ g asteroidal fragments as Apollo objects. More quantitative study of this problem [25, 26] suggests a deficiency by a factor of 10 to 100 if one tries to explain the origin of the majority of the observed Apollo objects by these asteroidal mechanisms. This is in contrast to cometary sources which appear to provide an « excess of riches ». In any case, it is very likely that *some* of the observed Apollo objects are asteroidal fragments.

4. – Apollo-Amor objects.

Apollo objects are small (typically ~ 1 km radius) bodies with perihelion inside 1 AU, *i.e.* they cross the orbit of the Earth. Amor objects are similar bodies with perihelia between 1.0 and 1.3 AU. Dynamical studies of the orbits of these bodies [27] using the methods of Arnold [28] show that they represent a single steady-state population and should be discussed together. These bodies are frequently called « asteroids » because they do not display the coma and tail characteristic of comets. However, as discussed in previous sections, it is likely that many, if not most, of these bodies are extinct cometary cores. Therefore, designating them as « objects » avoids a possibly incorrect genetic identification.

About 28 Apollo and 16 Amor objects are known at present. The total number of such bodies greater than 1 km in diameter has been estimated to be ~ 1000 [7, 26].

In spite of their small size and number, these bodies are of major importance in solar-system studies. Quantitative calculations [7, 27, 29] show that impact of these objects is the principal cause of the large ($\geqslant 1$ km diameter) craters on the Moon, Earth and the other terrestrial planets, with the possible exception of Mars. These bodies will be fragmented as they traverse the portion of their orbits which lies in the asteroid belt. The fragments will continue to move in similar Earth-crossing orbits and will collide with the Earth as meteoroids and presumably as meteorites. Calculation of the expected terrestrial flux of meteorites from Apollo objects shows it to be compatible with the terrestrial flux of chondrites ($\sim 10^8$ g/y) [26], and, therefore, these objects are strong *prima facie* meteorite sources and require no special mechanism to transport their fragments into Earth-crossing orbit. Some of these objects have been suggested as relatively accessible sources of raw materials for construction in space [30].

Unlike the Oort-cloud comets and the main-belt asteroids, the orbits of these objects are not stable on the time scale of the solar system. As first shown by ÖPIK [31], and extended by subsequent work [7, 32, 33], these objects will be removed in 10^7 to 10^8 years as a result of planetary impacts and perturbation into hyperbolic solar-system escape orbits. In order for their presence to be maintained, it is necessary that one or more sources exist to replace those objects removed from the population. Quantitative studies [27] show that an injection rate of Apollo-Amor bodies (larger than 1 km diameter) of $\sim 15/10^6$ y is required to maintain the observed number in a steady state. Observational evidence that such a steady state is in fact maintained is provided by the lunar and terrestrial cratering record. The younger lunar maria have recorded the integrated flux of crater-forming Apollo objects since the end of major mare volcanism about $3.2 \cdot 10^9$ y ago. Various terrestrial surfaces (*e.g.* portions of the Canadian Shield and the Eastern U.S.A.) provide a similar record for the last $600 \cdot 10^6$ y [27, 29, 34, 35]. Within about a factor of two, the average impact rate over these very different time periods is the same. If anything, the recent terrestrial crater frequency suggests a possible *increase* in the recent impact rate, rather than the decay of a primordial population of crater-forming Apollo objects.

Since at least some meteorites are likely to be fragments of Apollo-Amor objects, it is of importance to know what type of meteorite is most plausibly identified with these objects. Multiband spectrophotometry has been completed for only two of these bodies, 1685 Toro and 433 Eros. More limited photometry, radiometric measurements, polarization and radar observations have been carried out for a few others. These data show that with 3 exceptions (*e.g.* Betulia) their surfaces resemble S asteroids much more closely than C asteroids. It must be remembered, however, that these statistics are biased against C asteroids because they have lower albedos and hence are fainter than a S asteroid of the same radius. There is also some suggestion that they are more similar to ordinary chondrites than typical main-belt S asteroids.

Because these objects are in a steady state, a continuing source of supply is required. In the previous sections it was pointed out that it is most likely that both comets and asteroids contribute to the Apollo-Amor population, and that the contribution from comets appears to be the larger. This suggests the syllogism:

1) Apollos come from comets,

2) ordinary chondrites come from Apollos,

3) therefore ordinary chondrites come from comets.

Many workers will agree with 1) or 2), or even both, but few will accept 3). This is in part due to a strong traditional association of meteorites with asteroids and in part for more valid reasons. Chondritic meteorites have had a complex chemical history, involving loss of volatile and moderately volatile elements and high-temperature (~ 1000 K) metamorphism. Such a history is not generally associated with the low-temperature icy material known to exist in comets. On the other hand, the discovery of ^{26}Mg from the decay of extinct ^{26}Al in at least some meteorites [36] introduces the possibility of heating and melting of even small parent bodies at any heliocentric distance. Following such heating and metamorphism, a mantle of icy material appropriate to the distance may be deposited upon the core material, or mixed with it to form an icy breccia.

Probably the greatest difficulty for a cometary origin for chondrites is the evidence that gas-rich chondrites have been surface regolith [14, 37]. In this regard it is crucial to understand whether these are relatively recent breccias or if they accumulated early in solar-system history. Evidence for both of these conclusions exists at present, but does not yet appear to be conclusive in nature.

There is a possibility that orbital data on Apollo objects and meteorites may contribute to our understanding of the sources of these bodies. This can be obtained from telescopic discovery of Apollos, radiant and time-of-fall data on meteorites [38], identification of photographic fireballs with meteorites on the basis of their physical characteristics [39, 40] and inference of meteorite velocities from ablation studies using charged-particle tracks [41]. However, this will not be a simple matter, as at least to a first approximation the distribution of Apollo and meteorite orbits appears to be independent of their original source [27, 42].

In conclusion, we have learned much in recent years concerning small bodies in the solar system and their interrelationships. Nevertheless, there are still large gaps in our understanding, which preclude a clear understanding of these relationships. Because of the relevance of these bodies to the formation and early history of the solar system, efforts directed toward closing these gaps will represent an important contribution to many areas of science.

REFERENCES

[1]   J. H. OORT: *Bull. Astron. Inst. Neth.*, **11**, 91 (1950).
[2]   F. L. WHIPPLE: *Comets, Asteroids, Meteorites*, edited by A. H. DELSEMME (Toledo, O., 1977), p. 25.
[3]   A. H. DELSEMME: *Comets, Asteroids, Meteorites*, edited by A. H. DELSEMME (Toledo, O., 1977), p. 453.
[4]   B. G. MARSDEN: *Annu. Rev. Astron. Astrophys.*, **12**, 1 (1974).
[5]   P. R. WEISSMAN: unpublished Ph. D. dissertation, University of California (Los Angeles, Cal., 1978).
[6]   Z. SEKANINA: *Astron. J.*, **74**, 1223 (1969).
[7]   E. M. SHOEMAKER: *Impact and Explosion Cratering*, edited by D. J. RODDY, R. O. PEPIN and R. B. MERRILL (New York, N. Y., 1977), p. 617.
[8]   D. E. BROWNLEE, R. S. RAJAN and D. A. TOMANDL: *Comets, Asteroids, Meteorites*, edited by A. H. DELSEMME (Toledo, O., 1977), p. 137.
[9]   G. T. SILL and L. L. WILKENING: *Icarus*, **33**, 13 (1978).
[10] G. W. WETHERILL: *Proceedings of the Soviet-American Conference on Cosmochemistry of the Moon and Planets* (Moscow, 1974), p. 411.
[11] E. J. ÖPIK: *Armagh Observ. Contr.*, **53J** (1965).
[12] M. J. GAFFEY and T. B. MC CORD: *Space Sci. Rev.*, **21**, 555 (1978).
[13] C. R. CHAPMAN: *Comets, Asteroids, Meteorites*, edited by A. H. DELSEMME (Toledo, O., 1977), p. 137.
[14] E. ANDERS: *Asteroids, an Exploration Assessment*, edited by D. MORRISON and W. C. WELLS (Washington, D. C., 1978), p. 57.
[15] G. W. WETHERILL: *J. Geophys. Res.*, **72**, 2429 (1967).
[16] J. W. DOHNANYI: *J. Geophys. Res.*, **74**, 2531 (1969).
[17] J. G. WILLIAMS: unpublished Ph. D. dissertation, University of California (Los Angeles, Cal., 1969).
[18] D. BROUWER and A. J. J. VAN WOERKOM: *Astron. Papers U.S. Naval Obs. Naut. Almanac Office*, **13**, part II, 85 (1950).
[19] P. D. ZIMMERMAN and G. W. WETHERILL: *Science*, **182**, 51 (1973).
[20] J. G. WILLIAMS: *Eos*, **54**, 233 (1973).
[21] G. W. WETHERILL: *Fragmentation of Asteroids and Delivery of Fragments to Earth*, edited by A. H. DELSEMME (Lyon, 1977), p. 283.
[22] G. W. WETHERILL and J. G. WILLIAMS: *Origin and Distribution of the Elements, Second Symposium*, edited by L. H. AHRENS (Oxford, 1979), p. 19.
[23] H. SCHOLL and C. FROESCHLÉ: *Comets, Asteroids, Meteorites*, edited by A. H. DELSEMME (Toledo, O., 1977), p. 587.
[24] J. T. WASSON: *Meteorites* (New York, N. Y., 1974).
[25] E. J. ÖPIK: *Adv. Astron. Astrophys.*, **2**, 219 (1963).
[26] G. W. WETHERILL: *Geochim. Cosmochim. Acta*, **40**, 1297 (1976).
[27] G. W. WETHERILL: *Icarus*, **37**, 96 (1979).

[28] J. R. ARNOLD: *Astrophys. J.*, **141**, 1536 (1965).

[29] R. A. F. GRIEVE and M. R. DENCE: *Icarus*, **38**, 230 (1979).

[30] B. O'LEARY: *Science*, **197**, 363 (1977).

[31] E. J. ÖPIK: *Proc. R. Ir. Acad. Sect. A*, **54**, 165 (1951).

[32] E. ANDERS and J. R. ARNOLD: *Science*, **149**, 1494 (1965).

[33] G. W. WETHERILL and J. G. WILLIAMS: *J. Geophys. Res.*, **73**, 635 (1968).

[34] E. M. SHOEMAKER, R. J. HACKMAN and R. E. EGGLETON: *Adv. Astronaut. Sci.* **8**, 70 (1963).

[35] G. NEUKUM, G. KÖNIG, H. FECHTIG and D. STÖRZER: *Proceedings of the VI Lunar Science Conference* (New York, N. Y., 1975), p. 2597.

[36] T. LEE, D. A. PAPANASTASSIOU and G. J. WASSERBURG: *Astrophys. J.*, **211**, 107 (1977).

[37] R. S. RAJAN: *Geochim. Cosmochim. Acta*, **38**, 777 (1974).

[38] G. W. WETHERILL: *Science*, **159**, 79 (1968).

[39] Z. CEPLECHA and R. E. MC CROSKY: *J. Geophys. Res.*, **81**, 6257 (1976).

[40] D. O. REVELLE and G. W. WETHERILL: *Meteoritics*, **13**, 611 (1978).

[41] R. S. RAJAN, D. O. REVELLE and G. W. WETHERILL: *Metheoritics*, **43**, 604 (1978).

[42] G. W. WETHERILL: *Asteroids, an Exploration Assessment*, edited by D. MORRISON and W. C. WELLS (Washington, D. C., 1978).

Evolution of Small Bodies in the Inner Solar System.

G. W. Wetherill

Carnegie Institution of Washington, Department of Terrestrial Magnetism
5241 Broad Branch Road, N.W., Washington, D. C. 20015

1. – Introduction.

For various practical reasons it is necessary to reduce solar-system investigations into component parts, both in space and time. Thus we speak of lunar science, meteor astronomy, the study of Jupiter, and meteoritics, as if they were separate sciences. Similarly, the accumulation of the planets, their early heavy bombardment history and the present interplanetary flux of cratering projectiles are usually discussed separately.

In doing so it is easy to lose sight of the fact that the solar system is a system, and our understanding of individual bodies and events must be fitted together in a harmonious way. The present paper represents a limited attempt to accomplish a small part of this task. I will try to trace a plausible course of events for the history of the residual bodies which remained in the inner solar system at the time when the accumulation of the Earth and Venus was nearly complete. This discussion is not intended to be in any way a definitive treatment of this subject. Rather it is meant as an example of the possibilities for integrated treatments of solar-system processes in time and space.

At the time the inner planets were 98 % grown, there still remained in their vicinity at least $2 \cdot 10^{26}$ g of unaccumulated material. Although this represents but a small part of the mass of the terrestrial planets, it represents a mass about 100 times that of the present asteroid belt, confined to a smaller volume of space.

The general framework of this discussion of the subsequent evolution of this material derives from those theories in which planetary growth is accomplished by the sweeping up of smaller bodies (*e.g.* [1-7]). This was accomplished about 100 million years after the beginning of planetesimal growth in the solar system. If the time scale of the solar system is fixed to the uranium decay constants of Jaffey *et al.* [8], the age of the earliest planetesimals is $4.55 \cdot 10^9$ y. (See discussion by Wetherill [9].) Under these assumptions, the growth of the Earth was nearly complete $4.45 \cdot 10^9$ y ago, which is, therefore, chosen as the starting point of the present discussion. The size distribution adopted is that found earlier [10], in which the largest nonsurviving body formed in the Earth-Moon-Venus region was $\sim 10^{26}$ g, *i.e.* greater than the mass of the Moon.

This work also showed that the r.m.s. velocities of the residual swarm were $(6 \div 10)$ km/s. This same result has also been calculated in an independent manner using a stochastic simulation model for the growth of the planets [11]. These sizes and velocities are significantly greater than those found by SA-FRONOV [12], because the more recent work included the effects of the planetes-imals crossing the orbits of both Earth and Venus.

By the time the growth of the Earth was nearly complete, most of the bodies larger than $\sim 10^{24}$ g in mass which had not impacted a planetary embryo had probably been disrupted by approaches within the Roche limit of a planet [10, 13]. Approaches within 3 planetary radii are eight times as probable as planetary impact, which must have occurred in order for the planets to grow. Therefore, approaches within the Roche limit were not im-probable or *ad hoc* occurrences as might be thought, but were an inevitable consequence of the same dynamical processes which caused planetary growth and cratering, and for a given planetesimal may easily be calculated to have occurred every few million years during the latter $\frac{2}{3}$ of the Earth's growth. The uncertainty requiring the statement that the larger bodies had only « probably » been disrupted arises from the uncertainty that disruption will actually occur during the ~ 1 hour required for the body to traverse the region in which it is unstable with respect to tidal disruption. If this disruption actually can occur, and on a time scale comparable to the characteristic time for collision with a planetary embryo ($\sim 10^7$ years), then the residual swarm was likely to have contained many bodies $\sim 10^{23}$ g in mass resulting from disruption of larger bodies, and a collisional steady-state size distribution of smaller bodies.

The actual problem of the evolution of this residual swarm is too complex to calculate with any degree of precision. The collisional steady-state size distribution will deviate from the theoretical value found by DOHNANYI [14] for the present asteroid belt as a consequence of even a somewhat enhanced solar wind being unable to clear the region of the finest comminutional debris [15]. Rather than being swept away, this material will be accumulated by the planets and Moon and any residual bodies large enough to be in an accretional state. The population of this fine material will probably build up. The increased collisional dissipation associated with the large cross-section of these small bodies may be expected to reduce their mean velocity to a value where the enhanced gravitational cross-section of the planets causes the rate of their planetary capture to come to equilibrium with their rate of production. Thus the velocity distribution of the swarm will be size dependent. At the high-mass end, the size spectrum will be sensitive to the uncertainties in the prob-ability of Roche-limit destruction following an encounter with a planet at a given distance, and to the uncertainty in the size distribution of the fragments produced by a Roche-limit breakup. Electromagnetic effects and gas drag resulting from collisional volatilization of the planetesimals could be important.

In complex problems such as this, the approach can be taken of investigating idealized problems which are more tractable. This commonly permits conceptual insight into the more difficult problems of the real solar system. Given the quantitative results of the idealized problem, it is frequently possible to semi-quantitatively consider the consequences of various deviations from ideality. An assemblage of solutions of these idealized problems can also be used as cornerstones in formulating one's thoughts concerning approaches to more difficult calculations. This is analogous to use of an idealized lunar « magma ocean » by lunar petrologists in considering the earliest chemical evolution of the Moon, rather than a probably more realistic model including heterogeneity resulting from the size and velocity dispersion of the accumulating planetesimals [12, 16]. In both cases the possibly questionable merit of the model arises from its being amenable to quantitative discussion and solution, rather than from arguments based on Ockham's razor, or on similar dull tools.

2. – The idealized problem.

The simplified problem identified here is that of calculating the orbital evolution of bodies with initial geocentric velocities of 8 km/s subject to gravitational perturbations by the terrestrial planets with their present masses and orbits. Although the evolution of the osculating elliptical orbit of a planet-crossing body can be of a deterministic nature on a short time scale of $\sim 10^3$ years, on longer time scales of $> 10^5$ years it is known that very different perturbed orbits can result from small differences in the geometry of close planetary encounters [17]. This effect precludes a deterministic approach to long-term orbital evolution, and a probabilistic approach is required. An analytical method of this kind has been presented by ÖPIK [17, 18], but does not provide the information required for this investigation because of complexities introduced by the nonlinear effects of multiple planet-crossing. Instead, the Monte Carlo iteration of Öpik's basic equation described by ARNOLD [19] is used. Some minor differences between the program used by ARNOLD and that used in this work have been discussed [20, 21]. In addition, a fairly major modification results from the use of a procedure for including the effects of large secular amplitudes in eccentricity in the vicinity of the secular resonance ν_6 as discussed elsewhere [22]. Monte Carlo studies of the evolution of orbits similar to those considered here have been presented previously (e.g. [23, 24]). These earlier calculations showed that material in an Earth-crossing orbit is removed by planetary impact or solar-system escape with a typical « half-life » of $\sim 2 \cdot 10^7$ years. A longer-lived « tail » to this orbital evolution was also found, but the number of Monte Carlo runs was insufficient to permit statistically significant discussions of this effect. This has been accomplished herein by a combination of brute force (700 runs), increasing the information yield from each run by properly using the information content of close encounters as well as impacts [13]. In addition, it was shown that all of the Earth impacts which occurred after $750 \cdot 10^6$ y had evolved through a stage in which their perihelion was beyond 1.5 AU. This fact permits use of these evolved orbits as starting orbits, and thereby improves the statistical significance for the low impact rates at these longer times. Thirty runs were made for each of 31 of these Mars-crossing orbits generated by the initial 700 runs. This calculation extends the orbital evolution to times when the initial impact rate has decayed to less than 10^{-4} of its initial value (fig. 1). It was also found

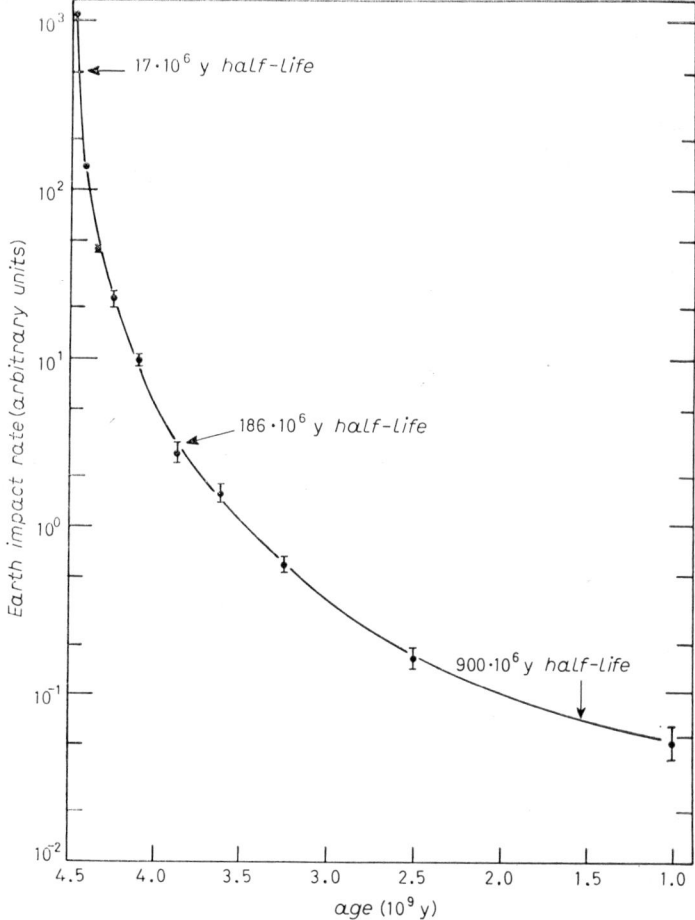

Fig. 1. – Calculated Earth impact rate as a function of time for a residual Earth-crosser with initial elements $a = 0.90$, $e = 0.27$, $i = 5.7°$, $V_G = 7.9$ km/s. The initial rapid decay in impact rate is followed by a long-lived « tail » as a consequence of bodies being stored in Mars-crossing orbits and later returned to Earth-crossing.

that this result is insensitive to the choice of initial Earth-crossing orbit by studying the randomization of such initial orbits and by calculation of the evolution of other initial orbits using a smaller number of runs (*e.g.* fig. 2).

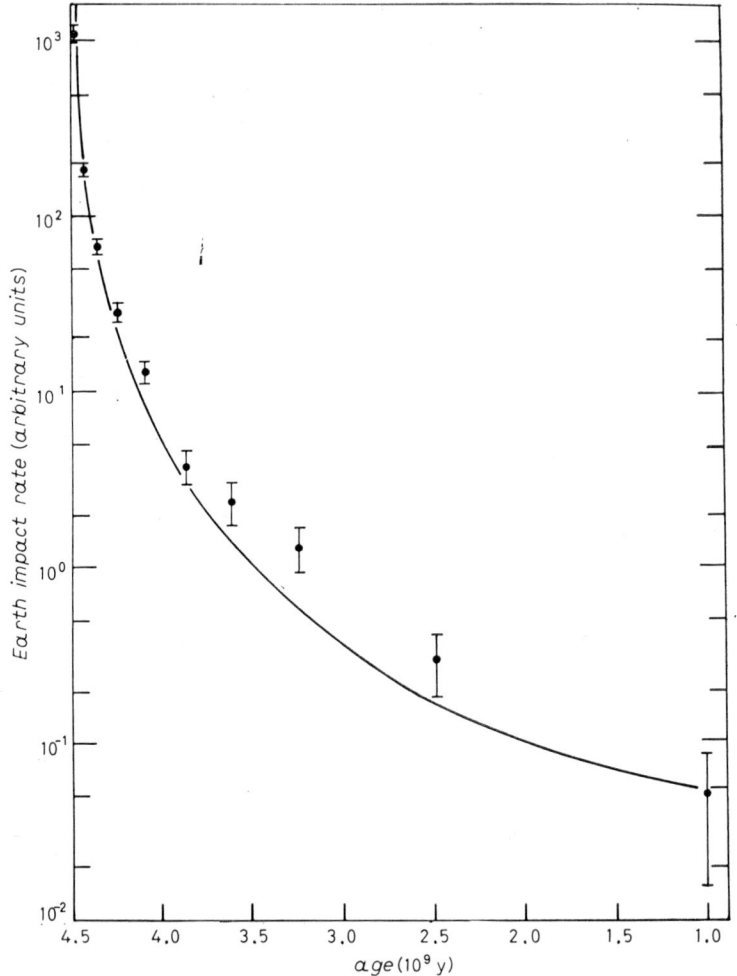

Fig. 2. – Calculated impact rate for a residual Earth-crosser with initial elements $a = 1.1$, $e = 0.27$, $i = 5.7°$, $V_G = 8.5$ km/s. In this case the initial orbit is principally exterior to that of the Earth, in contrast to the case shown in fig. 1, where the orbit is initially principally interior. The general features of the decay in impact rate are insensitive to differences of this kind. ——— curve fitted to points in fig. 1.

In every case it is found that the initial rapid decrease $((15 \div 20) \cdot 10^6$ y half-life) in impact rate slows after $\sim 100 \cdot 10^6$ y. By $4 \cdot 10^9$ y ago the impact rate is decreasing with a longer half-life of 150 to $200 \cdot 10^6$ y and at later times the half-life has increased to $\sim 1000 \cdot 10^6$ y. This initial half-life is essentially the same as the value of $16 \cdot 10^6$ y found by summing the probabilities of Earth and Venus impact calculated from the Öpik collision formula [17].

Examination of the Monte Carlo histories of those bodies calculated to impact after more than $100 \cdot 10^6$ y shows that the « tail » in fig. 1 owes its origin to the combined effects of Mars' perturbations and of the ν_6 secular resonance. As a result of Earth and Venus perturbations, the semi-major axis of the residual body is sometimes in the vicinity of 2 AU, which is near the position of the resonance for low inclinations. Under these circumstances secular perturbations cause large-amplitude (*e.g.* ± 0.2) oscillations in eccentricity. When the eccentricity is near minimum, the body is often only a Mars-crosser. A perturbation following a close encounter to Mars can then change the semi-major axis away from the resonance and « freeze in » the low value of eccentricity associated with pure Mars-crossing. The body then evolves as a Mars-crosser, which includes the small but significant probability of evolving into an orbit with perihelion near Mars, and even to the long-lived situation where crossing of Mars at perihelion only occurs during the maximum of the oscillations in eccentricity. The sequence of events found is shown schematically in fig. 3. At any stage in the evolution depicted there and described above, the chain of events may be reversed and random walk back to simple Earth-crossing can occur, which will then usually be followed by Earth or Venus impact on a $\sim 20 \cdot 10^6$ y time scale. As an ever smaller fraction of the Monte Carlo histories evolve into more distant and shallow Mars-crossing orbits, the time required to evolve back into Earth-crossing becomes longer, causing the half-life for Earth impact to increase, and thereby produce the long-lived tail seen in fig. 1. Although the ν_6 resonance played an important role in these calculations and probably in the real solar system as well, a similar tail is found to result from simple Mars' perturbations in calculations where the effect of the secular resonance is completely ignored. This results from the fact that, although the resonance facilitates the « stealing » of the body by Mars, it also facilitates the transfer of the body back into Earth-crossing. The principal effect of the resonance is, therefore, to speed up the evolution of an equilibrium between a population of Mars-crossing and Earth-crossing orbits, *i.e.* to behave as a « dynamic catalyst ». With the resonance, shallow Mars-crossers can evolve from Earth-crossers on a time scale of $\sim 50 \cdot 10^6$ y rather than requiring the hundreds of million years for the accumulation of weak Mars perturbations to effect the necessary long random walk. This property does not depend on the details of the exact position of the resonance, on the amplitudes of eccentricity as a function of the difference between the actual semi-major axis and the resonant value, nor even on whether only one or a number of resonances are introduced.

The time dependence for Earth (and Moon) impacts calculated for this idealized problem is similar to that used by some workers who have used crater statistics to obtain a time scale for early-solar-system history (*e.g.* [25, 26]).

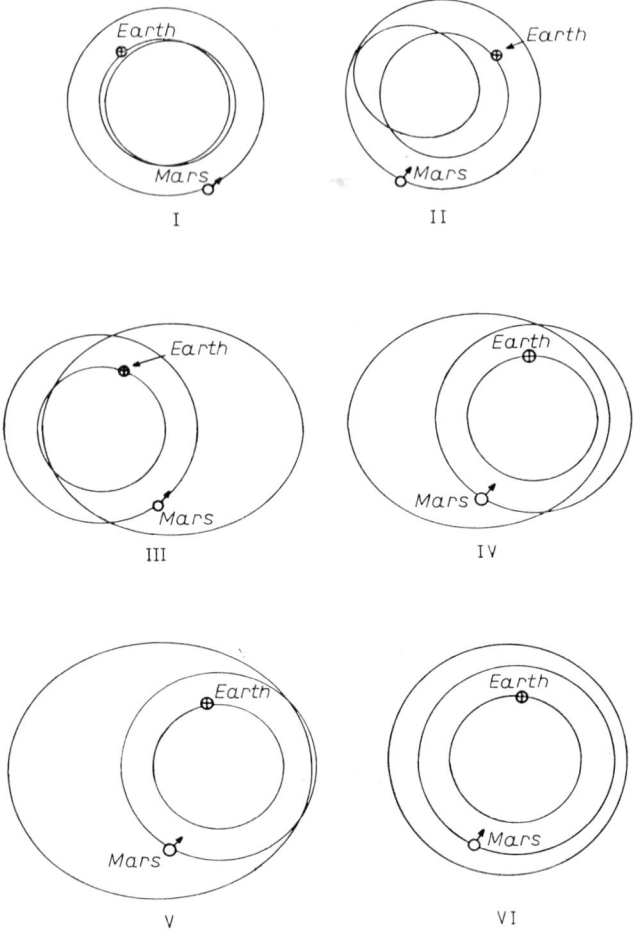

Fig. 3. – Schematic representation of the type of orbital evolution found to be responsible for the long-lived «tail» in fig. 1 and 2: I, low-velocity Earth-crosser; II, higher-velocity Earth-crosser; III, Earth and Mars crosser; IV, deep Mars-crosser; V, shallow Mars-crosser; VI, «implanted» asteroid noncrosser. A residual Earth-crosser is perturbed by Earth and Venus into a Mars-crossing orbit. The combined effects of Mars perturbations and those of the major planets, expressed in the secular resonance ν_6, result in evolution all the way to very shallow and occasional Mars-crossers in a small fraction of the cases.

At first sight it appears to suffer from the drawback that it provides no explanation for the gross difference between the abundance of annealed impact breccias and radiometric age resetting found on the youngest (3.9 to $4.0 \cdot 10^9$ y old) lunar highland surfaces when compared to the oldest ($\sim 3.8 \cdot 10^9$ y) mare surfaces [27]. This problem is sometimes dismissed by statements to the effect that the breccias and resetting can be attributed to the formation of one or more large basins and are irrelevant to the discussion. As will be discussed later, it is quite possible that basin formation is indeed the major cause of the $\sim 4 \cdot 10^9$ y impact metamorphism. Nevertheless, a primary task of an adequate theory for the flux of impacting bodies is to provide a natural explanation for the basin-forming projectiles, which constitute the greatest part of the mass flux in this time period. To ignore these principal bodies while concentrating on the smaller mass of lesser projectiles seems hardly appropriate, particularly in view of the possibility that many of these smaller projectiles are simply fragmentation products of the basin-forming bodies. In order to progress in this discussion, it is necessary to investigate differences between the idealized problem considered here and probable actual conditions in the early solar system.

3. – The effect of mutual collisions in the residual planetesimal swarm.

A principal difference between the real and idealized conditions in the early solar system is that, at the densities and velocities of the residual swarm, collisions between the bodies will be frequent and extensive fragmentation of these objects will occur. This can be easily seen by recognizing that the $\sim 10^{26}$ g remaining after Earth and Venus are 98 % formed is about 50 times the mass of the present asteroid belt. Furthermore, the Öpik collision formula shows that the probability of collision per unit time is inversely proportional to the semi-major axes of both colliding bodies and to the $\frac{3}{2}$ power of the distance of the collision from the Sun. Therefore, a change in these dimensions from 2.5 to 1 AU causes the collision rate to increase by an additional factor of $2.5^{\frac{3}{2}} = 25$. The mean collision velocity is about 1.6 times larger than that in the asteroid belt, resulting in greater fragmentation upon collision for a given mass. For a conventional mass distribution, the effect of this is to increase the impact rate by about a factor of 2. Because of these factors, lifetimes against collisional fragmentation will be reduced by a factor of ~ 2500 relative to the present asteroid belt. Bodies which can survive for 10^{10} years in the present asteroid belt will be destroyed by collision in $4 \cdot 10^6$ years. Since this is short compared to the $\sim 10^8$ year time scale for the orbital evolution under discussion, the only bodies which can survive under these conditions are those so large that their gravitational energy inhibits their destruction. A real calculation of the minimum mass necessary to avoid destruction is difficult, as it depends on uncertainties associated with the size distribution following Roche-limit destruction, on the probability of Roche-limit destruction itself and on the fraction of the available impact energy converted into kinetic energy of the ejecta. Some rough impression of this minimum mass can be obtained by simple scaling to the present asteroid belt as discussed above and comparison with previously calculated asteroid lifetimes [14, 28, 29]. For example, the shortest-lived values given by WETHERILL [28], table VII, are in accord with current data based on cratering experiments. From these data the lifetime of a 100 km (10^{22} g) radius body will be

$$(1) \qquad \sim \left(\frac{10^7}{50}\right)^{0.4} \cdot 30 \simeq 4000 \cdot 10^6 \text{ y} .$$

When reduced by the factor of 2500 given above, a lifetime of $\sim 2 \cdot 10^6$ y is obtained. Therefore, even with the uncertainties of this calculation, it is plausible that bodies smaller than 10^{21} g would be destroyed. The increase in gravitational energy with the fifth power of the radius leads to significantly greater mean lives, i.e. $\sim 30 \cdot 10^6$ y for 10^{24} g bodies, thus permitting most of them to survive for $> 50 \cdot 10^6$ y in view of the rapid initial decline in the mass of the swarm. A reasonable estimate of the smallest mass with a high probability of survival thus appears to be within an order of magnitude of 10^{23} g, corresponding to a radius of 190 km for a density of 3.5 g/cm³. For the Earth- and Venus-crossing populations, the collisional loss of smaller bodies will be replaced by production of fragments following Roche-limit destruction. However, for bodies perturbed into purely Mars-crossing orbits, Roche-limit destruction will not be adequate to maintain the steady-state population of $\leqslant 10^{22}$ g bodies. The Monte Carlo results show that times of $\sim 50 \cdot 10^6$ y are typically required for the perihelion of the Mars-crossers responsible for the long-lived tail to evolve out of the densely populated zone of Earth residua, i.e. to ~ 1.5 AU. Therefore, it is probably incorrect to consider the evolution shown in fig. 1 applicable to bodies $\leqslant 10^{21}$ g in mass; the population of smaller bodies will decay on the much shorter time scale associated with sweep-up by Earth and Venus.

The importance and even the existence of the « tail », therefore, depends on the number of these large bodies surviving to the final stages of the accumulation of the Earth. The accretion model described previously [10] leads to the expectation that there should be $\geqslant 10^3$ bodies with mass $\sim 10^{23}$ g and about 100 larger bodies with masses up to $\sim 10^{26}$ g at the time the Earth and Venus embryos became completely dominant. A Roche-limit lifetime of $\sim 10^7$ years then implies that most of the larger bodies will likely be destroyed during the 75 to $100 \cdot 10^6$ y required for the completion of accumulation. However, until the late stages of accumulation they will probably be sufficiently numerous to maintain a steady-state population of 10^{22} to 10^{23} g bodies, of which there could still be several hundred at the time Earth and Venus were 98% complete. Most of these would be swept up or disrupted during the next 100 million years, but fig. 2 shows that a few would be expected to return to Earth-crossing during the time between 4.3 and $3.9 \cdot 10^9$ y ago, relatively unaccompanied by smaller bodies.

4. – Contribution of the residual swarm to the $4.0 \cdot 10^9$ y late heavy bombardment of the Moon.

A principal task of any theory which attempts to describe the history of interplanetary bodies in the solar system is to explain the rapid decrease in extent of lunar breccia formation and resetting of radiometric ages between 4.0 and $3.8 \cdot 10^9$ y ago. A general discussion of this problem has been presented [30]. There it is pointed out that there are two « storage places » in which bodies can be preserved for several hundred million years and still have a reasonably high probability of returning to Earth-crossing. These are the inner edge of the asteroid belt and the outermost solar system, the region of Uranus and Neptune. In this section it will be shown that this present discussion may point the way to the removal of major difficulties associated with asteroidal storage. This discussion should not be interpreted as implying that the outer-solar-system source should be abandoned or down-graded in possible importance. It might seem reasonable to do this on the grounds that it is far away and consequently that projectiles from this region are far-fetched. This reasoning is probably not valid; a principal source of interplanetary material in the present solar system is the even more distant Oort cloud of comets. It is likely that this cloud had its origin in the outer solar system during the time period under discussion here, and a cometary cloud at 20 AU should affect the inner solar system more than one at the present 10^4 to 10^5 AU, as discussed in more detail elsewhere [13, 30].

In earlier work it was pointed out that, if the size distribution of the projectile population responsible for the late heavy bombardment were such that almost all the mass were concentrated in a few large bodies, then stochastic fluctuations in the return of these bodies to Earth-crossing would produce peaks in the mass flux on the Moon and terrestrial planets. Furthermore, if these bodies were $\sim 10^{23}$ g in mass, Roche-limit disruption following close approaches to Earth and Venus would produce simultaneous basin-forming heavy bombardments on all of the terrestrial planets as well as the Moon. These disrupted fragments would be removed, primarily by planetary impact, on a time scale of less than $100 \cdot 10^6$ y.

The delayed transfer of one or more $\sim 10^{23}$ g residual bodies into Earth-crossing between 4.3 and $3.9 \cdot 10^9$ y ago, unaccompanied by the usual collisional-size spectrum of smaller bodies, as discussed in the previous section, meets the requirements for the late heavy bombardment described above. The absence of basin-forming events after $\sim 3.9 \cdot 10^9$ y would then be simply a consequence of losing the last one of a few large bodies by that time, as would be expected from the results shown in fig. 1.

Whether or not this can explain quantitatively the extent of age resetting is not clear at present. In lunar mare and terrestrial cratering, the radiometric ages of most of the crater ejecta is not reset; it is simply thrown out with little heating or other alteration. Under these circumstances repetitive impacts are required before the age of any given fragment has a high probability of being reset. If the lunar resetting at 3.9 to $4.0 \cdot 10^9$ y is to be attributed primarily to basin-forming events, as is suggested by the relatively light cratering of the oldest post-Imbrium light plains, then this would imply that many basins were formed within about $100 \cdot 10^6$ y. The evidence on this question is not clear. Radiometric data on Apollo 17 breccias [31-33], when combined with geological evidence for their being related to the Serenitatis event (*e.g.* ref. [34]), argues in favor of all six of the near-side circular mare basins being formed within about $100 \cdot 10^6$ y. Even if this is true, it is not clear that this would represent sufficient repetitive bombardment and it may be necessary to also postulate that to some extent the $\sim 4 \cdot 10^9$ y ages found in basin ejecta were set by excavation of material previously at an elevated temperature. The smaller bodies formed following Roche-limit breakup will also cause some resetting, but the number and consequently the effect of these cannot be large, unless somehow the Imbrium and Orientale projectiles impacted the Moon near the very end of the late heavy bombardment. In any case this hypothesis wherein the late heavy bombardment is associated with the Earth-Mars-Earth transfer mechanism presented in this paper may explain the observed timing, magnitude and size distribution of the bombardment, and the uncertainties associated with the details of the resetting mechanisms will exist for any hypotheses consistent with these observations.

A possible objection to this Earth-Mars-Earth explanation of the late heavy bombardment is that a similar swarm of Mars residua has been ignored. If the growth of Mars were similar to that of Earth and Venus, a residual Mars-crossing population of $> 10^{25}$ g should have remained at $4 \cdot 10^9$ y and would have produced a greater flux of Earth and Moon projectiles than the residual Earth material returned from the vicinity of Mars. Roche-limit destruction by Mars would be less frequent. Some of the large bodies in this swarm should have been in very long-lived shallow Mars-crossing orbits, and complete absence of lunar basin-forming impacts younger than $\sim 3.9 \cdot 10^9$ y would be unexpected. This is one aspect of a previously recognized problem regarding Mars: the time scale for its growth from a zone containing only the mass of the present planet is $\sim 10^9$ years. This problem led SAFRONOV [12] and WEIDENSCHILLING [35] to propose that Mars represented an « aborted » Earth- or Venus-sized planet, which grew for about 10^7 years in a region of particle density comparable to the Earth or Venus zones, which was then swept nearly clear of planetesimals by the massive flux of high-velocity bodies expected to be generated in the zone of Jupiter as has been discussed by numerous authors [12, 15, 36-39]. This has been designated the « early heavy bombardment » [30] as distinct from the much less intense « late heavy bombardment » responsible for the formation of the breccias in the lunar highlands and which produced the circular mare basins and persisted for about $500 \cdot 10^6$ y after the formation of the Moon.

It is too early to say whether or not this is the correct solution to the problem of the time scale for the growth of Mars, and for the related questions of the small size of Mars and the presence of only minor planets between Mars and Jupiter. If this is correct, then the residual Mars swarm was destroyed long before the transfer to Mars-crossing of the Earth and Venus residua $\sim 10^8$ years later, and the Mars swarm would not have participated in the late heavy bombardment of the inner solar system.

5. – Transfer of residual Earth planetesimals to the asteroid belt and the origin of the sources of differentiated meteorites.

In addition to the long-lived tail in fig. 2, the same Monte Carlo calculations also produced the result that 0.4% of the original bodies will survive for the entire $4.5 \cdot 10^9$ y of solar-system history. This is similar to the fraction found to impact Earth and Venus during the 100 million year period centered at $4 \cdot 10^9$ y (0.3%). Therefore, if one or more bodies of mass $10^{23 \pm 1}$ g were disrupted and impacted the inner solar system to produce the late heavy bombardment during this time period, a similar population of residual Earth planetesimals should still exist. The exact number and size of these bodies are uncertain, if for no other reason than this information is in the « noise » resulting from the stochastic nature of this problem. However, it seems likely there should be some survivors. Even if the last 10^{23} g body was spent in producing the late heavy bombardment, there should have been some 10^{21} or 10^{22} g bodies in addition. Although the population transferred to Mars-crossing orbits was greatly depleted in small bodies, it does not seem plausible that this depletion went so far that the *number* of small bodies was actually less than the number of larger objects, but merely that the total *mass* of objects smaller than a given object was less than the mass of this single object.

The calculated distribution of semi-major axes and inclinations of the surviving bodies is shown in fig. 4. The eccentricities of these objects is such that they barely cross the aphelion of Mars, and do so only at the maximum in their secular oscillations in eccentricity. Close encounters with Mars are thus rare, and it is this fact which permitted their survival.

There are a number of observed asteroids in orbits of this kind, but, with the exception of 313 Chaldaea, the larger objects of this group, such as 753 Tiflis, have radii $\leqslant 10$ km, corresponding to a mass of about 10^{19} g. It is possible, but not particularly likely, that these bodies are the survivors under discussion, as their collisional lifetime in the present asteroid belt is probably less than the age of the solar system, and at least some of these objects should be collision fragments of larger asteroids which lie safely beyond Mars. If the

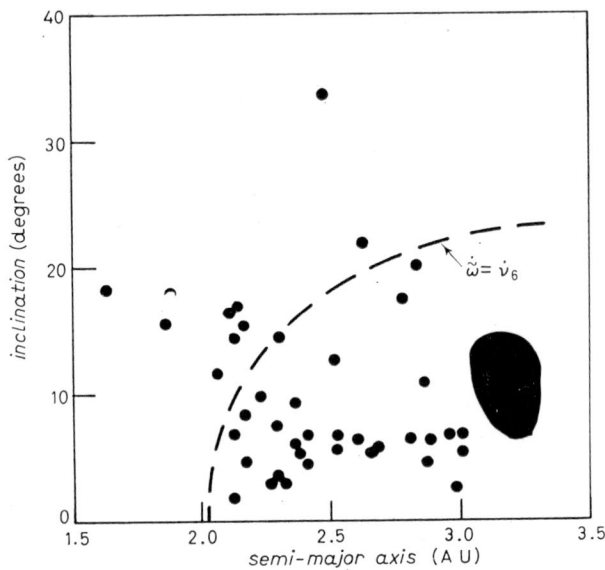

Fig. 4. – Distribution of semi-major axes and inclinations of initial Earth-crossers calculated to survive for the age of the solar system.

radiometric albedo of 0.014 reported for 313 Chaldaea by HANSEN [40] is correct, then the radius of this body is 80 km, and is in the mass range of plausible large surviving objects. However, this low albedo also suggest that Chaldaea is a carbonaceous object, which would make it all the more interesting if it were indeed a survivor of Earth's swarm, but this composition hardly constitutes evidence for association of this body with Earth material.

There are some large asteroids in the vicinity of the ν_6 resonance which at present can approach within 0.02 to 0.06 AU of Mars during the phase in their secular perturbations when their eccentricities are near maximum and their orbits are favorably oriented with respect to the orbit of Mars (J. G. WILLIAMS; private communication (1975)). These include 7 Iris (105 km), 8 Flora (76 km) and 18 Melpomene (75 km) (fig. 5). It has been previously argued [22, 41] that these bodies are very likely to be sources of differentiated silicate and iron meteorites at the present time. It was shown that their collision ejecta are sufficiently abundant and of high enough velocity to become Mars-crossing by the combined effect of ν_6 and Mars perturbations. The reflectance spectra of these bodies (S-type, ref. [41]) suggests a silicate-metal mixture appropriate to the composition of these meteorites. Could these bodies be residual Earth planetesimals? Transfer of these bodies from very shallow Mars-crossing orbits to their present almost-crossing orbits could occur if the position of ν_6 and possibly other resonances were not entirely stable during the first several hundred million years of solar-system history. One possibility, which requires quantitative verification, is that this might result from the growth of Uranus

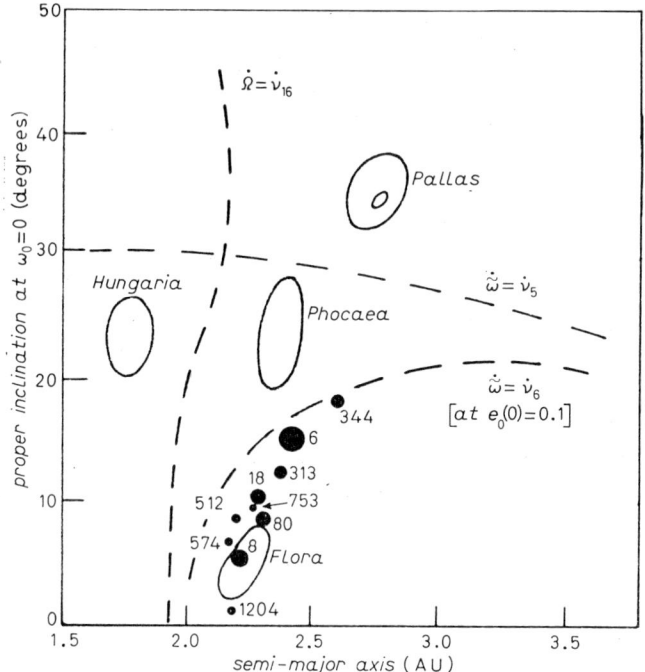

Fig. 5. – Observed distribution of large asteroids in the vicinity of the ν_6 resonance, many of which are believed to be important sources of iron and differentiated stone meteorites [22, 42].

and Neptune on a long time scale [13, 30, 43], which caused the mass and possibly the position of Jupiter and Saturn to change slightly by accumulation and ejection of the massive cloud (10^{29} g?) of residual Uranus and Neptune planetesimals. Thus the amplitude of the oscillations in the eccentricity of these bodies may at one time have been high enough for them to have been Mars-crossing, but by now the maximum eccentricities have decreased by about 0.02 and Mars encounters are no longer possible. Although the theoretical basis for these speculations is minimal, even more extreme speculations are necessary in order to explain the present distribution of asteroid orbits. It seems likely that, as our understanding of these very difficult problems increases, our understanding of the relationships between residual planetesimals of varied origins and the present asteroids should be improved to the point where more intelligent discussion of this matter will be appropriate.

6. – Return home of Earth's swarm, or origin of differentiated meteorites.

Meteorites may be divided into two principal classes: undifferentiated and differentiated. The undifferentiated meteorites, also known as chondrites, have preserved to a remarkable degree the relative proportions of the non-volatile elements in the early solar system, but differ from one another in the extent to which the volatile-element abundances have been preserved (*e.g.* ref. [44]). The differentiated meteorites, which include the irons, stony-irons and achondrites, have undergone further chemical fractionation, usually as a consequence of crystal-liquid fractionation, and in several ways are analogous to igneous rocks from the Earth and Moon. Oxygen isotope studies [45] show that these two classes of meteorites are unlikely to be genetically related, and their principal common property may simply be that they had orbits which intersected that of the Earth, even though the complex dynamic processes which led to their being in these orbits were quite different.

Most of our understanding of processes occurring in the primordial solar system has been obtained by studies of these two classes of meteorites. The chondrites have preserved relics of those processes which would be destroyed by magmatic differentiation, whereas the differentiated meteorites record the existence and nature of these processes in the early solar system, and are relevant to our understanding of the planetesimals which accumulated to form the Earth and Moon.

It is generally thought that the differentiated meteorites are of asteroidal origin. The principal basis for this, other than preconceived views regarding the chemical history of comets, has been the long $(\sim (10^8 \div 10^9)\text{ y})$ exposure ages of iron meteorites, which are very difficult to reconcile with the expected dynamic lifetime $(\sim (10^6 \div 10^7)\text{ y})$ of cometary material in the inner solar system [23]. More recently, strong evidence for an asteroidal origin of a major class of differentiated meteorites, the polymict-brecciated basaltic achondrites, or howardites, has been obtained from their similarity to lunar regolith breccias [46].

This section will describe a dynamic theory for the derivation of differentiated meteorites from the inner edge of the asteroid belt, presented by WE-THERILL and WILLIAMS [42]. At least some of these asteroids may be implanted members of Earth's swarm, as discussed in the previous section.

The theory is essentially fundamentally that of Arnold [19] in which relatively low-velocity collision ejecta (*i.e.* < 1 km/s) from asteroids with perihelion near Mars achieve the large acceleration (~ 5 km/s) necessary to become Earth-crossing by the random-walk accumulation of Mars perturbations. The present work extends that of Arnold by using cratering studies of Gault, Shoemaker and Moore [47] and O'Keefe and Ahrens [48] to estimate the yield of meteorites from observed asteroids. It also provides a way to circumvent a difficulty in obtaining achondrites by the Mars perturbation mechanism: the long time previously required to reorient a meteorite's orbit from perihelion near Mars to aphelion near Mars. This resulted in a minimum time of $\sim 5 \cdot 10^8$ y for reaching Earth-crossing [24], by which time collisional destruction of meteorite-size silicate bodies is nearly inevitable [14, 28, 49]. This minimum delay is found to disappear when the effects of the secular resonance ν_6 [50, 51] are included. The importance of this resonance in the production of asteroidal meteorites was emphasized by WILLIAMS [52]. The present theory differs from that of Williams in that Earth-crossing primarily results from the non-linear combination of these resonant Jupiter and Saturn secular perturbations with those resulting from close encounters to Mars. In Williams' theory resonant perturbations led directly to Earth-crossing, which can occur, but less frequently.

For most low-inclination orbits the secular perturbations are of small amplitude and cannot have much effect on the subsequent evolution of asteroidal collision debris. However, the forced oscillations can reach high amplitudes in the vicinity of surfaces in a, e, i space for which the rates of precession of the nodes and arguments of perihelion are equal to eigenfrequencies occurring in the general theory of planetary secular perturbations [50]. One of these resonant surfaces, ν_6, plays a major role in the present treatment.

6'1. Ejection of collision debris into Mars-crossing orbits. – Estimates of cratering yields have been calculated for large asteroids in the vicinity of Mars and the ν_6 resonance by use of the asteroidal collision theory [28] and experimental and theoretical cratering data [47, 48]. The calculated yield from the more important sources, as well as from several less productive objects in deeper Mars-crossing orbits, are given in table I. Because of the concentration of asteroidal mass in the larger bodies, the total yield will be similar to that obtained from the largest bodies. Large statistical fluctuations can be expected in the yields from these objects, because the observed mass-frequency relationship in the asteroid belt results in dominance of the yield by a few large

TABLE I. – *Observed asteroids most productive of ejecta in Mars-crossing orbits (together with representative smaller bodies).*

Object	Spectral class	Diameter (km)	Present perihelion (AU)	Δ_{Mars} (AU)	$\Delta_{resonance}$ (AU)	Calculated yield (g/y)
6 Hebe	S	197	1.94	0.141	0.13	$4.60 \cdot 10^9$
8 Flora	S	150	1.86	0.024	0.16	$1.96 \cdot 10^9$
18 Melpomene	S	141	1.80	0.023	0.16	$2.96 \cdot 10^9$
80 Sappho	$S?$	95	1.84	0.095	0.14	$6.91 \cdot 10^8$
313 Chaldea	$C?$	40?	1.95	-0.039	0.19	$1.28 \cdot 10^9$
344 Desiderata	?	58	1.79	0.015	0.17	$2.75 \cdot 10^9$
512 Taurinensis	?	16	1.64	-0.145	0.10	$7.93 \cdot 10^7$
574 Reginhild	?	7	1.71	-0.118	0.19	$1.13 \cdot 10^7$
1204 Renzia	?	8	1.60	-0.127	0.23	$1.39 \cdot 10^7$

bodies and because most of the ejected mass derives from rare large collisions on these bodies.

In the calculation of these yields, collision debris were assumed to be ejected in random directions and at random phases of the secular perturbation phases of their sources. Ejection velocities from 25 m/s up to 800 m/s were used, weighted in accordance with the cratering data referred to above. The mass yields were calculated by the methods used previously [24, 53]. Although many of the sources are in orbits which can never make close encounters to Mars, some of the ejecta will be in orbits for which such encounters are possible. This arises in part because of changes in semi-major axis and eccentricity resulting directly from their ejection velocity following collision. More important, if ejection occurs near the minimum in the secular oscillation in eccentricity of the source object, and the semi-major axis is changed to be closer to the resonant value, then the increased amplitude in the amplitude of the oscillations in eccentricity will cause the maximum eccentricity of the fragment to be larger. The combined effects of the decreased semi-major axis and the increased eccentricity can thereby decrease the perihelion of the fragment sufficiently to produce Mars-crossing. It is sometimes supposed that mere approach to the resonant value of the semi-major axis is sufficient to achieve

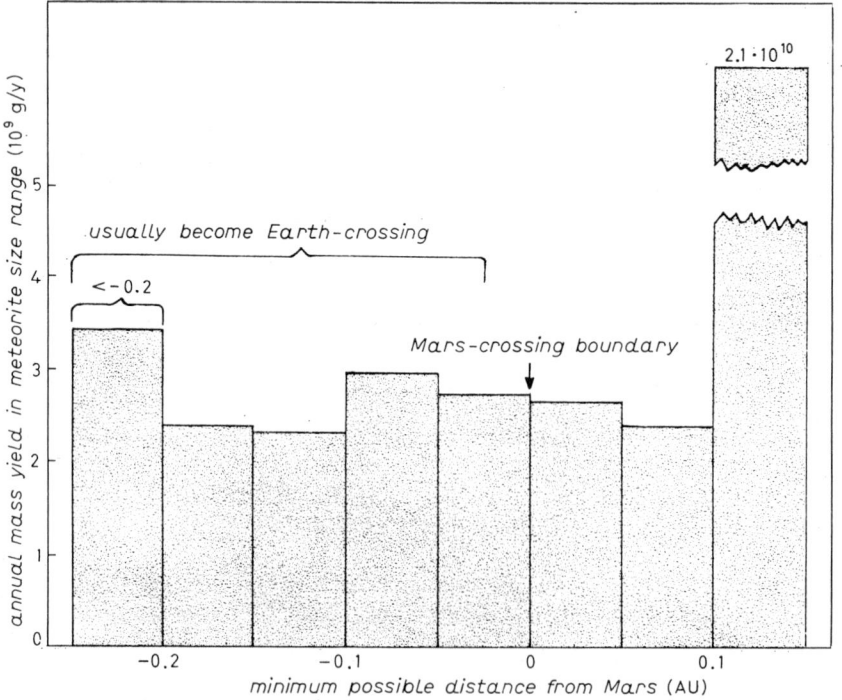

Fig. 6. – Calculated annual yield of meteoritic fragments from asteroids as a function of the minimum possible encounter distance from Mars ($\varDelta_{\text{Mars}}$).

this reduction in perihelion. This is not the case; ejection near the maximum phase in eccentricity will cause the increased amplitude to produce a smaller minimum eccentricity, leaving the maximum eccentricity unchanged. In any case, with random ejecta, a calculable fraction will be ejected at phases favorable for the production of small perihelia, and the remainder will simply contribute to the ordinary collisional steady-state population of small bodies in the asteroid belt.

The results of these ejecta calculations are given in fig. 6. Material having negative values of the « minimum possible distance from Mars » can make close encounters to Mars and undergo further orbital evolution as a consequence of Mars-crossing. This is described in the next subsection. The total annual yield of these Mars-approaching fragments in the meteorite-size range (100 g to 10^6 g) is calculated to be $\sim 10^{10}$ g/y. The total flux of differentiated meteorites on the Earth is $\sim 10^7$ to 10^8 g/y, so, if this material is transferred to Earth-crossing orbits with reasonable efficiency and with little collisional destruction during the time needed to complete this transfer, then this material will be adequate in quantity to provide the terrestrial flux of differentiated meteorites. Of course, it is not known if these asteroids actually correspond in composition to these meteorites. Spectrophotometric data, however, show that several of the principal sources listed in table I correspond to S-type asteroids, which by comparison with terrestrial materials of geochemically plausible composition are believed to represent a mixture of metallic iron with silicates of composition similar to those of the achondrites [41].

6`2. Subsequent orbital evolution of the fragmented material. – It was shown in the previous section that a significant quantity of debris from the inner edge of the asteroid belt will be placed in orbits which at least some of the time will have perihelia within the orbit of Mars, and the nodes of which will occasionally intersect the orbit of Mars. Close approaches to Mars will then occur at a calculable rate, and the further orbital evolution can be followed by the Monte Carlo method described previously [22] which includes the effect of the ν_6 resonance.

These calculations have been carried out for four starting orbits with different values of the initial minimum possible distance from Mars (Δ_{Mars}) and the initial distance of the semi-major axis from the resonant value ($\Delta_{resonance}$). The calculated subsequent evolution of the orbits of these bodies illustrates well the « synergistic mechanism » qualitatively proposed earlier [54]. Close approaches to Mars are initially rare (*i.e.* once in $\sim 10^7$ years), requiring proper orientation of the orbits of Mars and the asteroid as well as favorable values of the forced-oscillation phases. Perturbations following these rare close approaches will often random-walk the semi-major axis closer to the resonance, causing a slight increase in maximum eccentricity. The increase in maximum eccentricity will only be small at first, because the perturbation occurs near the phase at which the eccentricity is at a maximum. However, this small increase in eccentricity decreases the time between close approaches to $\sim 10^6$ years, and permits close approaches at phases further from maximum eccentricity. Random walk toward the resonance can then occur on a more rapid time scale, the eccentricity can build up sufficiently to produce Earth-crossing at the maximum phase in eccentricity. Earth perturbations will then often reduce the semi-major axis well below the resonant value, and Earth-crossing will be « frozen in » with the eccentricity at least temporarily fixed near the maximum value. Of course, Earth perturbations can also random-walk the semi-major axis back into resonance, and there is then a chance to lose the fragment temporarily to pure Mars-crossing. This chain of events is

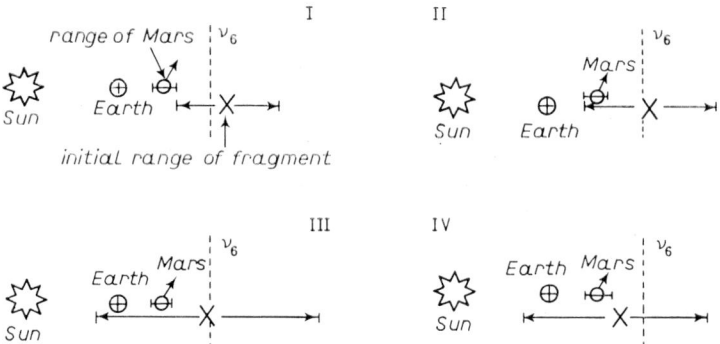

Fig. 7. – Schematic description of the combined effect of secular, Mars and Earth perturbations in transferring a fragment from I) an occasional very shallow Mars-crosser to II) an orbit with perihelion frequently well inside Mars orbit (Mars perturbations decrease a of fragment, decreasing distance to resonance; increase in eccentricity causes deeper Mars-crossing) to III) an orbit with semi-major axis near the resonant value and semi-major axis large enough to produce Earth-crossing (more frequent Mars perturbations decrease a to near-resonance value, increasing e to produce deep Mars-crossing or even Earth-crossing) to IV) a relatively stable Earth-crossing orbit with semi-major axis inside the resonant value (Earth perturbations decrease a, remove fragment from resonance and stabilize Earth-crossing).

shown schematically in fig. 7. The results of actual calculations, in which these events occurred as the spontaneous result of the interaction of these secular, Mars and Earth perturbations, are given in fig. 8 to 12.

Figure 8 shows that the most probable time of first Earth-crossing is within the first 100 million years, but that the mean time to Earth-crossing is considerably longer, about $600 \cdot 10^6$ y. The results for all four starting orbits are combined in fig. 8. The two starting orbits with the more negative values of $\varDelta_{\text{Mars}}$ (C and D) contribute about twice as much to the peak at less than $100 \cdot 10^6$ y than the two very shallow starting orbits A and B. Starting orbits A and B contribute 80% of the tail of late Earth-crossers $> 2000 \cdot 10^6$ y. How-

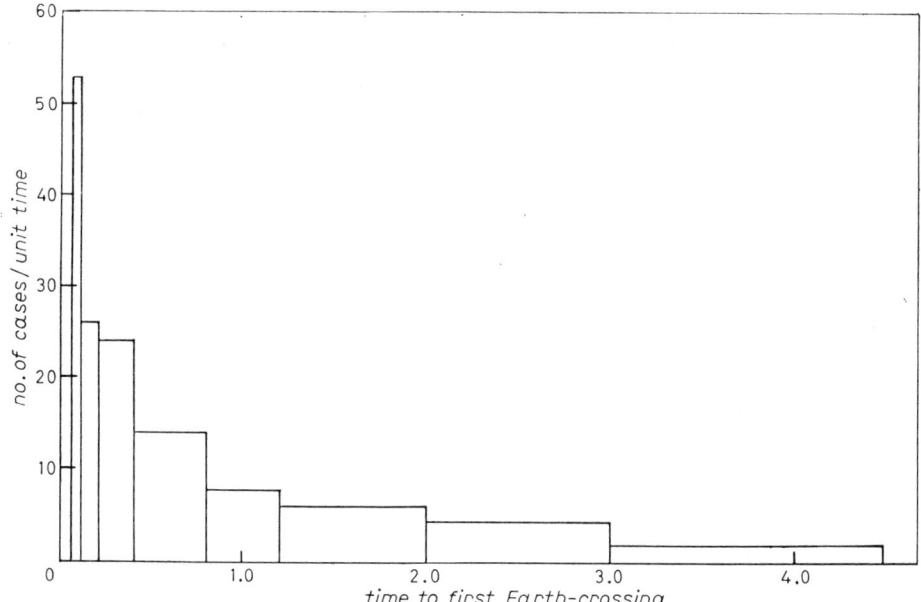

Fig. 8. – Calculated distribution of times at which fragments in the following initial orbits first become Earth-crossers (four Mars-grazers, combined data):

Orbit	Present perihelion (AU)	Δ_{Mars} (AU)	$\Delta_{\text{resonance}}$ (AU)
A	1.96	-0.015	$+0.15$
B	1.79	-0.05	$+0.16$
C	1.60	-0.13	$+0.22$
D	1.64	-0.15	$+0.08$

ever, since the distribution of Δ_{Mars} for the ejecta shown in fig. 6 is rather uniform over a wide range in Δ_{Mars}, the combined data shown in fig. 8 give a good idea of the actual times required to achieve Earth-crossing, and show that these times are distributed over a wide range of values.

The calculated times of Earth impact are shown separately for each of the starting orbits in fig. 9-12. The very shallow starting orbit A (fig. 9) contributes Earth impacts over the entire history of the solar system, with some preference for times between 100 and $2500 \cdot 10^6$ y. Slightly deeper values of Δ_{Mars} (fig. 10) cause a shift toward smaller impact times, whereas the even deeper initial orbits C and D (fig. 11 and 12) will primarily generate impacts within $1000 \cdot 10^6$ y, and some impacts within the first $100 \cdot 10^6$ y are to be expected.

These calculated impact times may be compared with the measured cosmic-ray exposure ages for iron meteorites shown in fig. 13 [55]. The typical exposure ages of $\leqslant 10^9$ years are in agreement with the results of the calculations. However, it is unlikely that the scarcity of exposure ages of $> 10^9$ years can be entirely explained on a dynamic basis, as the calculations predict 5 to 10 times as many of these long exposure ages as are actually found. It is likely that collisional fragmentation will begin to cause reduction of the number of stone

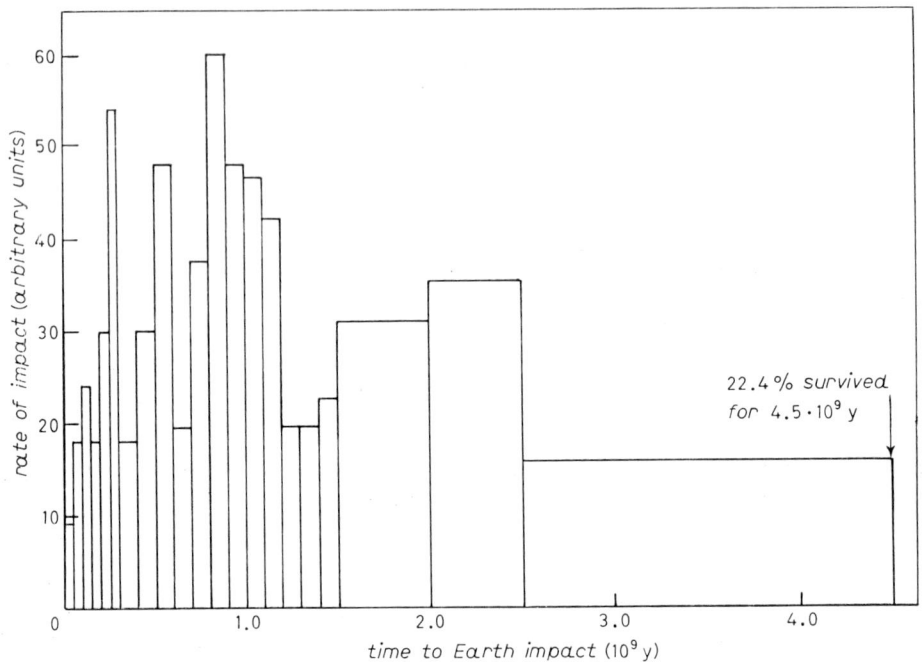

Fig. 9. – Calculated distribution of times at which fragments impact the Earth for an initial orbit which is a very shallow Mars-crosser. The initial elements correspond to those of a real asteroid 1480 Aunus (initial orbit A: $a = 2.203$, $e = 0.111$, $i = 4.3°$, $\Delta_{\mathrm{Mars}} = -0.015$ AU, $\Delta_{\mathrm{resonance}} = +0.15$ AU), but the result should be taken to represent that of a fragment of any object ejected into a similar orbit; Earth impact probability $= 10.1\%$.

meteorites after $\sim 2 \cdot 10^7$ years of exposure [56] and comparison of the tensile strength of iron and silicate rock [28] suggests that the number of irons should be similarly reduced after $(500 \div 1000) \cdot 10^6$ y. Therefore, it is plausible that the reduction in observed exposure ages above $1000 \cdot 10^6$ y is at least partially produced by collisional destruction. It also must be recognized that the observed distribution is strongly affected by a few large fragmentation events. Particularly remarkable is the grouping of 17 of the 18 known type IIIAB irons [57, 58] at $(650 \pm 100) \cdot 10^6$ y. This is to be expected from the observed distribution of the asteroids, in which the total mass is strongly concentrated in the largest bodies. However, this precludes the possibility of a detailed agreement between the results of the present calculations and the observed exposure ages. The peaks produced by the infrequent but dominant collisions are random phenomena and are in principle unpredictable. Other measurements [59-61] show that a larger number of iron objects with these short exposure ages exist, and, furthermore, their existence is predicted by the present theory.

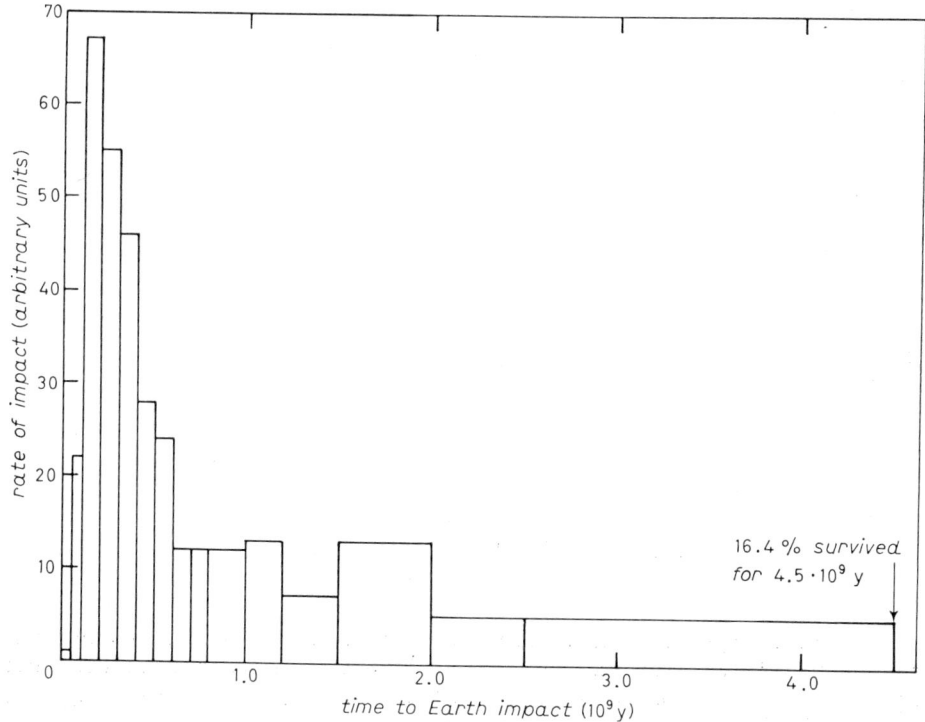

Fig. 10. – Calculated distribution of times of Earth impact for a slightly more negative initial value of Δ_{Mars} (initial orbit B (228 Agathe): $a = 2.201$, $e = 0.183$, $i = 3.0°$, $\Delta_{\text{Mars}} = -0.049$ AU, $\Delta_{\text{resonance}} = +0.16$ AU). A peak at $\sim 300 \cdot 10^6$ y now appears in the distribution, but longer times are still probable; Earth impact probability $= 10.4\%$.

In contrast, the achondrites have short exposure ages, mostly less than $20 \cdot 10^6$ y, but ranging up to $100 \cdot 10^6$ y for the large enstatite achondrite Norton County. If these are derived from the asteroids under consideration here, then some mechanism must exist to eliminate the older ages. Collisional destruction of the fragments is the most plausible mechanism to accomplish this, and the age limit of $\sim 100 \cdot 10^6$ y is consistent with the collisional lifetime of small silicate bodies, as discussed earlier in this paper. This same collisional lifetime will also eliminate any long-lived «tail» in the exposure age distribution of chondrites. It is likely that the chondrite exposure ages are also dynamically limited, as their distribution of radiants and times of fall imply a short lifetime with regard to perturbation by Earth and Venus into Jupiter-crossing [45, 62, 63].

A plausible theory for a meteorite source must also provide a quantitatively adequate source. As shown in fig. 8-12, an Earth impact rate of very nearly 10 % is calculated for all four starting orbits. It is not known what the con-

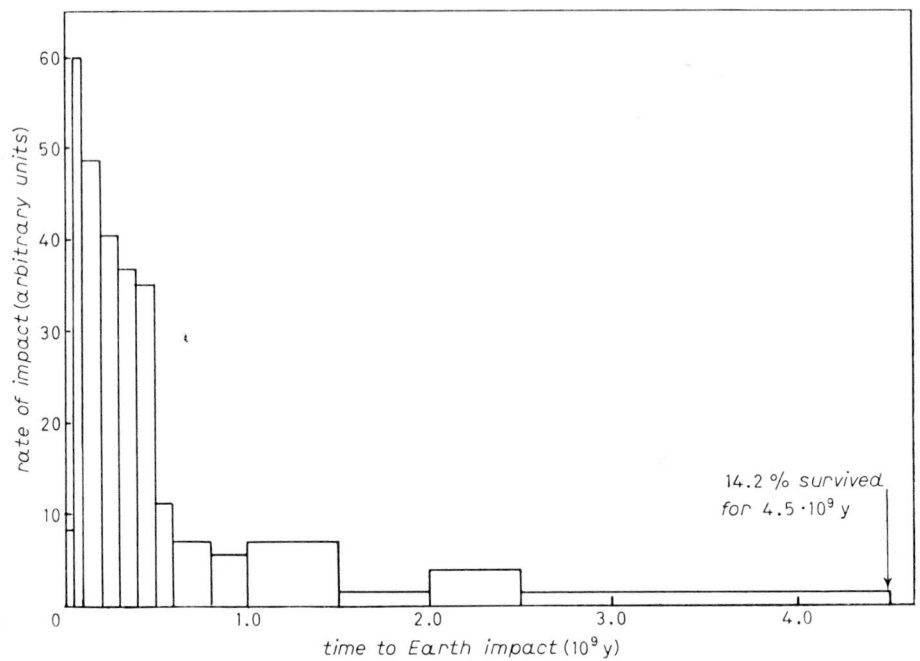

Fig. 11. – Earth impact times for an initial orbit with a fairly large negative value of Δ_{Mars} initially rather distant from the resonance, and with a very low inclination (initial orbit C (1204 Renzia): $a = 2.263$, $e = 0.2937$, $i = 1.9°$, $\Delta_{\text{Mars}} = -0.13$ AU, $\Delta_{\text{resonance}} = +0.22$ AU); Earth impact probability = 9.6%.

centration of iron objects in the surface regions of these asteroids is. The Earth's metallic core represents about $\frac{1}{3}$ of its total mass. If the asteroid ejecta contained this proportion of iron objects, the Earth impact flux of $(100 \div 10^6)$ g iron meteorites would be $\sim 3 \cdot 10^8$ g/y. The flux of chondrites has been estimated to be $\sim 10^8$ g/y. The rate of observed iron falls relative to chondrites is $\sim 7\%$, corresponding to a flux of $7 \cdot 10^6$ g/y. Therefore, the present source is adequate even if the ejected matter is depleted in iron objects by a factor of ~ 50. The rate of fall of achondrites is similar to that of irons. As discussed earlier, the number of these bodies will be greatly depleted by collisions. The Monte Carlo calculations predict that $\sim 1\%$ of the impacts will occur on the Earth during the first $25 \cdot 10^6$ y, leading to an expected terrestrial flux of $\sim 10^7$ g/y, in agreement with observation.

In order for a stone meteorite to survive passage through the atmosphere, it is necessary that the fragments enter the atmosphere at less than about

22 km/s. The calculated geocentric velocity distribution for starting orbit D is shown in fig. 14. A geocentric velocity of 19 km/s corresponds to an entry velocity of 22 km/s when the Earth's gravitational acceleration is included. The great majority of the calculated impacts are of lower velocity, and atmos-

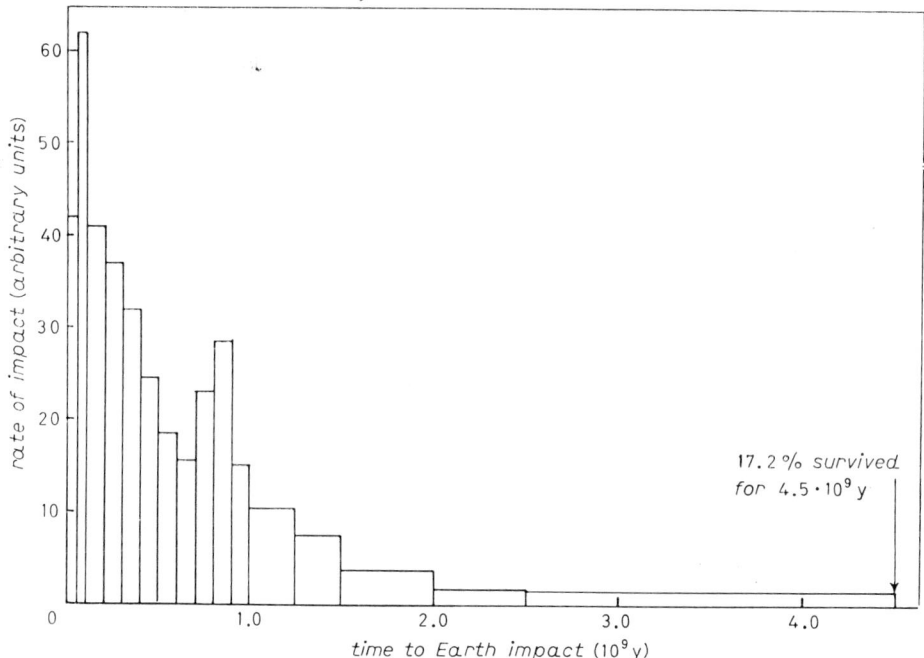

Fig. 12. – Earth impact times for an initial orbit with a fairly large negative value of Δ_{Mars} and initially rather close to the resonance (initial orbit D (512 Taurinensis): $a = 2.190$, $e = 0.254$, $i = 8.7°$, $\Delta_{\mathrm{Mars}} = -0.15$ AU, $\Delta_{\mathrm{resonance}} = +0.08$ AU). Most of the impacts occur within $1000 \cdot 10^6$ y. The larger number of $\sim 800 \cdot 10^6$ y impact times, compared to fig. 7, is a result of the higher inclination. The larger value of impact times $< 50 \cdot 10^6$ y is the result of the smaller $\Delta_{\mathrm{resonance}}$. Earth impact probability $= 10.1\%$.

pheric destruction would not be expected to decrease the impact flux significantly.

Figure 14 also gives the calculated distribution of the elongation of the true geocentric radiant (λ). As discussed elsewhere [45, 63], this distribution determines the distribution of the expected time of day of the meteorite falls. A significant excess of afternoon falls, as observed for chondrites, requires a marked concentration of radiants with $\lambda < 90°$. For the case shown, typical of all four calculated, radiants of $\leqslant 90°$ are in excess by a factor of ~ 1.6. However, the previous work cited showed that a much larger excess, *i.e.* a factor of ~ 5, is required to produce the observed A.M./P.M. asymmetry. This results from the effects of the Earth's gravitational field, the inclination of the Earth's

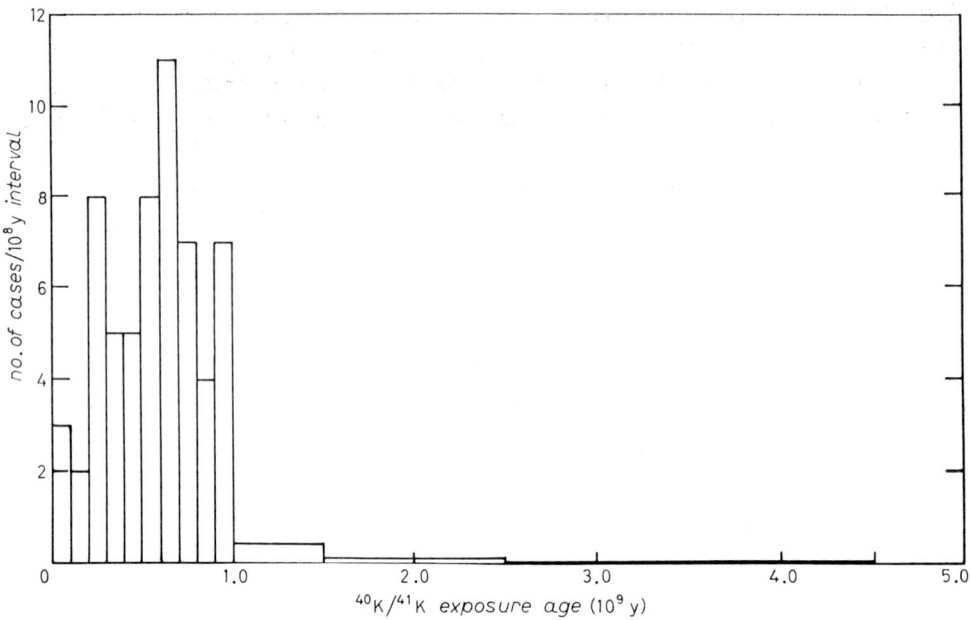

Fig. 13. – Observed exposure ages of iron meteorites (data from ref. [55]).

axis and, most important, that meteorites with radiants $< 90°$ still have a high probability of impacting in the morning hemisphere, unless the elongation of the radiant is near $0°$. In fact, observed A.M. meteorite falls for which radiants are known almost all have radiants less than $90°$ [62, 63]. Therefore, no significant A.M./P.M. asymmetry is expected for these bodies, and none is found. Velocity and impact rate for Venus, Earth and Mars are summarized in tables II and III.

It is concluded that, when the combined effects of secular, Mars and Earth perturbations are considered, the low-inclination asteroidal bodies near the inner edge of the asteroid belt seem to provide an adequate source of differentiated meteorites. This does not preclude the possibility that other sources may also be important. However, none is known at present which will provide

TABLE II. – *Fraction impacting the terrestrial planets.*

Starting orbit	Percent		
	Venus	Earth	Mars
A	9.8	10.1	12.1
B	9.4	10.4	12.2
C	9.9	9.6	14.5
D	10.5	10.1	9.5

TABLE III. – R.m.s. *velocity of impacts* (km/s).

Starting orbit	Venus	Earth	Mars
A	19.2	15.1	7.3
B	22.2	16.8	7.7
C	22.0	16.3	7.35
D	20.2	15.9	8.60

so great a yield and be consistent with the observed iron meteorite exposure ages. A more complete theoretical treatment of this problem would also include the effects of the ν_{16} resonance (fig. 5). This effect is currently being studied. The principal effect of this resonance is to change the amplitude of oscillations

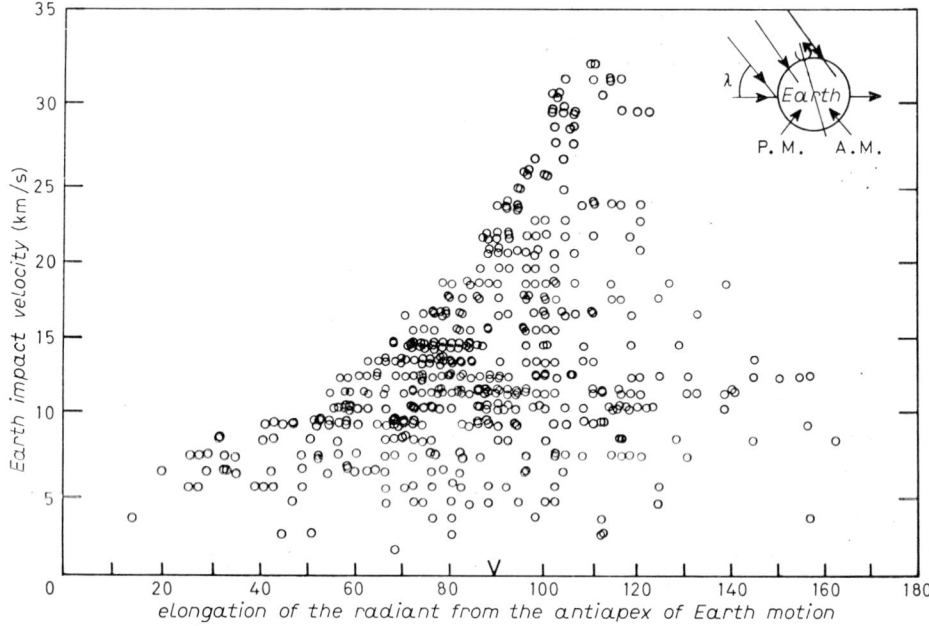

Fig. 14. – Distribution of geocentric velocities *vs.* elongation of the radiant from the antiapex of Earth's motion. Although the radiants are predominantly $< 90°$, the excess of small radiants is not sufficient to produce a pronounced afternoon excess. The initial orbit is that of orbit D (fig. 8) but the distribution is insensitive to the choice of initial orbit.

in inclination rather than those of eccentricity, and, therefore, does not directly change the perihelion distance of fragments. Although conclusions must be tentative at this stage, it does not appear likely that this resonance will have major consequence for the starting orbits considered here.

7. – Concluding remark.

The journey of this material from Earth planetesimals to differentiated meteorites has been a long one, both in space and time. In its entirety it may not even have occurred. However, meteorites and meteoroids are fragments of small bodies which were present in the early solar system and similar journeys must have occurred. To understand them is an important goal of planetary studies.

REFERENCES

[1]   T. C. CHAMBERLIN: *Carnegie Inst. Washington Yearb.*, **3**, 195 (1904).
[2]   V. S. SAFRONOV: *Vopr. Kosmog.*, **7**, 63 (1958).
[3]   V. S. SAFRONOV: *Vopr. Kosmog.*, **7**, 59 (1960).
[4]   A. H. MARCUS: *Icarus*, **4**, 267 (1965).
[5]   A. H. MARCUS: *Icarus*, **11**, 76 (1969).
[6]   S. J. WEIDENSCHILLING: *Icarus*, **22**, 426 (1974).
[7]   W. K. HARTMANN: *Icarus*, **24**, 504 (1975).
[8]   A. H. JAFFEY, K. F. FLYNN, L. E. GLENDENIN, W. C. BENTLY and A. M. EASLING: *Phys. Rev. C*, **4**, 1889 (1971).
[9]   G. W. WETHERILL: *Annu. Rev. Nucl. Sci.*, **25**, 283 (1975).
[10] G. W. WETHERILL: *Proceedings of the VIII Lunar Science Conference* (1976) p. 3245.
[11] G. W. WETHERILL: *Protostars and Planets*, edited by T. GEHRELS (Tucson, Ariz., 1979), p. 545.
[12] V. S. SAFRONOV: *Evolution of the Protoplanetary Cloud and Formation of the Earth and Planets* (Moscow, 1969) (translated for NASA and NSF by Israel Program for Scientific Translations (1972)).
[13] V. S. SAFRONOV: *Vopr. Kosmog.*, **7**, 59 (1960).
[14] J. W. DOHNANYI: *J. Geophys. Res.*, **74**, 2531 (1969).
[15] F. L. WHIPPLE: *Mem. Soc. Sci. Liège*, Ser. 6, **9**, 101 (1976).
[16] G. W. WETHERILL: *Conference on Origin of Mare Basalts and Their Implications for Lunar Evolution*, Lunar Science Institute, Contr. 234 (Houston, Tex., 1975), p. 184.
[17] E. J. ÖPIK: *Proc. R. Irish Acad. Sect. A*, **54**, 165 (1951).
[18] E. J. ÖPIK: *Interplanetary Encounters* (New York, N. Y., 1976).
[19] J. R. ARNOLD: *Astrophys. J.*, **141**, 1536 (1965).
[20] G. W. WETHERILL: *Origin and Distribution of the Elements*, edited by L. H. AHRENS (Oxford, 1968), p. 1178.
[21] G. W. WETHERILL: *Meteorite Research*, edited by P. MILLMAN (Dordrecht, 1969), p. 573.
[22] G. W. WETHERILL: *Comets, Asteroids, Meteorites*, edited by A. H. DELSEMME (Toledo, O., 1976), p. 283.
[23] E. ANDERS and J. R. ARNOLD: *Science*, **149**, 1494 (1965).
[24] G. W. WETHERILL and J. G. WILLIAMS: *J. Geophys. Res.*, **73**, 635 (1968).
[25] W. K. HARTMANN: *Astrophys. Space Sci.*, **17**, 48 (1972).
[26] G. NEUKUM and D. V. WISE: *Science*, **194**, 1381 (1976).

[27] F. Tera, D. A. Papanastassiou and G. J. Wasserburg: *Earth Planet. Sci. Lett.*, **22**, 1 (1974).
[28] G. W. Wetherill: *J. Geophys. Res.*, **72**, 2429 (1967).
[29] C. R. Chapman and D. R. Davis: *Science*, **190**, 553 (1975).
[30] G. W. Wetherill: *Proceedings of the VI Lunar Science Conference* (1975), p. 1539.
[31] W. Compston, J. J. Foster and C. M. Gray: *Proceedings of the VIII Lunar Science Conference* (Houston, Tex., 1977), p. 199.
[32] G. Turner and P. H. Cadogan: *Proceedings of the VI Lunar Science Conference* (1975), p. 1509.
[33] E. K. Jessburger, J. C. Huneke, F. A. Podosek and G. J. Wasserburg: *Proceedings of the V Lunar Science Conference* (1974), p. 1419.
[34] V. S. Reed and E. W. Wolfe: *Proceedings of the VI Lunar Science Conference* (1975), p. 2443.
[35] S. J. Weidenschilling: *Icarus*, **26**, 361 (1975).
[36] G. W. Wetherill: *Tectonophysics*, **13**, 31 (1972).
[37] W. M. Kaula and P. E. Bigeleisen: *Icarus*, **25**, 18 (1975).
[38] S. J. Weidenschilling: *Astron. J.*, **80**, 145 (1975).
[39] W. H. Ip: *Comets, Asteroids, Meteorites*, edited by A. W. Delsemme (Toledo, O., 1976), p. 485.
[40] O. L. Hansen: *Astron. J.*, **81**, 74 (1976).
[41] C. R. Chapman, D. Morrison and B. Zellner: *Icarus*, **25**, 104 (1975).
[42] G. W. Wetherill and J. G. Williams: *Proceedings of the II International Conference on Origin and Abundance of the Elements*, edited by L. H. Ahrens (Oxford, 1978), p. 19.
[43] V. S. Safronov: *Proceedings of the Soviet-American Conference on the Cosmochemistry of the Moon and Planets* (Moscow, 1975), p. 624.
[44] E. Anders: *Geochim. Cosmochim. Acta*, **35**, 516 (1971).
[45] R. N. Clayton, N. Onuma and T. K. Mayeda: *Earth Planet. Sci. Lett.*, **30**, 10 (1976).
[46] R. S. Rajan: *Geochim Cosmochim. Acta*, **38**, 777 (1974).
[47] D. E. Gault, E. M. Shoemaker and H. J. Moore: *NASA Tech. Note 1767*, 39 (1963).
[48] J. D. O'Keefe and T. J. Ahrens: *Proceedings of the VII Lunar Science Conference* (1976), p. 3007.
[49] D. G. Gault: *Meteoritics*, **4**, 177 (1969).
[50] J. G. Williams: *Secular perturbations in the solar system* (unpublished Ph. D. Thesis, UCLA, 1969).
[51] J. G. Williams: *Physical Studies of Minor Planets*, edited by T. Gehrels (NASA, 1971), p. 267.
[52] J. G. Williams: *Eos*, **54**, 233 (1973).
[53] G. W. Wetherill: *Geochim. Cosmochim. Acta*, **40**, 1297 (1976).
[54] G. W. Wetherill: *Annu. Rev. Earth Planet. Sci.*, **2**, 303 (1974).
[55] H. Voshage: *Z. Naturforsch. Teil A*, **22**, 477 (1967).
[56] G. F. Herzog and E. Anders: *Geochim. Cosmochim. Acta*, **35**, 239 (1971).
[57] J. T. Wasson and J. Kimberlin: *Geochim. Cosmochim. Acta*, **31**, 2065 (1967).
[58] J. T. Wasson: *Meteorites*, Chap. XI (New York, N. Y., 1974).
[59] E. Vilcsek and H. Wänke: *Radioactive Dating* (Vienna, 1963), p. 381.
[60] J. C. Cobb: *J. Geophys. Res.*, **72**, 1329 (1967).
[61] F. Begemann, E. Vilcsek, L. E. Nyquist and P. Signer: *Earth Planet. Sci. Lett.*, **9**, 317 (1970).
[62] G. W. Wetherill: *Science*, **159**, 79 (1968).
[63] G. W. Wetherill: *J. Geophys. Res.*, **74**, 4402 (1969).

Molecules and Grains in Interstellar Space.

N. C. WICKRAMASINGHE

University College - Cardiff, U.K.
Department of Applied Mathematics and Astronomy - Cardiff, U.K.

1. - Interstellar molecules.

1˙1. *Introduction.* – Optical detections of the radicals CH, CH⁺, CN, CN⁺ are now nearly four decades old. Yet an understanding of how even these simple molecules form under astronomical conditions has scarcely begun. For instance, there is a major problem that has persisted for over a generation to explain the observed ratio CH⁺/CH~ 1. Although the formation rate of CH⁺ is rapid through the reaction

$$C^+ + H \rightarrow CH^+ + h\nu$$

with a rate constant of $\sim 10^{-17}$ cm³ s⁻¹, the destruction of CH⁺ proceeds too fast through the reactions

$$CH^+ + H_2 \rightarrow CH_2^+ + H$$

and

$$CH^+ + e \rightarrow C + H \,,$$

the former having a rate constant of $\sim 10^{-9}$ cm³ s⁻¹ and the latter a rate constant of $\sim 10^{-7}$ cm³ s⁻¹. The fact that the theoretical expectation CH⁺/CH$\ll 1$ is not borne out by the observational data is already quite disturbing for the theory of molecule formation in the gas phase in interstellar clouds. DAL-GARNO [1] has recently reviewed the subject of interstellar CH, CH⁺, and we refer the reader to this review for more details.

Theoretical arguments for the presence of molecules such as H_2 have been discussed by astronomers for more than a generation. But direct evidence of any molecules, apart from CH, CH⁺, CN, CN⁺, is relatively quite recent. New data on interstellar molecules have come from radio astronomy and later from millimetre wave as well as ultraviolet astronomy. Radio astronomers entered the field with their discoveries of OH masers in the 1960's. This entry

TABLE I. – *Observed interstellar molecules in dense molecular clouds.*

Inorganic		Organic
H_2 hydrogen	diatomic	CH methylidyne
OH hydroxyl		CH^+ methylidyne ion
SiO silicon monoxide		CN cyanogen
SiS silicon sulfide		CO carbon monoxide
NS nitrogen sulfide		CS carbon monosulfide
SO sulfur monoxide		
H_2O water	triatomic	CCH ethynal
N_2H^+		HCN hydrogen cyanide
H_2S hydrogen sulfide		HNC hydrogen isocyanide
SO_2 sulfur dioxide		HCO^+ formyl ion
		HCO formyl
		OCS carbonyl sulfide
		HNO nitroxyl
NH_2 ammonia	4-atomic	H_2CO formaldehyde
		HNCO isocyanic acid
		H_2CS thioformaldehyde
		C_3N cyanoethynyl
	5-atomic	H_2CHN methanimine
		H_2NCN cyanamide
		HCOOH formic acid
		HC_3N cyanoacetylene
		H_2C_2O ketan
	6-atomic	CH_3OH methanol
		CH_3CN cyanomethane
		$HCONH_2$ formamide
	7-atomic	CH_3NH_2 methylamine
		CH_3C_2N methylacetylene
		$HCOCH_3$ acetaldehyde
		H_2CCHCN vinyl cyanide
		HC_5N cyanodiacetylene
	8-atomic	$HCOOCH_3$ methyl formate
	9-atomic	$(CH_3)_2O$ dimethyl ether
		C_2H_5OH ethanol
		HC_7N cyanotriacetylene
		C_2H_5CN ethyl cyanide
	11-atomic	HC_9N cyano-octa-tetrayne

opened the flood-gates for a spate of further discoveries of large numbers of organic and inorganic molecules. Today the list of molecules known to exist in interstellar space includes some forty chemical species of widely ranging degrees of complexity. This list (table I) continues to expand, as more and more transitions are predicted, and more molecules are searched for and found. The limits of discovery in this field are still impossible to define, and are most probably almost infinite!

The molecules listed in table I are not uniformly distributed throughout the interstellar medium. The thickness of the interstellar medium in H I is typically ~ 200 pc in the solar neighbourhood. The mean hydrogen density is $\sim 10^{-24}$ g cm^{-3}, but large density fluctuations are represented by the now quite familiar cloud structure of interstellar space. Cloud densities range from ~ 10 atoms cm^{-3} to $(10^4 \div 10^6)$ cm^{-3}. The higher densities are appropriate to condensations in the larger cloud complexes which have come to be known as giant molecular clouds [2]. Such condensations which make for a large fraction of all interstellar matter have been extensively mapped in CO. It is in these condensations also that most of the complex organic molecules listed in table I are to be found. Giant molecular clouds are known to be sites of active star formation with the presence of large numbers of infra-red sources, as well as O and B type stars.

The dominant nuclear species everywhere in the interstellar medium is of course hydrogen, and the disposition of this species (whether it is atomic, ionized or molecular) is likely to play a controlling role in the gas phase chemistry in any given region. Ionized hydrogen H II can exist in appreciable quantity only in the immediate vicinity of O and B stars. Within H II regions the main constituents are ions and the more refractory kinds of solid particles.

In H I regions, which are opaque to radiation at wavelengths shortward of 912 Å, ionizing electromagnetic radiation is mainly absent. The dominant ionizing process operative in these regions is thus due to cosmic rays, particularly at energies less than 300 MeV, with a smaller contribution from soft X-rays. There are uncertainties in the estimates of low-energy cosmic-ray fluxes, which in turn lead to uncertainties in ionization rates. Although cosmic-ray fluxes at high energies are fairly well determined, the low-energy tail of the cosmic-ray spectrum is still quite uncertain. Many indirect arguments, including the idea that cosmic rays must maintain the heating of clouds to gas kinetic temperatures ~ 50 K, give a cosmic-ray ionization rate [3]

$$\xi_{\text{CR}} \simeq 10^{-17} \text{ s}^{-1} \text{ per atom} .$$

An intricate theory of interstellar chemistry hangs on this rather uncertain number.

1‵2. *Molecular hydrogen, carbon monoxide.* – Grain surface effects are thought the to be important in the production of at least the one molecular species

H_2 [4]. Hydrogen atoms incident on a grain could recombine thus:

$$H + H + grain \rightarrow H_2 + grain .$$

The excess energy of a newly formed H_2 molecule goes in part towards heating the grain and in part to eject the molecule off the grain surface. Limiting conditions involve the values of the grain temperature and the grain opacity. The grain temperature has to be sufficiently low (say $T_g < 12$ K) to permit long enough retention times of incident molecules, and a grain optical depth of $\tau \sim 1 \div 2$ at visual wavelengths appears necessary for shielding the H_2 molecules against dissociative ultraviolet radiation. These conditions are satisfied within almost all « dense » clouds, and the hydrogen will then be converted into molecular form on a very short time scale. These expectations are borne out by the recent ultraviolet detections of H_2 in molecular clouds. As a rough rule of thumb, we could assume that all hydrogen in relatively cold clouds becomes converted into H_2 when the gas density exceeds about 1000 cm^{-3}.

A similar situation is relevant for the molecule CO. For clouds with gas densities upwards of 1000 cm^{-3} about 10% of all the interstellar C appears to be tied up in the form CO. This is due in part to the strong binding of CO (11 eV), in part to the long radiative lifetime (10^{10} s) and in part to a very efficient mechanism of formation for CO. Unlike for the case of H_2, the formation of CO on grain surfaces is likely to be unimportant, both on account of the lower densities of C and O and also because recombinations of absorbed atoms with H are likely to occur in all cases of interest. CO could be formed most efficiently in mass flows from stars where thermodynamic concentrations prevail, or through the destruction of grains of an appropriate composition. For instance, for the break-up of grains comprised mainly of H_2CO, CO would be expected to be a dominant component, because of the weak binding of H_2 and CO in H_2CO.

1'3. *General remarks on molecule formation.* – The interstellar medium, even in its relatively denser phases, is grossly nonthermodynamic. The black-sphere temperature, grain temperature, the temperature of the dilute interstellar radiation field and the gas kinetic temperature are all unequal. Densities of gas are in general so low that 3-atom encounters play a totally negligible role in molecular association. That is to say, reaction rates for processes of the type

$$A + B + C \rightarrow AB + C$$

are negligibly small.

Under standard cloud conditions, where the interstellar radiation field is essentially unattenuated, dissociative lifetimes of gas phase molecules range within (100 ÷ 1000) years. These time scales are considerably shorter than the gravitational-collapse times $\sim (G\varrho)^{-\frac{1}{2}}$ of clouds. Thus the continual presence of molecules in clouds implies an ongoing, *in situ* formation process in these clouds.

If $k = \langle \sigma v \rangle$ is the rate constant for a radiative association

$$A + B \to AB + h\nu,$$

we have a formation rate

$$\frac{\mathrm{d}n(AB)}{\mathrm{d}t} = kn(A)\,n(B),$$

where $n(X)$ denotes the concentration of species X and t is the time. For a species AB whose dissociative lifetime under interstellar conditions is τ, we have a steady-state concentration

$$n(AB) = k\tau n(A)n(B).$$

For the case when A and B are both neutral species (*e.g.* C, H) the typical rate constant k is $\sim 10^{-18}$ cm^3 s^{-1}. Setting $n(\mathrm{H}) \simeq 10$, $n(\mathrm{C}) \simeq 10^{-2}$ for « standard » cloud conditions, we get $n(\mathrm{CH}) \simeq 10^{-10}$ cm^{-3} with $\tau \sim 10^{10}$ s. This is 10^2 lower than the observed density $\sim 10^{-8}$ cm^{-3}.

1`4. Ion-molecule chemistry. – Conditions are more favourable for molecule formation when one of the reacting species A or B is charged. These are known as charge exchange reactions, examples of which are

$$\mathrm{H}_2^+ + \mathrm{H}_2 \;\to \mathrm{H}_3^+ + \mathrm{H} \qquad (k = 2 \cdot 10^{-9} \text{ cm}^3 \text{ s}^{-1}),$$

$$\mathrm{O}^+ + \mathrm{H}_2 \;\to \mathrm{OH}^+ + \mathrm{H} \qquad (k = 10^{-9} \text{ cm}^3 \text{ s}^{-1}),$$

$$\mathrm{CO}^+ + \mathrm{H}_2 \to \mathrm{HCO}^+ + \mathrm{H} \qquad (k = 1.4 \cdot 10^{-9} \text{ cm}^3 \text{ s}^{-1}).$$

(These reactions are in fact a combination of charge exchange and chemical exchange. Chemical exchanges between neutrals of the type $AB + C \to \to AC + B$ already have rate constants $k \sim 5 \cdot 10^{-11}$ cm^3 s^{-1}. Chemical-exchange reactions between neutrals, however, require activation energies of $\sim (0.2 \div 0.3)$ eV, so hot gas or fast grains are required to promote reactions [5].) On account of the greatly increased rate constants that are relevant when charged species are involved, ion-molecule chemistry in interstellar space looks at first sight promising. Indeed, successes of this theory have been considerable with regard to correct predictions of *some* of the abundances of the smaller organic and inorganic molecules and radicals.

The starting point of charge exchange reaction chains involves ionization of the abundant species H_2 and He (see, for example, ref. [6]). We require first a knowledge of the cosmic-ray ionization rate, which is set somewhat arbitrarily in the range

$$\xi \simeq (10^{-17} \div 10^{-18}) \text{ s}^{-1} \text{ per molecule}.$$

H_2 and He ionization proceed thus

$$H_2 + CR \nearrow \quad H_2^+ + e + CR^1 \qquad (k = 0.95\xi),$$
$$\searrow \quad H^+ + H + e + CR^1 \qquad (k = 0.05\xi),$$

$$He + CR \rightarrow He^+ + e + CR^1 \qquad (k = \xi),$$

where CR, CR^1 are initial and final cosmic-ray energies. The next step in the argument involves the *assumption* that most of the C is in the form CO, as is indicated by the millimetre wave observations. (The formation of CO could involve reactions of the type $CH^+ + O \rightarrow CO + H^+$, $OH + C \rightarrow CO + H^+$, or it could involve pre-existence in mass flows from stars.) A complex network of reactions is considered, including

$$H_2^+ + H_2 \rightarrow H_3^+ + H,$$
$$He^+ + H_2 \rightarrow H_2^+ + He \rightarrow H + H^+ + He,$$
$$CO + H_3^+ \rightarrow HCO^+ + H_2,$$
$$N_2 + H_3^+ \rightarrow N_2H^+ + H,$$
$$N_2H^+ + CO \rightarrow HCO^+ + N_2.$$

Although this theory, with its many free parameters, has no difficulty in explaining the abundances of the smaller molecular species, it fails conspicuously to account for larger molecules—molecules which are even as complex as H_2CO. Grain surface reactions have been discussed as a possible way out, but such schemes are immediately self-defeating. If complex molecules can form through grain surface reactions, at low grain temperatures they would condense onto grains and not evaporate back into the gas phase to any significant degree.

1'5. *The synthesis of larger molecules.* – One of the most striking features of the astronomical data is that the abundances of the larger molecular species, where they could be determined, are in general comparable to the abundances of the smaller radicals such as CH, CH^+. This property is difficult to explain in terms of purely gas phase reaction networks operating at low densities. One would generally expect to obtain diminishing abundances for larger and larger molecules, instead of the situation that is actually observed.

A solution of this problem could involve a radically different approach to molecule formation in space. Sir Fred HOYLE and the present author [7] suggested a mechanism involving the ultraviolet-light processing of grain mantles which could lead to the evolution of stable polymeric compositions of such mantles. The smaller radicals and molecules observed in interstellar clouds could then be considered as break-up fragments of the polymeric mantles and the conditions of break-up could produce the equality of abundances as observed.

GOLDANSKII [8] has also discussed the elaboration of organic molecules in the solid state at very low temperatures. The argument is that interstellar gas phase reactions yield a hybrid mix of low-molecular-weight organic molecules which condense on grains. Polymerization reactions are then supposed to occur in the solid phase by quantum-mechanical tunnelling between adjacent molecules. GOLDANSKII has argued that at low temperatures appropriate to grains an entropy factor $Q + TS$ becomes unimportant to an extent that slightly endothermic reactions become weakly exothermic, so that polymerization reactions can proceed spontaneously. GOLDANSKII substantiates his claims with experimental results on the polymerization of H_2CO under laboratory conditions.

Sir Fred HOYLE and the author [9] have also considered the possibility of organic polymers condensing in the mass flows from stars. Although thermodynamic conditions are likely to prevail throughout mass flows from cool giant stars, the situation is found to be different for massive early type stars of mass $> 50 \, M_\odot$. These stars are pulsationally unstable when they reach the main sequence, and they have an initial mass loss driven essentially by the nuclear energy generated in the star. This mass loss could be as high as $(10^{-3} \div 10^{-2}) \, M_\odot \, y^{-1}$. At the stellar photosphere where a typical temperature is $T \sim 50\,000$ K all the hydrogen is ionized and there are no molecules. However, recombination to H I occurs further out in the mass flow where the temperature falls to ~ 5000 K. At this «level» in the flow an effective low-temperature photosphere becomes established, and thermodynamic conditions persist to a good approximation. For solar abundances of the elements we have $CO/C \sim 10^{-6}$ in the equilibrium $C + O \rightarrow CO$. The next step in the argument is that recombination of $C + O \rightarrow CO$ further out in the flows occurs more slowly than the condensation of much of the excess free carbon into carbonaceous polymers. Rapid exponential growth of polymers interspersed by fragmentation leads to a logic similar to biological replication. We have argued that polysaccharides are the main product in such a condensation process.

1'6. *Diffuse interstellar bands.* – In addition to the well-identified atomic and molecular lines, stellar spectra are known to have a number of diffuse unidentified absorption bands at well-defined wavelengths. These features are characterized by half-widths ranging from $(100 \div 50)$ Å for the broadest band to less than 5 Å for the narrowest unindentified lines. These widths contrast strikingly with widths of several hundredths of an ångström for atomic or molecular lines. Since their original discovery over forty years ago, some thirty or so bands have been listed. Table II gives the half-widths and central wavelengths of some of the strongest and best-studied features.

TABLE II.

Central wavelength (Å)	Half-width (Å)
4430	28
4760	30
4890	30
5776	16
5778	4
5780	4
5797	3
6180 (6176)	$25 \div 40$
6203	3
6270	4
6284	4
6614	2

Of these bands the one centred at $\lambda = 4430$ Å is by far the strongest in intensity and also the best documented. The extinction at the band centre is $\sim (0.1 \div 0.2)$ mag/kpc.

The correlation of band strengths one with another and also with the general visual extinction of starlight has indicated a possible connection with grains.

TABLE III.

Laboratory data		Astronomical data	
Central wavelength (Å)	Range (Å)	Central wavelength (Å)	Range (Å)
6663	$1 \div 2$	6661	$1 \div 2$
6614	$1 \div 2$	6614	$1 \div 2$
6289	$1 \div 2$	6284	4
6284	$1 \div 2$		
6174	14	6175	~ 20
4428	40	4428	~ 30

Simple molecules, perhaps in grain matrices, have been considered as possible candidates, but without success. Some years ago JOHNSON [10, 11] obtained data on the absorptions produced by a magnesium porphyrin synthesised in the laboratory. Comparisons between the strongest absorptions in the laboratory molecule with the strongest of the diffuse interstellar features are shown in table III.

These correspondences strongly point to the general correctness of Johnson's claim, though the *precise* molecule or molecules responsible for these bands may not be the very particular porphyrin considered by JOHNSON.

2. – Interstellar grains.

2˙1. *Introductory remarks*. – For nearly half a century astronomers have been debating the composition of interstellar grains. There have been a welter of new observations, several theories and a succession of changing fashions— but no convergent solution has emerged so far [12]. All the indications are that we have consistently followed a wrong path.

Observations of interstellar extinction and polarization at visual wavelengths provide perhaps the most stringent constraints on permissible grain models. To produce the observed average visual extinction of 2 mag/kpc, we require about $(2 \div 5)\%$ by mass of the interstellar medium to be in the form of grains which have typical dimensions of several tenths of a micron. This implies that only the CNO elements could contribute significantly to the mass of grains. Indeed a large fraction $\sim 80\%$ of all the interstellar CNO has to be in the form of grains, and this is consistent with the observed gas phase depletions of these elements as inferred from recent far-ultraviolet spectra of stars. Further, from the data on interstellar polarization—both linear as well as circular—we know that the grains causing visual extinction must be asymmetric in their shape (probably rodlike) and have predominantly dielectric properties [13].

A great deal has been written in recent years concerning the possible role of silicates and metallic oxides. But these models, if they are meant to represent the bulk of grains, are in flagrant inconsistency with the most fundamental cosmic-abundance data. Mg, Si, Fe are cosmically under-abundant by *at least* a factor of 5 to account for the observed scattering of starlight at visual wavelengths. If the bulk of grains were made up of these elements, it would be impossible to understand how solar abundances are what they are known to be.

2˙2. C, N, O *composition of the dust*. – Thus it seems that only C, N, O could contribute significantly to the scattering of starlight at visual wavelengths. And here there are two broad divisions in types of material that are possible:

1) a mixture of inorganic ices, dominated by H_2O ice;

2) a mixture of organic solids: the C, N, O elements occurring in molecular combinations with H that have an organic character.

The first of these possibilities has been discussed and scrutinised for the longest time, starting with the early work by VAN DE HULST in the 1940's. But it was evident right from the start that ices alone could not fit the bill. Figure 1 shows a comparison of the best-fitting calculation for the extinction properties of ice spheres (without impurities) and observations. This calculation is for a size distribution of ice spheres that VAN DE HULST believed could arise from

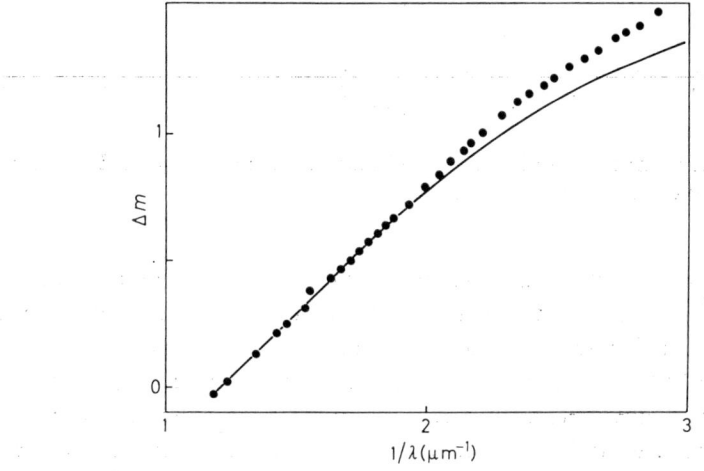

Fig. 1. – Theoretical extinction curve for a size distribution of pure ice grains compared with Nandy's observations (•) for stars in the Cygnus direction. The curve is for ice grains with refractive index $n = 1.33$ with a mean radius $r_1 = 0.34$ μm. Normalization is to $\Delta m = 0.409$ at $\lambda^{-1} = 1.62$ μm^{-1}, $\Delta m = 0.726$ at $\lambda^{-1} = 1.94$ μm^{-1}.

a balance between formation and destruction processes in interstellar clouds. The points are observations due to NANDY over the visual wavelength region. The disagreement seen here is endemic to all *pure* ice grain models. It is, however, possible to obtain significantly better fits by introducing an arbitrarily high value (~ 0.1) for the absorption coefficient, and attributing this to metal impurities. (If the absorption is due to Fe impurities, we require three times more Fe in grains than is allowed for by the observed Fe/O ratio.) Alternatively, one could introduce graphite or metal cores into the ice grains to obtain better fits to the extinction data. But both these « refinements » to the ice grain theory are *ad hoc*, arbitrary, and cannot be justified. Even if we allowed ourselves the liberty of such *ad hoc* refinements, there are many severe problems. Firstly, there are several parameters in the model which need to be arbitrarily fitted, and, secondly, the agreement with the data, even for optimal choice of parameters, is not perfect.

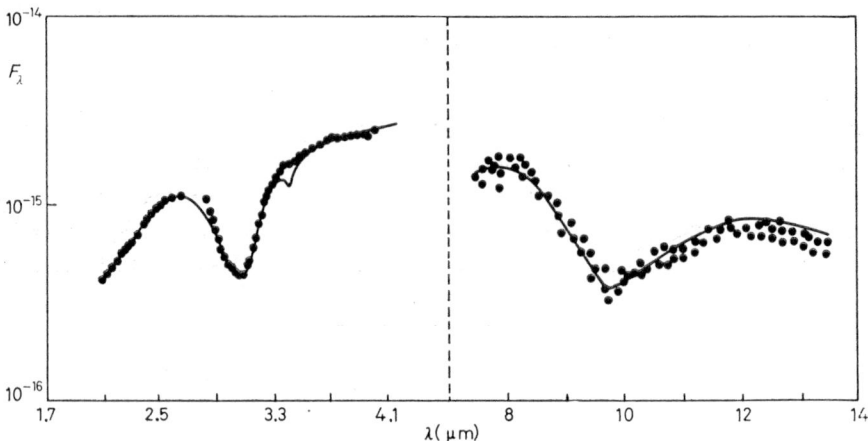

Fig. 2. – Calculated infra-red emission for cellulose in a mass flow model (solid curve) compared with data for the BN object (points).

There are several advantages to be gained from an organic composition of the dust. A significant difference from water-ice grains results from the fact that organic polymers are more refractory and can withstand *in vacuo* temperatures of 700 K or more. The infra-red emission from galactic sources could be explained quite readily if the polymers that condense in mass flows from young stars are similar to cellulose [14]. Figures 2 and 3 give examples of the fits between observed infra-red spectra and calculations based on data. for cellulose. The close agreement seen here was considered by Sir Fred HOYLE and the author to provide a strong case for an identification of a polysaccharide very similar to cellulose. Although mixtures of ices and silicates in varying proportions could explain some aspects of the astronomical observations, such mixtures fail to give a satisfactory overall agreement.

One of the most striking features of the interstellar extinction data is their uniformity across much of the Galaxy. Indeed, it has recently been found that the interstellar extinction curve for Magellanic-cloud stars is also similar to the galactic-extinction curve. To reproduce even the gross structure of this extinction curve at visual wavelengths, for any particular type of grain and for a given size distribution function, we need a size parameter to remain constant to within 10 % across much of the Galaxy. It has always been a difficulty to understand compositional invariance and size invariance of grains to such an amazingly high degree of precision. Such an invariance could scarcely be achieved through purely inorganic processes involving random growth and destruction under the highly variable conditions of interstellar space.

2˙3. *The case for biological grains.* – In the rest of this article an outline will be given of a solution to these problems, recently proposed by Sir Fred

HOYLE and the author [15]. This solution, which is somewhat radical, involves the widespread occurrence of terrestrial-style microbiology.

We first note that the most inevitable logic for replicating a whole range of complex molecules, and for reproducing structures of almost invariable size,

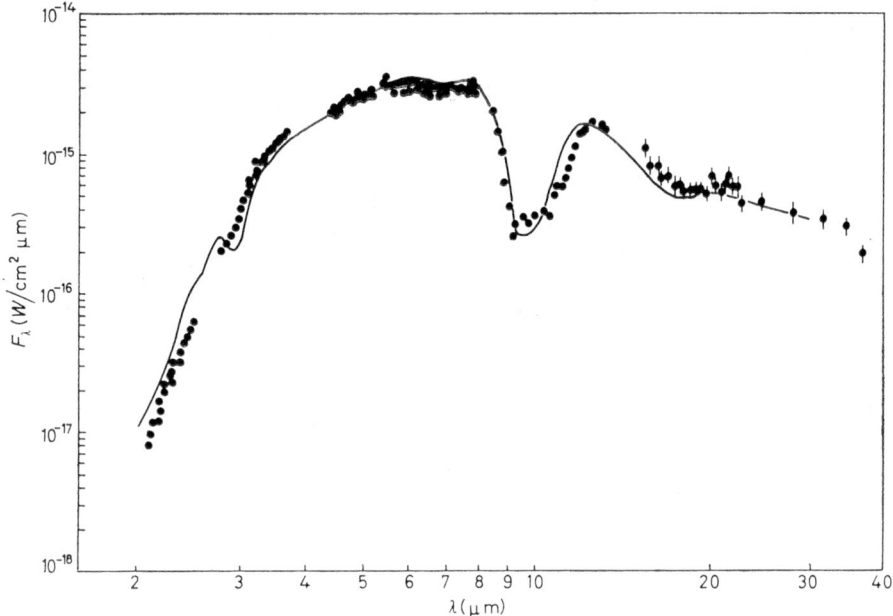

Fig. 3. – Same as in fig. 2 but with data for the source OH 26.5 + 06.

is biology itself. Bacteria have the right sizes and shapes to be *prima facie* candidates for grains. Under terrestrial conditions a single bacterium of mass $\sim 10^{-12}$ g supplied only with H_2, N_2, CO_2, Mg and phosphate ions and visual light could convert all the CNO in the Galaxy into identical bacteria in one week. Under astronomical conditions, where bacterial particles can be propelled around the Galaxy by radiation pressure, limitations to growth arise only from limited access to molecular feedstock. We estimate a regeneration time for all the 10^{40} g of interstellar bacteria to be $(10^8 \div 10^9)$ years, most of the regeneration occurring not in planetary or cometary conditions, but in protostellar or circumstellar situations with $n_{H_2} \sim (10^8 \div 10^{10})$ cm^{-3}, $T \sim (100 \div 300)$ K.

Figure 4 shows the observed size distribution of a representative sample of spore-forming bacteria (cross-sectional diameters of rods and spheres). Terrestrial bacteria are bound by *rigid* outer cell walls comprised mainly of polysaccharides and lipoproteins, with an interior that is comprised of a rich variety of biochemicals and water. Water makes up about 80% by volume of a bacterium. We estimate that interstellar-cloud conditions lead to evaporation of

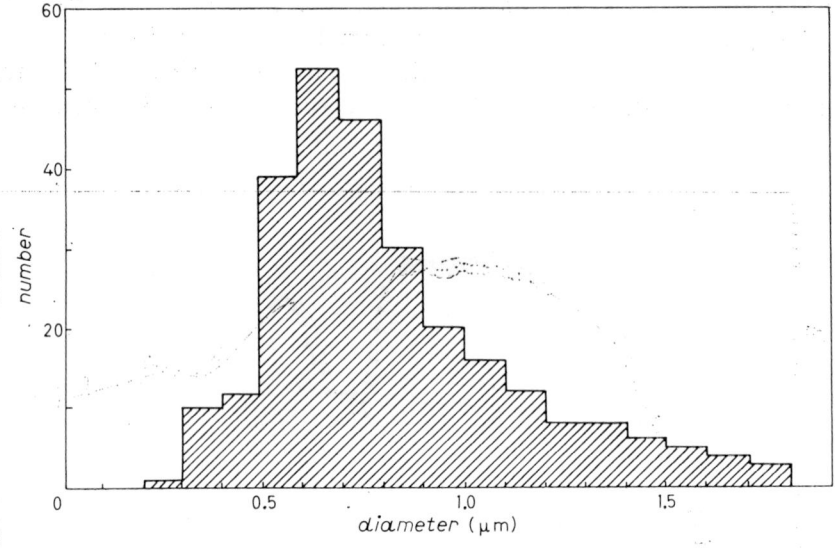

Fig. 4. – Size distribution of spore-forming terrestrial bacteria.

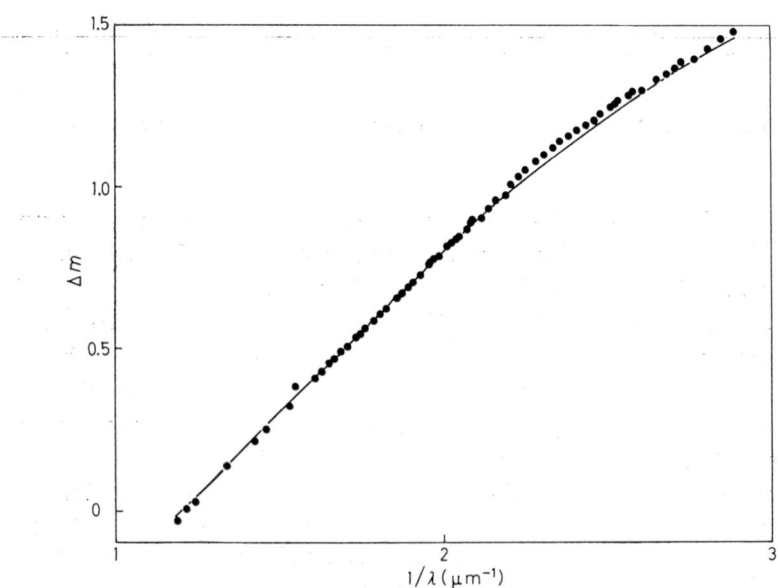

Fig. 5. – Calculated extinction curve for the size distribution of bacteria prescribed by the histogram in fig. 4, assumed to be freeze-dried under interstellar conditions. Normalization is to $\Delta m = 0.409$ at $\lambda^{-1} = 1.62 \, \mu m^{-1}$, $\Delta m = 0.726$ at $\lambda^{-1} = 1.94 \, \mu m^{-1}$. Points are the mean extinction data of Nandy for the Cygnus region.

water to 60 % by volume, assuming the bulk of interstellar C is tied up as bacteria. Thus the mean refractive index of a freeze-dried bacterium is made up of 20 % organic material with $m = 1.5$, 20 % water with $m = 1.3$ and 60 % vacuum with $m = 1$, giving a mean value

$$\overline{m} = 1 + 0.2 \times 0.5 + 0.2 \times 0.3 = 1.16 \,.$$

Figure 5 shows the normalized extinction properties of bacteria thus freeze-dried under interstellar conditions. If one accepts the laboratory data on size distribution as being representative, there are *no* free parameters left here. The calculated curve and consequently this agreement with the astronomical data are then unique. The fit is a consequence of the size distribution $\propto da/a^3$ and the property that $n \simeq 1.16$ for freeze-dried conditions. The agreement is better than for any previously considered model. The excess absorption in the (4000÷5000) Å wave band also turns out to have a profound significance. Photosynthetic bacteria and algae (blue-green algae) have pigments made up of a carotenoid-chlorophyll complex that absorbs over the wave band (4000÷ ÷5000) Å and at (6200÷7000) Å. This is shown in fig. 6. Contributions of various components to the absorption of a blue-green algae are shown here.

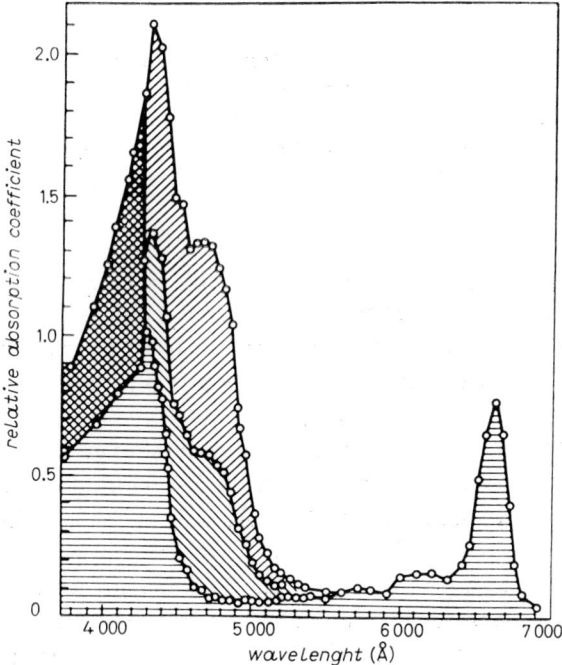

Fig. 6. – The contribution of different molecules to the pigment absorption in the planktonic diatom Nitzchia Clostesium: ▤ chlorophyll *A*, ▨ total carotenoids, ▨ fucoxanthin, ▨ carotenoids other than fucoxanthin.

Figure 7 shows the agreement between the astronomical excess *over and above*
the pure scattering curve due to bacteria and the prediction for algal-type
pigments. The observed excess absorption amounting to ~ 0.1 mag/kpc im-
plies a pigment mass density $\varrho \approx 10^{-28}$ g cm^{-3}, $\sim 1\%$ of the mass of bacteria.

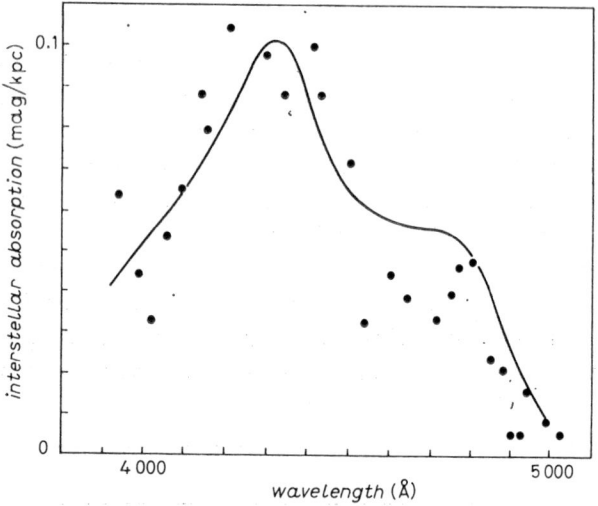

Fig. 7. – Points represent the excess interstellar extinction over the wave band
$(4000 \div 5000)$ Å differenced against our theoretical calculation for scattering by freeze-
dried bacteria (fig. 5). Normalization is to $\Delta m = 0.1$ mag at 4300 Å. Curve is the
expected absorption excess by phytoplankton pigments taken from fig. 6 with normalization to $\Delta m = 0$ at $\lambda = 5100$ Å, $\Delta m = 0.1$ at $\lambda = 4300$ Å.

Viruses and bacteriophages can be thought of as genetic units from which
bacteria themselves are assembled. Some free viral component is also to be
expected in the interstellar medium. Viruses have average radii $a \sim 0.02$ μm
and, since they are basically dielectric ($n \approx 1.5$), they can be important in
producing scattering at $\lambda < 1500$ Å.

Figure 8 shows the interstellar-extinction data over the wave band 1 μm$\div$
$\div 1200$ Å. We first note that the upturn in UV could be due to viruses. To
produce the observed extinction at 1200 Å, only $\sim 5\%$ of CNO is required in
the form of viruses. Secondly, we note that the interstellar-extinction data
show a pronounced hump at $\lambda \simeq 2200$ Å. This feature is usually attributed to
graphite, but the requirement for spherical graphite poses a difficulty. An
organic explanation seems preferable. Figure 9 shows the distribution of absorption peaks for over 185 representative biochemical chromophores involving
conjugated double bonds which we expect to find in bacterial cells. The solid
curve is a calculated average absorption, seen to peak strongly on $\lambda = 2200$ Å.

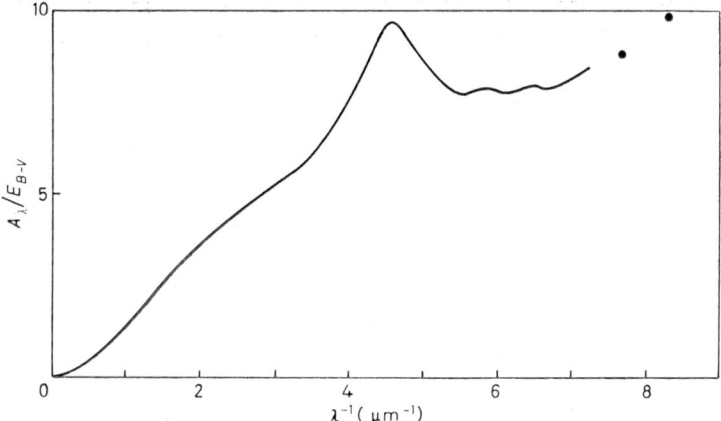

Fig. 8. – Mean interstellar-extinction curve.

The $\lambda = 2200\,\text{Å}$ interstellar feature is thus seen to be readily explained on the basis of biochemical chromophores which we expect to find in bacteria and viruses. The minor ripples at $\lambda^{-1} = 5.8$, $6.5\ \mu\text{m}^{-1}$ are also most likely due to biological chromophores involving single bond σ-electrons in saturated hydrocarbons.

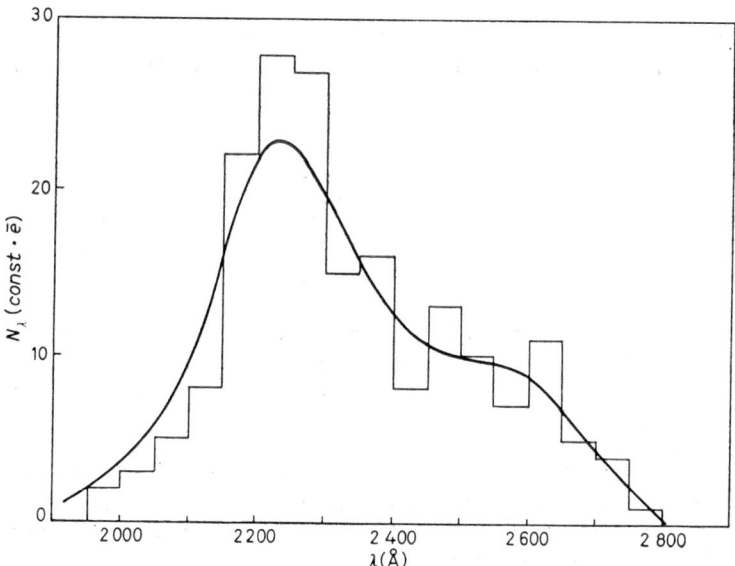

Fig. 9. – Histogram shows distribution of main absorption peaks for naturally occurring chromophores in 186 molecules. Solid curve is the computed average absorption curve due to this ensemble.

Figure 10 shows the gross structure of the interstellar-extinction curve decomposed into bacterial and viral scattering components *plus* an excess absorption

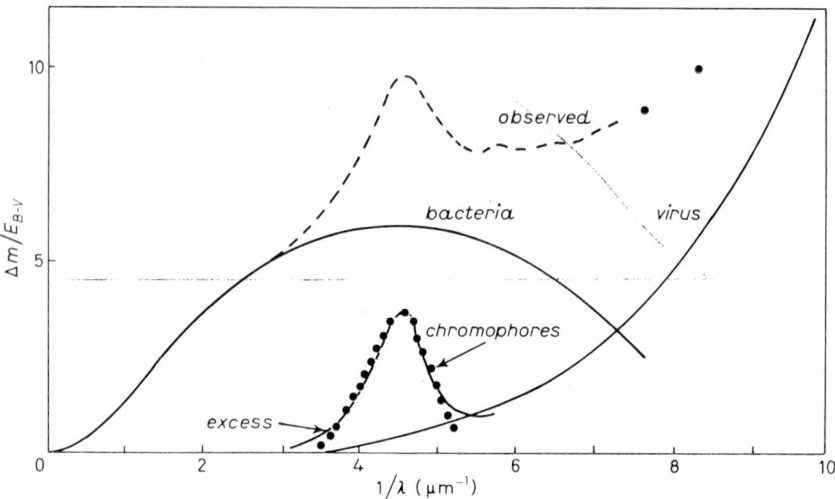

Fig. 10. – The observed interstellar-extinction curve decomposed into bacterial and viral scattering components and an excess absorption calculated for curve in fig. 9 is shown to be in good agreement with the excess absorption as shown.

in the middle ultraviolet. Chromopheric absorption in biomolecules could fit this excess remarkably well with only 10 % of CNO in the Galaxy in the form of these molecules.

REFERENCES

[1] A. DALGARNO: in *Atomic Processes and Applications*, edited by P. G. BURKE and B. L. MOISEIWITSCH (Amsterdam, 1976).
[2] P. M. SOLOMON: in *Giant Molecular Clouds*, edited by P. M. SOLOMON and M. G. EDMUNDS (Oxford, 1980).
[3] L. SPITZER: *Diffuse Matter in Space* (New York, N. Y., 1967).
[4] P. M. SOLOMON and N. C. WICKRAMASINGHE: *Astrophys. J.*, **158**, 449 (1969).
[5] T. P. STECHER and D. A. WILLIAMS: *Astrophys. J.*, **146**, 88 (1966).
[6] W. D. WATSON: in *Atomic and Molecular Physics and the Interstellar Matter*, edited by R. BALIAN (Amsterdam, 1975).
[7] F. HOYLE and N. C. WICKRAMASINGHE: *Nature (London)*, **264**, 45 (1976).
[8] V. I. GOLDANSKII: *Nature (London)*, **279**, 109 (1979).
[9] F. HOYLE and N. C. WICKRAMASINGHE: *Nature (London)*, **270**, 701 (1977).
[10] F. M. JOHNSON: *Ann. N. Y. Acad. Sci.*, **194**, 3 (1971).

[11] F. M. JOHNSON, D. T. BAILEY and P. A. WEGNER: in *Interstellar Dust and Related Topics*, edited by H. C. VAN DE HULST and J. M. GREENBERG (San Francisco, Cal., 1973).

[12] N. C. WICKRAMASINGHE: *Interstellar Grains* (London, 1967).

[13] P. G. MARTIN: *Cosmic Dust* (Oxford, 1979).

[14] F. HOYLE and N. C. WICKRAMASINGHE: *Astrophys. Space Sci.*, **53**, 489 (1978).

[15] F. HOYLE and N. C. WICKRAMASINGHE: *Astrophys. Space Sci.*, **66**, 77 (1979).

Condensation in the Early Solar System.

J. R. ARNOLD

Department of Chemistry, B-017, University of California - San Diego, La Jolla, Cal. 92093

The subject of this lecture is the stage of the formation of the solar system in which gases condense to solids, and the initial stage of growth of these solids, to a size of at most 1 cm. It is traditional among chemists (but not astrophysicists) to locate this stage in the preplanetary disk, moving around the protosun. Is this bit of traditional wisdom correct? I believe there is at least one strong reason to think so. It is best summarized by an idea of Suess which I paraphrase here: « The most remarkable fact about our samples of the Earth, Moon and meteorites is that they all contain all known stable nuclides, and that the isotopic pattern of each element is the same everywhere, except for well-understood processes ». Since SUESS said this, exciting discoveries have been made of isotopic anomalies, which WASSERBURG has discussed in other lectures here. But we must remember that the places where those anomalies occur are rare and hard to find. Only the rare gases, which are really rare in our samples, are a partial exception.

It seems to me very plausible, though of course not certain, that an isotopic mixing so complete in every detail was achieved in the gas phase, that is at a temperature high enough that all the elements we study behave alike. At any rate, in what follows I choose, mainly for this reason, to adopt the traditional view stated above. We can then proceed in a very simple way. Consider a planet, for example the Earth. Its mass of $6 \cdot 10^{27}$ g must once have been spread over a broad zone. We can, for illustration, say it was the zone from 0.85 to 1.2 astronomical units, half the distance to Venus and to Mars, which occupy the neighboring zones. Moreover, if the material was not yet fractionated, there was much H_2, He and other elements with solar abundances to match the silicates and irons of the Earth. A reasonable guess might be 10^{30} g (more, of course, if some solids were lost later). The surface density is then

$$\varrho_* = \frac{m}{\pi(R_2^2 - R_1^2)} = \frac{10^{30}}{\pi(1.5 \cdot 10^{13})^2 \, (1.2^2 - 0.85^2)} \sim 10^3 \text{ g/cm}^2 \, ,$$

coincidentally the same as the Earth's atmosphere. If we take then a temperature T on the order of 10^3 degrees, consider that the gas is mostly H_2, and derive the z-component of the gravitational acceleration of the disk, we obtain, following many authors, a scale height for the gas of $\sim 10^{12}$ cm. The gas density at the midplane was then

$$10^3/10^{12} = 10^{-9}\ \text{g/cm}^3$$

and

$$\varrho \sim 4 \cdot 10^{-5}\ \text{atm}\ .$$

This simple-minded calculation has some important consequences:

1) At this pressure, far above that of even a dense interstellar cloud, we are in a regime of neutral gas, where classical physical chemistry is valid, and where thermal (but not necessarily chemical) equilibrium prevails.

2) The pressure is below the triple point of nearly all substances. Hence the equilibrium condensates are solids, not liquids.

3) At such pressures, gas discharge develops easily and high electrical fields cannot exist.

We must note, however, some reservations. If the pressure is only a little lower, say 10^{-6} atm, serious departures from equilibrium begin to appear. For example, as noted by ARRHENIUS and DE [1], grain temperature is lower than gas temperature. Also, the thermodynamically favored condensed phase does not always occur. There is good reason to believe that complex solids, especially silicates, may form glassy condensates, essentially supercooled liquids, rather than the equilibrium phases which are crystalline.

What are the particle sizes of the condensed solids? At least at the beginning they must be small, $\leqslant 10^{-4}$ cm. The physics of such particles is a well-known subject, but in general not to physicists. It has been the province of the engineer. I know now of two very useful books, by FUKS [2] and by FRIEDLANDER [3], written by and for engineers. They should be studied by anyone who hopes to make progress in this field.

If a gas of solar composition, in the pressure range of $(10^{-5} \div 10^{-3})$ atm, cools down from a temperature above 2000 K (where no solids are stable), a temperature will be reached where solids begin to condense. For a simple atomic species, such as Fe, the equilibrium vapor pressure is a function of temperature, of the following form

$$P_v = P^* \exp\left[-\Delta H/RT\right]\ .$$

Here P^* is a collection of constants, with units of pressure, which are more or less temperature independent. The value of P_v must change very rapidly with T; the old chemists' rule of thumb is a change of a factor of 2 in $(10 \div 20)$ K.

The degree of « supersaturation » P/P_v is the critical parameter in the theory of nucleation. SALPETER [4] gives a clear discussion of this theory. For our purposes we can reduce this to a simple notion: the competition of nucleation and growth. When cooling is rapid, P/P_v can reach values > 10. Then the number of nuclei formed quickly becomes extremely large. The remaining Fe vapor condenses rapidly on these numerous nuclei; the result is many small grains.

When the cooling rate is slow, the first nuclei have time to form leaving a large pool of remaining vapor. The rate of nucleation remains low until the vapor is depleted; the final particles are few and large. The overall time for the cooling is controlled by macroscopic conditions—the evolutionary track of the protosun. This time is certainly $> 10^3$ years, and probably much greater. Hence there has been a tendency to believe that the slow-cooling case was the applicable one [5]. I think not.

The conditions under which a slow, steady cooling can occur in Nature are well known. They occur in the deep interior of planetary bodies. That is, the thermal inertia must be very large and the thermal path length separating the sample from sources and sinks must be long. The conditions in the postulated solar nebula are the opposite: very low thermal inertia, and direct or near radiative paths to the protosun and to deep space. The appearance or growth of solid grains increases the opacity, providing a direct thermal instability. In fact, except that they are a more extreme case, the conditions are those which produce weather at the Earth's surface, with a time scale on the order of one or a few days. The cooling, according to this argument, is steady only in the long-term average. The short term should be dominated by fluctuations, at least in the temperature band where condensation is important. Condensible substances must have condensed and re-vaporized many times before the last cooling.

From this point of view one expects nucleation to dominate, and hence a small particle size. Rough scaling calculations suggest a size less than 10 micrometers, and very likely less than 1 μm. The number of particles to be expected can be calculated if the composition and the mean particle size are known. At 1 μm it is on the order of 1 particle/cm³, at 0.1 μm around 10^3, for a solar composition. These particles will move through the gas with the Brownian motion. They will adhere at substantially every collision. The result will be agglomeration [2, 3, 6, 7] according to the simple rate equation

$$\frac{dN}{dt} = - kN^2 ,$$

where N is the particle density. The rate constant k depends only weakly on particle radius in the range $r = 10^{-7}$ to 10^{-3} cm; thus it is similar at various stages of the process [2]. Its value is $\sim 10^{-10}$ to 10^{-9} in CGS units. The process changes character when the particles grow large enough to be affected by gravitational settling and inertial effects [5, 8]. It is not obvious how to fix this limit; perhaps it is in the range of 0.1 cm or so. Meanwhile, the agglomerates need not remain unaltered after sticking. At temperatures where the vapor pressure is significant, the solid-state diffusion constants for most substances are appreciable. Thus recrystallization and even reaction between diverse substances are quite likely to produce grains larger than the original condensates. Other forces, especially electrostatic ones, may in some cases speed up the agglomeration process.

In summary, it appears likely, though not certain, that the bulk of planetary solids accessible to us last condensed from the vapor after the formation of a protoplanetary disk at a pressure high enough to assure a neutral gas and the applicability of physical chemical principles. The behavior of the resulting fine grains can be studied, however, mainly in the engineering literature. Arguments are given that the resulting gas-solid clouds are thermally unstable on a fairly short time scale. The attractiveness of this last result, if true, is that it brings the condensation process into, or much closer to, a range accessible to direct study in the laboratory.

REFERENCES

[1] G. ARRHENIUS and B. R. DE: *Meteoritics*, **8**, 297 (1973).
[2] N. A. FUKS: *The Mechanics of Aerosols* (translation) (London, 1964).
[3] S. K. FRIEDLANDER: *Smoke, Dust and Haze: Fundamentals of Aerosol Behavior* (New York, N. Y., 1977).
[4] E. E. SALPETER: *Astrophys. J.*, **193**, 579 (1974).
[5] P. GOLDREICH and W. R. WARD: *Astrophys. J.*, **183**, 7051 (1973).
[6] J. R. ARNOLD: in *Comets, Asteroids, Meteorites — Interrelations, Evolution and Origins*, edited by A. H. DELSEMME (Toledo, O., 1977).
[7] R. J. P. LYONS: NASA Technical Note D1871 (1963).
[8] V. S. SAFRONOV: *Evolution of the protoplanetary cloud and formation of the Earth and the planets*, Nasa TT F-677, Va. (1972).

Isotopic Heterogeneities in the Solar System.

G. J. WASSERBURG, D. A. PAPANASTASSIOU and T. LEE

The Lunatic Asylum of the Charles Arms Laboratory
Division of Geological and Planetary Sciences (*)
California Institute of Technology - Pasadena, Cal. 91125

> The world was so recent that
> many things lacked names, and
> in order to indicate them it
> was necessary to point.
>
> Gabriel Garcia MARQUEZ - 1967

1. – Introduction.

It has now been established that the isotopic composition of many elements (O, Ne, Mg, Si, Ca, Kr, Sr, Xe, Ba, Nd, Sm) in some meteoritic materials is distinctly different from that in terrestrial samples. In addition, evidence has been found to indicate that very-short-lived radioactive nuclides (^{26}Al, ^{107}Pd, ^{129}I, ^{244}Pu) were present in the early solar system. These discoveries indicate that we now have the possibility of identifying and studying the specific astrophysical and cosmochemical processes which immediately preceded the formation of the Sun and the solar system. The solar system was formed from a mixture of gas and dust *about* 4.6 AE ago. In general, the sequence of formation of the planets and the Sun is not known. At 4.54 AE ago some small planetary objects melted and differentiated (see [1]). At about 4.45 AE ago the Earth and the Moon were involved in a major chemical differentiation process. The determination of these ages is fundamentally dependent on the existence of chemical fractionation between the radioactive parent and stable daughter elements. The ages are thus the times of chemical fractionation. This difference of 0.1 AE may represent differentiation processes within the parent planets or possibly the time required to aggregate the larger terrestrial planets from the smaller debris in the inner solar system. To correlate this aggregation process with the 0.1 AE time difference requires that substantial fractionation of volatile from involatile elements take place during the final accretion events. Intervals of 0.1 AE for accretion of the planets have been argued from theoretical con-

---

(*) Division contribution number 3222 (290B).

siderations [2, 3]. The post-accretional history of the terrestrial planets is marked by planetary differentiation processes and a decreasing intensity of bombardment by planetary debris. The late (post-accretional) bombardment history was indexed by an intense and well-defined event on the Moon at 3.9 AE which has been associated with the major cratering of the whole inner solar system (Mars, Earth-Moon, Venus and Mercury [4]). The subsequent planetary evolution has been governed by internal differentiation supported by heat sources. This planetary differentiation was most intense on the Earth, but has occurred to varying degrees on all of the terrestrial planets. The particular history of the Earth for the past 3.9 AE is still being unraveled by use of old and new techniques and theories. However, the very early history of the Earth is not known. The post-accretional histories of the other planets are just now being addressed in the framework of a broad-scale exploration of the solar system which involves the use of spacecraft. In searching for the clues about the very early history of the solar system just before 4.555 AE the keys have come from studies of meteorites. Meteorites which come from small planetary objects (less than ~ 100 km) provide relics of very ancient solar-system processes as they are less subject to the more complex internal evolution which continues to occur on larger planetary bodies.

The meteorites comprise a highly diverse group of objects that includes the results of melting and chemical-mechanical segregation on small planets and those materials (the chondrites) which appear to be aggregates of more primitive debris. The relative abundances of the chemical elements in the chondrites have been used as a basis for estimating the solar abundances (relative to Si). In particular, the type-I carbonaceous chondrites which are relatively rich in volatile elements have provided one of the precise measures of the relative abundances of the elements. These abundances compare very closely with those which can be reliably measured from spectroscopic studies of the solar atmosphere. Knowledge of the isotopic abundances and of the chemical abundances gives the abundances of the nuclear species in the « Sun » and has provided the fundamental data base for all theories of the formation of the elements. All the precise isotopic-abundance data have come from measurements of terrestrial samples, meteorites and, subsequent to the Apollo missions, lunar samples. Good-precision data on isotopic abundances for a few elements have been obtained for the Sun, and these are in reasonable agreement with the observations on terrestrial materials.

Some time ago, it was recognized that the observed abundances of the nuclear species in the Sun are primarily the result of nuclear reactions in a variety of environments inside different stars. The concept that the solar abundances were the average of a variety of nuclear proccesses in numerous stellar sources was a major advance in understanding the origin of the elements. Some of the abundant nuclear species such as ^{4}He, ^{12}C, ^{16}O, ^{20}Ne, ^{28}Si and ^{56}Fe are the direct product of hydrostatic nuclear burning and are related to the major energy sources governing stellar production. These species are produced under conditions which are reasonably well understood and are associated with well-defined stellar sites. Other nuclear species are considered to be produced by modification of the production pattern of hydrostatic burning during transient explosive phenomena upon the death of massive stars. These processes appear to be important, but have not been clearly understood in the context of realistic stellar environments. Often the abundances of minor nuclear species are not clearly constrained by our knowledge of stellar evolution and may be affected by several distinct burning events. The production of such nuclides is at present not well known and cannot be associated with particular sites within stars. The mechanisms which contribute to the formation of a particular nuclear species are thus complex and depend upon a thorough knowledge of the nuclear-reaction rates, time scales, and an understanding of the different possible stellar sites in which the appropriate reactions may go on. Furthermore, the details of the ejection of processed matter back into the interstellar medium and the recycling of these materials in new generations of stars are not well understood.

One of the most important results to come out of studies of the abundances of the nuclear species in the solar system, which are widespread through the population-I stars, is the classification of the nuclei into s, r and p types. This discussion follows the pioneer work of Burbidge, Burbidge, Fowler and Hoyle [5] and Cameron [6]. The s-type nuclei may be produced by a continuous chain of neutron captures starting with around $Z = 26$ (Fe) and extending up to ^{209}Bi. These nuclei are connected by neutron capture followed by β-decay. The rates of neutron capture are assumed to be sufficiently slow so that β-decays with a lifetime of ~ 10 years may take place without a significant opportunity for the nucleus capturing another neutron. The r-type nuclei (see fig. 1) are unshielded, neutron-rich nuclei and would be the result of the β-decay of very neutron-rich unstable precursor nuclei. The r nuclei also include U and Th and all of the elements heavier than ^{209}Bi. In fact, it is the existence of all of these long-lived heavier elements, that are isolated from other stable nuclear species in the chart of the nuclides, which most clearly separates the r and s nuclei and defines the need for such a classification. Extensive theoretical discussions on these processes have been given in the literature.

Some nuclear species may be considered as both s and r types (see fig. 1). The neutron-rich precursors of the r-type nuclei are required to form in a neutron-rich environment at high temperatures and densities and may only exist for a short time (~ 1 s). The astrophysical sites in which the r-type nuclei may be produced have never been clear, but the strongest association has been with supernova events. It is quite possible that the « classical r » process with $(n, \gamma) \leftrightarrows (\gamma, n)$ quasi-equilibrium is not the true means of producing « r » process nuclei [7]. The p-type nuclei are neutron-poor isotopes which cannot be reached by the s chain and are typically rather rare. These have

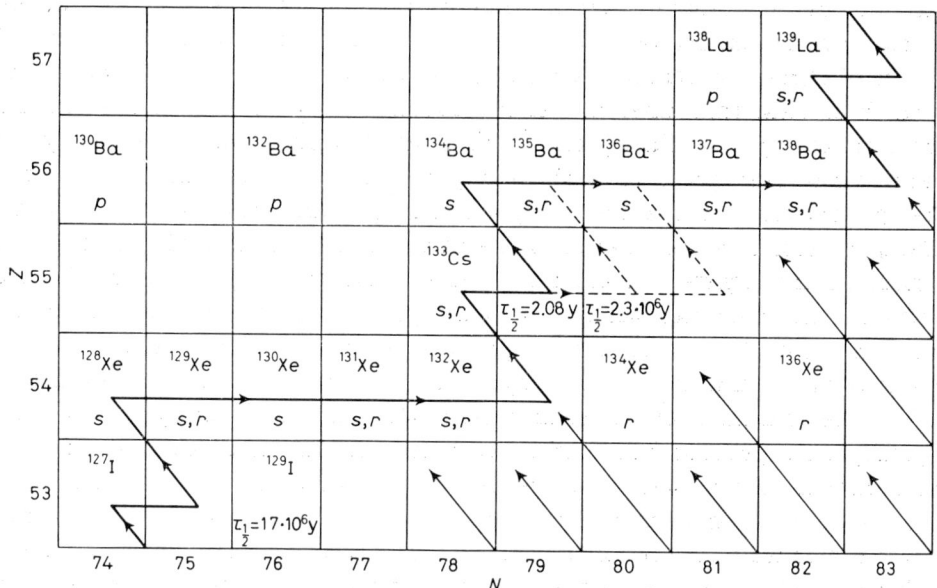

Fig. 1. – Chart of the nuclides showing the s-process path (heavy line) in the vicinity of Xe and Ba. Isotopes on the path are reached by neutron captures on a time scale slow compared to the half-lives for β-decay along the path. Isotopes on the proton-rich side of the path (*e.g.* ^{130}Ba, ^{132}Ba) are bypassed by the s process and require a separate, p process (p, γ or γ, n reactions). Isotopes on the neutron-rich side of the path are also bypassed by the s process and are produced by the successive β-decays of their more neutron-rich isobars produced by neutron captures on a time scale which is rapid compared to β-decays (r process). Note isotopes shielded from the r process and produced only by the s process (*e.g.* ^{134,136}Ba).

sometimes been considered as resulting from proton bombardment of pre-existing s and r nuclei or by (γ, n) reactions, but neither the nuclear processes nor the astrophysical sites associated with these processes have been clear [8].

There are a few other nuclei which cannot be classified in this manner, but none for $Z \geqslant 26$. The s, r, p classification has proven to be a basic one which is fundamental to any discussion of the formation of the elements and hence nuclei are called s, r, p « process ». However, the s, r and p processes referred to are the conceptual ones outlined above and are not as yet related to actual nuclear-astrophysical processes and sites. The solar-system abundances are believed to be averages of the production of the nuclear species through several different stellar sources and many cycles of processing and injection into the interstellar medium and reprocessing.

In order to understand the processes involved in generating the elements, it is necessary to utilize both astronomical observations which directly or indi-rectly indicate nucleosynthetic activity and to pursue, through theoretical considerations, the hints contained in the abundances of the nuclear species and in models of stellar evolution. Indeed, if it were possible to obtain samples of matter from distinctive stellar sources, then our understanding of the mechanisms of element generation could be greatly advanced. For over two decades it has been clear that different stellar components are present in the material which makes up the solar system. The possibility that some pre-solar-system stardust has been preserved until today has long been a matter of inquiry. The preservation of the pre- or proto-solar gas and dust requires that the mixing processes in the interstellar medium and in the formation of the solar system be incomplete. It further requires that the physical and chemical differentiation processes within the solar system over the past ~ 4.6 AE not erase the isotopic and chemical signatures of the precursor sources. Until recently, the dominant working hypothesis in the study of nucleosynthesis and of the formation of the solar system was that the solar system was isotopically homogeneous but chemically heterogeneous as a result of chemical fractionation. There was no evidence for the survival of identifiable components of pre-solar matter. From the discovery of ^{129}I present in meteorites it was concluded that the last injection of freshly synthesized nuclear material took place $\sim 1.6 \cdot 10^8$ y before the formation of the solar system. This viewpoint has now been changed as a result of a series of discoveries over the past few years. A review of isotopic effects by PODOSEK [9], which also contains a substantial discussion of anomalies in the « rare » gases, has recently appeared. This review, which was prepared before the more recent deluge of isotopic anomalies, should provide a complementary discussion to the present work.

2. – The magic of meteorites.

Meteorites are derived from relatively small planetary objects ($\leqslant 100$ km) which are thought to reside in the asteroidal belt and possibly some in the Oort cometary cloud. These bodies are subject to physical disruption and gravitational perturbations which cause them to collide with the Earth and other planets. Subsequent to disruption and breakup, they are known to survive typically for only a few million years and some as long as $\sim 5 \cdot 10^8$ y [10, 11]. The nature of the meteoritic debris which falls on the Earth shows it to be comprised of two distinctive classes. The first are aggregates of varied materials, often full of ovoidal pellets (chondrules) which appear to be droplets of some kind, imbedded in a fine-grained matrix. These meteorites, called chondrites, have strong similarities in relative chemical abundances to the Sun and may be considered as « primary » bodies. While they often show evidence of element migration on a local (microscopic) scale and some recrystallization due to heating, they do not show the typical results of major planetary differentiation processes (melting, gross loss of volatiles, outgassing, large-scale physical-chemical segregations). Some of the chondritic material has certainly resided and evolved within a planetary body. The chondrites, therefore, cannot be considered as cosmic virgins, but possibly as vestigial virgins. In general, these meteorites appear to preserve the characteristics of the earliest planetary material in the solar system. Some chondrites are found to contain chondrules and blebs of material rich in calcium and aluminum. Some of these inclusions are close in chemical composition to what would condense out of a hot cloud ($\sim$ 1800 K) of solar composition during cooling at a pressure of 10^{-3} bar. The major phases present in these so-called high-temperature Ca-Al inclusions are very similar to those calculated from thermodynamic considerations [12, 13] and are interpreted as direct condensates from a hot solar nebula. These inclusions often contain specks (or micrometer-size nuggets) of platinum group elements [14]. Some of the platinum group elements, Hf, Mo, W, are extremely refractory and may have been the earliest condensates to form. These could have been a soot which acted as the seed for later condensates. These chondrules and blebs of high-temperature phases may represent early rain or hail (seeded, nonequilibrium droplets or aggregates which precipitated from a somewhat supercooled medium) from local portions of the solar nebula which became hot and were then subject to cooling. Many of the inclusions show clear textural evidence of having crystallized from a liquid [15]. The most spectacular examples of such inclusions are in the Allende meteorite which fell on 8 February 1969 near Pueblito de Allende, in the vicinity of Hidalgo del Parral, in Mexico.

The other class of meteorites may be called « secondary » objects. These include igneous rocks which are the obvious result of early planetary melting processes: the iron meteorites, stony-iron meteorites and the achondrites (such as the eucrites, howardites). The achondrites are clearly the result of crystallization of silicate melts and represent lavas or agglomerates of pieces of lavas. The iron and stony-iron meteorites also appear to be the result of local planetary melting with accumulation of metallic Fe-Ni and FeS in small pods or pools in the meteorite parent bodies. These secondary objects are of critical importance in understanding the processes which took place in the early stages of planetary evolution. They are distinct from the « primary » objects in being removed from the preceding nebular evolution by additional stages of planetary evolution. Our discussion will include isotopic observations on both « primary » and « secondary » meteorites, but with an emphasis on primary objects which contain high-temperature Ca-Al-rich inclusions which could represent the earliest condensates [13, 16].

3. – Rules and regularities of isotopes.

The relative abundances of the isotopes of a given element may be changed by a variety of processes. Isotopic species when charged may be strongly separated through the effects of electric and magnetic fields depending on the ratio of charge to mass and the kinetic energy of the particles. For uncharged particles, the separation is usually not highly specific and for small separations is dependent on the masses of the isotopic species. Such separation is possible as a result of chemical equilibrium between different phases (*e.g.* solid-liquid-gas), from a host of kinetic effects and from settling in a gravitational potential. For kinetic effects the degree of separation depends on the extent to which the parent reservoir is depleted (*e.g.* 10 % loss of matter, 90 % loss, etc.). There may, of course, be shifts in isotopic abundance due to the laboratory procedures which are used in chemical separation or in the actual measurements. If we consider the shifts in relative isotopic abundance to be a function of mass for small effects, then we obtain $N_{M+1}/N_M \approx (1 + \alpha)(N_{M+1}/N_M)_0$, where α is the mass fractionation factor per mass unit difference and N_{M+1}/N_M is the ratio of the observed abundances relative to a standard (0). If this rule were followed, then all isotopic variations would be as indicated by curve *F-F* in fig. 2. This abundance trend is commonly observed both due to laboratory

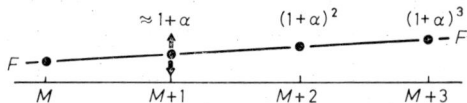

Fig. 2. – Isotope fractionation pattern for a uniform mass fractionation factor $1 + \alpha$ per unit mass difference. The vertical axis is the ratio $(N_{M'}/N_M)(N_{M'}/N_M)_0$, where $M' = M$, $M + 1$, $M + 2$, etc. For small fractionation effects, isotope ratios lie on a line *F-F*. Deviations of an isotope $M + 1$ (see arrows) from the linear array defined by at least three other isotopes would correspond to an excess or deficiency assignable to this isotope and indicative of nonlinear or nuclear effects.

effects and those which occur in Nature. Deviations of one isotope from this trend are not plausibly due to mass fractionation and are called nonlinear anomalies (see mass $M+1$ in fig. 2). These anomalies are presumed to be associated with nuclear effects. If, for example, only one isotope out of several is in excess relative to a fractionation trend, it may be considered to represent the addition of this isotope to the standard material. If it is deficient, then it may represent some special nuclear mechanism for the destruction of that isotope or the existence of material with a deficiency of that isotope relative to standard material. We may think of the average solar-system abundances as representing mixtures of isotopes from different sources. If some exotic isotope is added to average solar material, it will give an excess of that isotope as indicated (fig. 3). If, however, the solar values were originally made up of some average value to which an isotope had to be added in order to make up the *present* average value, then an isotope deficiency would be observable at that mass for some samples and the deficiency would give a lower limit to the number of such nuclei which had to be added to the whole solar system. More

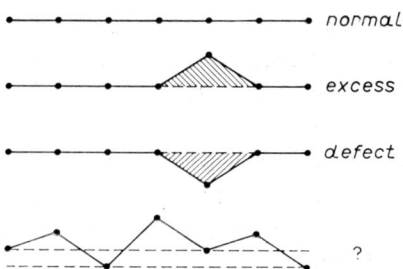

Fig. 3. – Schematic patterns of isotopic anomalies relative to terrestrial abundances. The positions of the dots in the top represent no fractional deviations from terrestrial values. An excess or a deficiency at a single isotope would be uniquely assignable. The second pattern corresponds to isotopes with terrestrial abundance but with one isotope in excess. For a more complex pattern (bottom) the identification of the isotopes which are anomalous depends on the normalization chosen, the baseline of the original unaltered abundances and possibly on other assumptions. In this case the identification of an anomaly is not unique.

generally, the solar system may have been heterogeneous in several isotopes of one element and it would no longer be possible to identify a simple excess or deficiency. We note that, for elements with only three naturally occurring isotopes, unless the effects are much larger than possible instrumental mass fractionation factors, it is not possible to determine which isotope is actually anomalous. In the simplest case where the solar values for an element resulted from mixtures of material from two distinctive nucleosynthetic sources, we would expect variations in the isotopic abundances in mixtures of different proportions (see fig. 4).

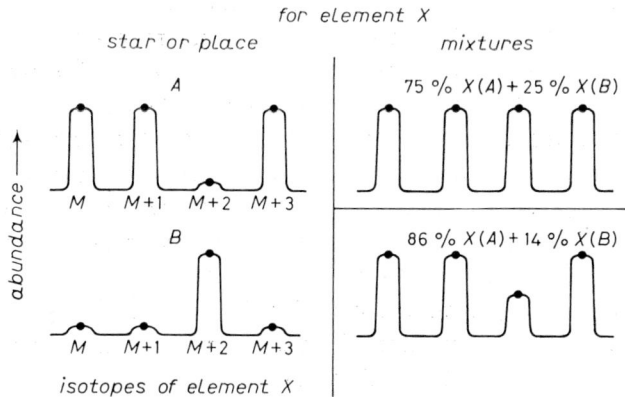

Fig. 4. – Mixing of distinct isotopic patterns, produced in stars (A, B), to produce different observed mixtures (right). The existence of at least two contributing isotope patterns and the distinction at mass $M + 2$ would be deduced from the mixtures on the right, although the compositions A and B on the left would not be uniquely determinable without a reliable nuclear-synthesis model.

As opposed to the more general isotopic anomalies, excesses are commonly observed for isotopes with long-lived radioactive parent isotopes which are known to exist today (*e.g.* ^{40}Ar, ^{87}Sr, ^{206}Pb, ^{207}Pb, ^{208}Pb, ^{143}Nd). Excesses which are not associated with existing parent isotopes may, however, result from the decay of parent radioactive isotopes which are now extinct (no longer found in the solar system). In order to attribute an excess of a given nuclide to an extinct radioactive nuclide, it is necessary to demonstrate that the excess nuclei have the same chemical distribution that the parent nuclide would have had under the circumstances when the object formed.

In addition to different nucleosynthetic mixes and the accumulation of decay products of radioactive nuclei, it is also necessary to consider isotopic effects produced by nuclear reactions in dispersed media such as dust or gas by cosmic rays, and the reactions in condensed matter by cosmic rays and reactions from emitted high-energy α's and fission of long-lived nuclei with the surrounding nuclei (*e.g.* (α, n) and (n, γ)).

4. – The oxygen problem.

Studies of different meteorites and of the Earth and Moon show that there exist small but distinct variations in the oxygen isotopic composition between solid planetary bodies in the solar system. This was first recognized by TAYLOR, DUKE, SILVER and EPSTEIN [17], who found small variations of ^{18}O/^{16}O between meteoritic and terrestrial rock samples. CLAYTON, GROSSMAN and MAYEDA [18] studied both the ^{18}O/^{16}O and ^{17}O/^{16}O abundances in inclusions in the Allende meteorite and showed that large variations existed in both ratios. These workers demonstrated that the correlation between the ^{18}O/^{16}O and ^{17}O/^{16}O effects was not due to mass-dependent isotopic fractionation, as previously assumed, but due to mixing of isotopically distinctive reservoirs. CLAYTON *et al.* [18] proposed a model in which interstellar dust grains containing essentially pure ^{16}O, without ^{18}O and ^{17}O, were mixed with normal solar-system material. In addition to observing variations between different objects, CLAYTON, ONUMA, GROSSMAN and MAYEDA [19] also reported large differences between coexisting mineral phases within individual inclusions from the Allende meteorite.

Let us consider the general patterns for the isotopic composition of oxygen using the three-isotope correlation diagram for oxygen in which ^{17}O/^{16}O is plotted against ^{18}O/^{16}O (fig. 5a)). The precise isotopic composition for oxygen in the Sun, which is the main oxygen reservoir of the solar system, is not in fact known, nor are there values for the other planets or their satellites. In fig. 5a) we have arbitrarily placed the solar value at a distinct point labeled ⊙. Reliable data exist for the Earth, the Moon, a wide variety of meteorites [20, 21] and limited data for the atmosphere of Mars [22] and are indicated schema-

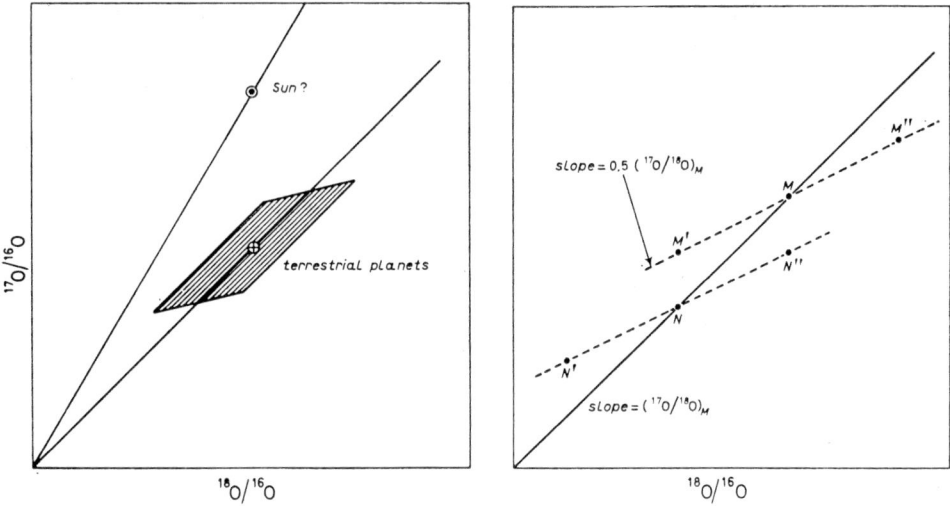

Fig. 5. – (left) Schematic representation of the oxygen isotopic composition for the terrestrial planets. The combination of distinct nucleosynthetic components and mass fractionation effects in principle results in the data lying in an area rather than on a line. The possibility that the Sun has a distinct composition from the terrestrial planets cannot be excluded. (right) Mass fractionation effects in oxygen superimposed on nuclear effects. Initial compositions M and N are distinct due to the presence of different proportions of ^{16}O; primed compositions (M', M'', etc.) would be obtained by mass fractionation effects (exaggerated) from their respective initial (M) compositions.

tically in fig. 5a). This represents an approximation to the average value for the terrestrial planets. If a reservoir M were subject to isotopic fractionation, producing new complementary reservoirs M' and M'', then they would lie approximately on a line $M'MM''$ with a slope of $\frac{1}{2}(^{17}O/^{18}O)_M$ for small fractionations (fig. 5b). If all solar-system reservoirs were simply the product of such processes and without extreme fractionations, then the mean solar-system value (including ⊙) and all solar-system samples would lie on such a line through the terrestrial value (⊕). The observations by CLAYTON et al. [18] showed that Allende inclusions were displaced from a fractionation line in the neighborhood of ⊕ in the direction of the origin by up to ∼ 40‰ (4%) of the length of the segment from the origin to ⊕. This clearly indicates the presence of at least two distinctive oxygen components. Samples of solar-system material will thus cover an area in the oxygen isotope diagram (see polygon in fig. 5a)) rather than a point or a line. It should be evident that if current solar-system reservoirs are mixtures of two components coupled with fractionation, no strict statement can be made about the oxygen composition in the Sun from samples of the terrestrial planets. Furthermore, comparisons of samples relative to an assumed « end member » may not fully reflect the displacement relative to the state of the solar system prior to admixing another component. It is

possible to consider the observed oxygen isotopic shifts as the result of *a*)
mixing an extraordinary material (*E*) consisting of nuclei containing essentially
pure ¹⁶O with normal uncontaminated (*U*) solar-system material (fig. 6*a*)),
or *b*) the result of mixing an extraordinary material (*E**) containing essentially

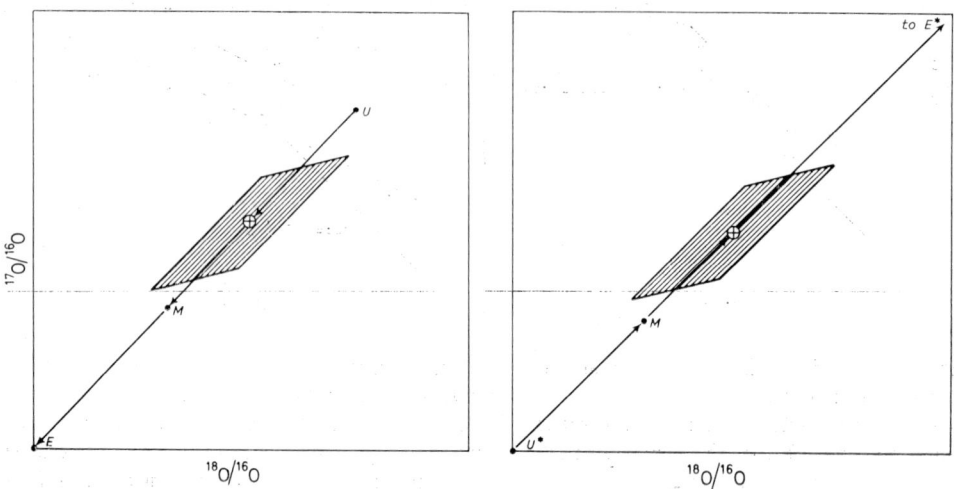

Fig. 6. – (left) Schematic mixing diagram for uncontaminated (*U*) and extraordinary (*E*)
reservoirs. The *E* reservoir is assumed to be ¹⁶O-rich. Contaminating reservoir *U* with *E*
causes the isotopic compositions to change in the direction of the arrows. (right) Same
diagram for the *U** and *E** model, where *E** corresponds to the addition of ¹⁷O and
¹⁸O in a fixed proportion.

pure ¹⁷O and ¹⁸O with the presently observed terrestrial ratio with originally
uncontaminated solar-system material (*U**), which is free of ¹⁸O and ¹⁷O (fig. 6*b*)).
    With regard to the model of addition of exotic pure ¹⁶O to uncontaminated
pre-solar material, it is possible that the *E* component is actually typical solar
oxygen in which the ¹⁷O and ¹⁸O were destroyed by heavy proton bombardment
by an active early Sun [23]. For the model in which the uncontaminated
solar-system material *U** contains essentially pure ¹⁶O and the extraordinary
contaminating material *E** contains most of the ¹⁷O and ¹⁸O in the present
solar system, the variations along the short arrow within the stippled region
about ⊕ in fig. 6 are considered to be residual fluctuations of a process which
is nearly complete. In contrast to the *E* model, the number of ¹⁸O nuclei which
must be added to produce the same isotopic shift for the *E** model is rather
small. If all of the ¹⁸O (and ¹⁷O) nuclei in a sample are from contamination
with an *E** reservoir, this constitutes only ∼ 1/400 of all the oxygen nuclei.
There is no apparent nucleosynthetic source for oxygen rich in ¹⁷O and ¹⁸O.
In addition, ¹⁷O and ¹⁸O seem to be widespread in the present interstellar
clouds with an abundance roughly near terrestrial (within ∼ 50 % [24]).

Thus there is no reason to believe that the proto-solar cloud was void of these species before the last injection. Furthermore, ^{18}O and ^{17}O are believed to be produced in two different sites rather than in a single source [25]. Therefore, this model seems unlikely.

We consider here the model in which the total oxygen of a sample is a mixture of extraordinary oxygen (pure ^{16}O) from reservoir E and an ordinary uncontaminated solar-system oxygen from reservoir U. The notation E follows Black's [26] characterization of extrasolar ^{22}Ne. We assume that U contains all the oxygen isotopes and with normal $^{17}O/^{18}O$ (~ 0.18), but with $^{16}O/^{18}O$ not specified. Equations for $^{16}O/^{17}O$ may be written which are completely analogous to those given below for $^{16}O/^{18}O$. Consider a mixture (M) made of E and U and which has not been subject to isotope fractionation. Then the fractional deviation (Δ) in per mil of $^{16}O/^{18}O$ in M *relative to the uncontaminated solar material* U may be conveniently written as

$$(1) \qquad \Delta(^{16}O/^{18}O)_{MU} \equiv \frac{[(^{16}O/^{18}O)_M - (^{16}O/^{18}O)_U] \cdot 10^3}{(^{16}O/^{18}O)_U} .$$

It follows that

$$\Delta(^{16}O/^{18}O)_{MU} = (^{16}O_{EM}/^{16}O_{UM}) \cdot 10^3 ,$$

where $^{16}O_{EM}$ and $^{16}O_{UM}$ are the numbers of extraordinary and of uncontaminated ^{16}O nuclei, repectively, in the mixture M. For measurements of M *relative to an arbitrary standard S* expressed as $\delta(16/18)_{MS}$ we have

$$(2) \qquad \Delta(^{16}O/^{18}O)_{MU} = \frac{(^{16}O/^{18}O)_S}{(^{16}O/^{18}O)_U} [\delta(^{16}O/^{18}O)_{MS} - \delta(^{16}O/^{18}O)_{US}] .$$

The usual form in which data are reported is $\delta(^{18}O/^{16}O)_{MS}$, which satisfies the identity

$$[1 + \delta(^{18}O/^{16}O)_{MS} \cdot 10^{-3}][1 + \delta(^{16}O/^{18}O)_{MS} \cdot 10^{-3}] = 1 .$$

Note in eq. (2) that $\Delta(^{16}O/^{18}O)_{MU}$ is not proportional to $\delta(^{16}O/^{18}O)_{MS}$ unless $\delta(^{16}O/^{18}O)_{US}$ is negligible. In general, $\Delta(^{16}O/^{18}O)_{MU}$ is not determined unless $(^{16}O/^{18}O)_U$ is known. The shift of ^{16}O relative to the « uncontaminated » value may in principle be quite large, although there is no evidence of wide differences (over $10\%_0$) between different *large* planetary objects [21]. However, the average solar-system value for $\Delta(^{16}O/^{18}O)_{MU}$ may not be adequately approximated by $\delta(^{16}O/^{18}O)_{MS}$ for any of the usual standards (*e.g.* « CCRS » [18]).

It is not yet possible to determine reliably the contribution of exotic oxygen to the whole solar system. The variations of bulk meteorite samples found by CLAYTON and MAYEDA [21] show a range of $\sim 6\%_0$. From the distinct oxygen isotopic composition of different classes of meteorites it is clear that the addition of an extraordinary component is not restricted to refractory inclusions in Allende, but represents a widespread phenomenon through at least all of the condensed material. The alternatives which appear to exist are the addition of ^{16}O (or of $^{18}O + ^{17}O$) to the solar system a) in small amounts and added only to the material which makes up the terrestrial bodies, or b) in large amounts to the whole solar system so as to significantly alter average uncontaminated solar-system abundances. If meteorites provide good sampling of the solar-system variability, then the amount of exotic oxygen added to the solar system must be at least $2 \cdot 10^{28}$ grams of O_E. If the addition is only to the condensed material in the terrestrial planets, this will be reduced by a factor of $\sim 10^3$.

In the above discussion we have concentrated on the variations in oxygen isotopic composition on a gross scale. However, it is important to recognize that there exist large variations between coexisting mineral phases within a single morphologically well-defined inclusion. CLAYTON et al. [19] showed that different intergrown phases associated with early condensation (melilite, spinel, fassaite) showed different oxygen isotopic compositions in each individual inclusion. The spinel showed the largest amount of O_E component, fassaite the next largest, and the melilite showed no significant amount of an O_E component. For example, it was commonly found that

$$\delta(^{18}O/^{16}O)_{\text{spinel}} - \delta(^{18}O/^{16}O)_{\text{melilite}} \approx -30\%_0 \ .$$

This variability suggested the presence of carrier grains which were derived from different nucleosynthetic sources and trapped in the inclusions. The question of whether each inclusion is an aggregate of grains from different sources (coarse stardust) or whether each inclusion and its constituent grains represent a single or homogenized source which was later altered is a fundamental issue which will be addressed in a later section. We note here that studies of other isotopic anomalies show the latter interpretation to be correct.

It is desirable also to explain other isotopic anomalies as part of the same addition phenomenon as found for oxygen. As first pointed out by BLACK [26] and confirmed and refined by EBERHARDT [27, 28], a component of possibly pure ^{22}Ne (called Ne-E) with $^{20}Ne/^{22}Ne < 1.3$ and $^{21}Ne/^{22}Ne < 0.015$ is present in the solar system (cf. fig. 7). This Ne-E was proposed by BLACK [26] to be exotic material from a distinctive nucleosynthetic source which was added to normal solar-system material. From measurements of solar wind [29] it is found that ^{22}Ne is present in the surface of the Sun with an abundance of $^{20}Ne/^{22}Ne \simeq 13.7$, while the terrestrial value of $(^{20}Ne/^{22}Ne)_{\oplus} = 9.8$. The Sun thus appears to be rich in ^{20}Ne, while the planets and meteorites are relatively rich in ^{22}Ne. The high ^{22}Ne abundance relative to ^{20}Ne in planetary gases as compared with the solar wind suggests that the added extraordinary Ne-E component is enriched in the dusty and condensable material making up the

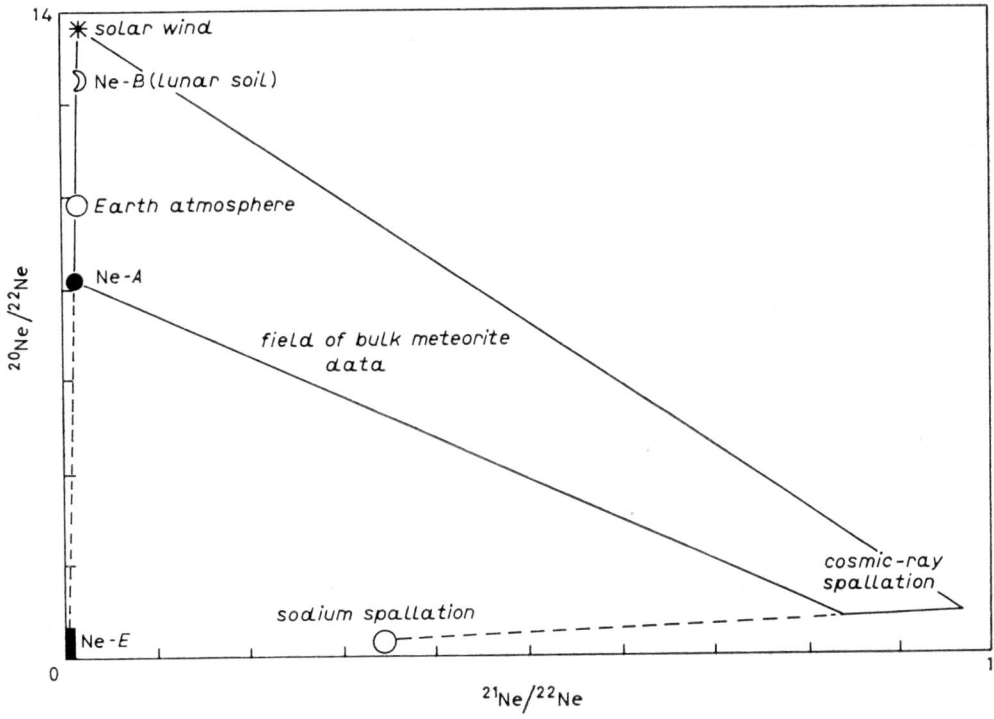

Fig. 7. – Neon isotope compositions. Note the very low $^{21}Ne/^{22}Ne$ values if spallation effects are excluded. Components corresponding to different proportions of ^{20}Ne and ^{22}Ne plot very near the ordinate and exhibit extreme variations. The limits on the Ne-E component found in meteorites are shown by the black area near the origin. The Earth's atmosphere and most meteorites are enriched in ^{22}Ne as compared to the Sun (from direct measurements on solar-wind Ne). The lunar Ne which is from implanted solar wind on lunar rocks is close to solar wind and altered by fractionation processes.

terrestrial planets. This suggests that a quantitative correlation should exist between Ne$_E$ and O$_E$. This is particularly appealing since ^{16}O and ^{20}Ne are produced in explosive C burning. No such correlation has been found, possibly due to the fact that there are no obvious mechanisms which would concentrate Ne in easily condensable materials or solids such as dust grains or in crystals which are high-temperature condensates. An estimate of exotic Ne contributions relative to exotic oxygen assuming that only the terrestrial-type materials were contaminated is thus not possible, because Ne is severely depleted in condensed objects.

It is possible that the variations in oxygen which are observed are not due to the addition of extraordinary material to the solar system, but rather due to differences in O between gas and dust. In this case the O isotopic variation is the result of incomplete exchange or mixing between the reservoirs of distinct isotopic character, *e.g.* gas and dust, in the proto-solar cloud. It is most consistent to associate the pure(r) ^{16}O component with the dust phase. The sizes of the two O reservoirs can be estimated using the present solar abundance if all the easily condensable elements such as Ca, Al, Mg, Si, Fe, etc. were in the dust. The fraction of total oxygen initially in the dust must be approximately $[(Mg+Fe+Si)/O]_\odot \sim 0.15$ for any reasonable mineralogical composition of the dust. So the solar-system oxygen was originally resident in two reservoirs which are different in size by only a factor of 7. Presumably, evaporation of the dust and gas-dust reactions would be the processes responsible for isotopic exchange between such reservoirs. A model of this type cannot explain the presence of short-lived nuclides nor can it explain anomalies in refractory elements, if the dust is treated as a homogeneous component. Therefore, this model may only be considered as a partial source of isotopic heterogeneity in a stratified solar nebula. If various objects in the solar system are formed as mixtures of refractory materials residing in grains and of more « volatile » material residing in the gas, then we would expect the size of isotopic anomalies to correlate with the « refractory »/« volatile » ratios. For the oxygen data such a progression of isotopic effects is not apparent, since the most volatile materials (carbonaceous chondrites) show intermediate effects between refractory Allende inclusions and ordinary meteorites [20, 21]. This indicates that for any model we require locally distinct vaporization-condensation histories and preservation of isotopic anomalies in parcels of gas during the high-temperature stage of the nebula as well as the addition to both gas and dust of freshly synthesized nuclides.

5. – FUN inclusions.

Magnesium and oxygen. One of the difficulties in understanding the observed isotopic anomalies has been the apparent lack of anomalies in refractory elements and the lack of correlation between effects in different elements except for inferred parent-daughter pairs. For some time, efforts to find a correlation between oxygen anomalies and the magnesium isotopic composition yielded no positive results [30]. In most cases the Mg isotopic composition was found to be same (and indistinguishable from terrestrial) in spinel, pyroxene and melilite from the same Ca-Al-rich inclusions where the oxygen was highly anomalous for the spinel and pyroxene but much closer to « normal » for the melilite. This could be due to an absence of any anomalies in Mg or to a much smaller exotic Mg component as compared with O_E. Correlations have been established for isotopic anomalies between the heavy rare gases (Kr, Xe), but they are not directly associated with any mineral phase nor are they shown to be correlated with oxygen. Correlations in isotopic anomalies for different nuclear species are extremely important as they should provide a signature of the nucleosynthetic processes which caused them.

The first samples to exhibit peculiar isotopic compositions for several elements were found in two inclusions, C-1 and EK-1-4-1. These inclusions are Ca-Al-rich chondrules from the Allende meteorite. They do not appear distinctive in terms of mineral composition or morphology from the typical high-temperature Ca-Al-rich inclusions. The original identification of the special nature of these samples came from the observation that they had 1) Mg isotopic compositions which were widely displaced from terrestrial values and could not be explained by decay of ^{26}Al, and 2) O isotopic compositions which did not lie on either the terrestrial fractionation line or on the O_E-O_N line [30-33]. The oxygen data showed that the coexisting phases in each inclusion had different isotopic compositions and that these could be explained either by the presence of a third component (different from O_E or O_N) or as the result of isotopic fractionation of material which was a mixture of O_E and O_N. The Mg data showed that all of the coexisting phases (both for macroscopic and microscopic samples) had the same Mg isotopic composition and that the isotopes were regularly enriched with increasing mass number. The enrichments were almost precisely what would obtain if they were the result of isotopic fractionation (see fig. 8). However, there were small, significant departures from any simple mass fractionation law which indicated that ^{26}Mg was deficient if ^{24}Mg and ^{25}Mg were originally terrestrial in abundance but subsequently shifted by fractionation. (Note that it is not possible in this case of three isotopes to distinguish between ^{24}Mg deficit, ^{25}Mg excess, or ^{26}Mg deficit.) Except for these isotopic peculiarities, the inclusions C-1 and EK-1-4-1 appeared to be indistinguishable, based on mineralogical and textural relationships, from the typical type-B Ca-Al inclusions [34]. The Mg data could be explained by a model in which each inclusion was made of peculiar material that had been isotopically homogenized in the inclusion. This could be produced by melting of an aggregate of isotopically heterogeneous matter (say, an interstellar dust ball) or by condensation of a liquid droplet from a supercooled homogeneous gas of isotopically peculiar matter. The crystals which then grew in the inclusion (up to 3 mm) would each have the same peculiar Mg isotopic composition. The isotopically peculiar material must have been subject to strong isotopic fractionation and contained a small component of exotic Mg. While the oxygen data for the bulk EK-1-4-1 and C-1 could be similarly interpreted (fractionation of material which initially was comprised of a given O_E-O_N mix), this model does not explain the differences in oxygen isotopic composition between the coexisting phases spinel, pyroxene and melilite (see fig. 9).

The interpretation of the Mg and O data by both WASSERBURG, LEE and PAPANASTASSIOU (WLP) [33] and CLAYTON and MAYEDA [31] is that these Ca-Al-rich chondrules were formed from a local parcel of gas and dust in the solar nebula which contained some exotic nuclei or nuclei with abundances somewhat different from the terrestrial values. The formation process involved

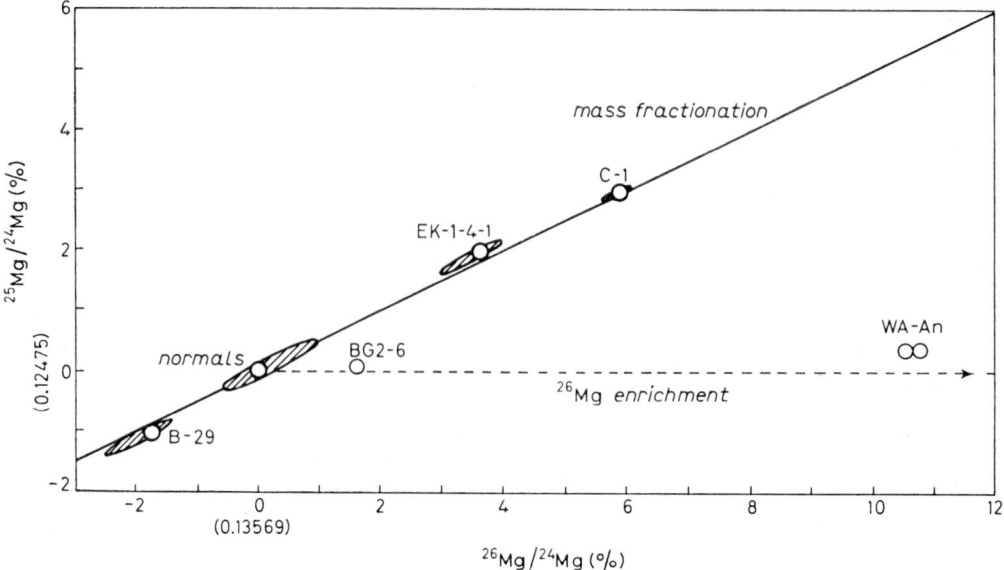

Fig. 8. – Mass fractionation effects for Mg. We show raw measured data as percent deviations from the mean $^{25}Mg/^{24}Mg$ and $^{26}Mg/^{24}Mg$ for normal Mg (terrestrial, lunar and chondritic). Shaded areas correspond to the observed maximum range of isotopic fractionation for each type of sample. The precision of the data is much better than the schematic display of the data range. FUN samples C-1 and EK-1-4-1 plot very near the expected correlation line for mass fractionation. Small but well-defined deviations for these samples on the upper side of the mass fractionation line are evidence for superimposed nuclear anomalies or a more complex law of mass fractionation. Sample B-29, a fine-grained Ca-Al-rich aggregate from Allende, shows a complementary fractionation pattern (depleted heavier isotopes). Effects due to pure ^{26}Mg enrichment move points to the right along a horizontal line (*e.g.* WA-An, anorthite from inclusion WA). (After [33].)

large isotopic fractionation, so that all of the primary mineral phases which grew in the chondrules were isotopically uniform but anomalous and fractionated. In a later stage, the minerals melilite and to some extent pyroxene were susceptible to a « back reaction » with cooler portions of the solar nebula, rich in volatiles and « normal » oxygen, thus causing a large shift in the oxygen composition of the melilite and only small shifts in the other phases. As Mg would be depleted in a volatile-rich, oxygen-rich (refractory-poor) medium, the melilites would maintain their original peculiar Mg isotopic composition, but shift in oxygen composition. This model is *ad hoc*, but constitutes an explanation coupling the high degree of alteration readily observed in the melilites and their nearly normal oxygen composition. The pyroxenes only show minor alteration. This interpretation follows the suggestion for oxygen proposed by BLANDER and FUCHS [15]. The fine-grained alteration products observed in the Ca-Al-rich inclusions have not been studied extensively, but

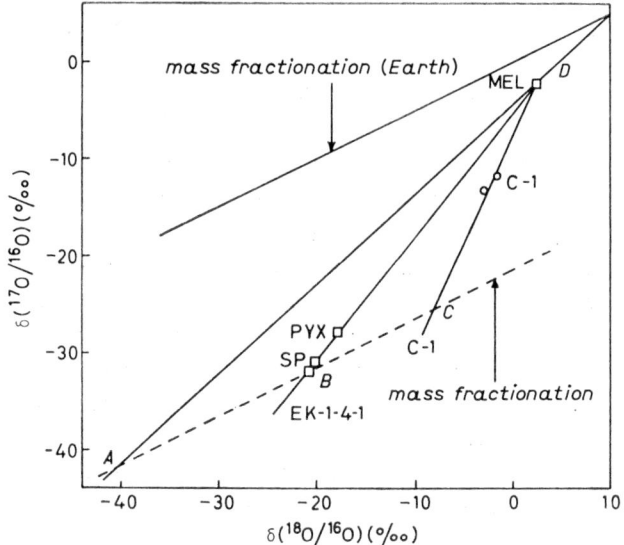

Fig. 9. – Oxygen isotope compositions for FUN inclusions EK-1-4-1 and C-1. Data for mineral separates of EK-1-4-1 (squares) are shown for spinel (SP), pyroxene (PYX) and melilite (MEL) and define line *BD*. Data on samples of the total inclusion C-1 are shown as circles. The data for both inclusions are far off line *AD* defined by oxygen measurements of total inclusions and mineral separates from all other Ca-Al-rich Allende inclusions. Note that EK-1-4-1 melilite is close to the terrestrial-data line and is consistent with a model in which FUN oxygen with a composition near point *B* has largely exchanged with oxygen of normal composition in the vicinity of *D*. If one assumes that the oxygen composition in EK-1-4-1 prior to exchange with normal oxygen is at point *B* (SP+PYX), then this oxygen composition could have been obtained from an initial composition *A* on the O_E-O_N mixing line (line *AD*), where line *AB*, with slope $\frac{1}{2}$ corresponding to mass fractionation, was simply drawn through point *B*. Displacement of FUN inclusions off the O_E-O_N line could also reflect distinct extraordinary oxygen compositions. The interpretation of displacement off line *AD* as due to mass fractionation was dependent on the observation of large fractionation effects for Mg in the same samples. (After [31].)

they are rich in Na, Cl and other volatile elements and contain phases which could only form at much lower temperatures than spinel, melilite and Ca-Mg-Ti-rich pyroxenes. These alteration products are seen to permeate the melilite crystals in veins and throughout their volume.

The general character of the isotopic shifts originally found for Mg and O for C-1 and EK-1-4-1 suggests two distinct effects: 1) isotopic fractionation (F) and 2) unidentified nuclear (UN) effects. This correlation caused WLP [33] to define these as FUN anomalies. As will be shown later, the existence of « nuclear » effects in many elements is apparently closely tied to the presence of mass fractionation. Subsequent to the identification of these two FUN

samples based on the Mg and O observations, it was discovered that a host
of other refractory elements were isotopically peculiar in these same inclusions.
The extent to which the nonlinear or nuclear effects are related or connected
to the fractionation observed in O and Mg is not clear. Clearly some elements
show no fractionation, but have nuclear anomalies. In all cases some care
must be taken in identifying deviations from a particular fractionation law
as reflecting true nuclear anomalies. For most samples which show small fractionation this is not important, but in some critical cases both the identification
and the magnitude of a nuclear anomaly will depend on the fractionation
law that is assumed. There is at present no satisfactory interpretation for
the apparent correlation between fractionation and nuclear anomalies. Having
identified a correlation between the peculiar isotopic patterns for both O and Mg
in the two FUN inclusions, it was of importance to ascertain whether any other
elements exhibited peculiar isotopic patterns. This search proved difficult
but successful and the resulting anomalies which have now been established
are discussed below in order of increasing atomic number.

Aluminum 26. A search was conducted for evidence of ^{26}Al in sample C-1
by ESAT, LEE, PAPANASTASSIOU and WASSERBURG [35]. No data have been
obtained on EK-1-4-1 because of the minute amount of material which remains.
These workers report that there is a suggestion of ^{26}Mg excess resulting from
^{26}Al decay in plagioclase crystals ($CaAl_2Si_2O_8$) in C-1, but with a ^{26}Al/^{27}Al ratio
which is a factor of seven less than found in typical Ca-Al-rich inclusions in
Allende (see general discussion on time scales). They further report the presence
of some Mg of terrestrial composition in impure plagioclase and attribute this
to a back-reaction process as outlined by WLP [33]. The effects of chemical
alteration subsequent to the formation of high-temperature condensates are
strongly emphasized by these authors. The strongest conclusion is that only
a small amount of ^{26}Al, and hence fewer late-stage nucleosynthetic products,
was included in the C-1 inclusion as compared with other materials (see section
on time scales for a fuller discussion of ^{26}Al).

Silicon. The presence of large isotopic-fractionation effects as observed
for Mg and inferred for O was substantiated by YEH and EPSTEIN [36] and
CLAYTON, MAYEDA and EPSTEIN [37], who found that Si was also strongly
fractionated in these two inclusions. The results show clear evidence of isotopic
fractionation with the heavier isotopes being enhanced similar to what was
found for Mg (see fig. 10). There is a close relationship between the degree
of fractionation for Si and Mg. Sample C-1 is the most fractionated and
EK-1-4-1 is less fractionated. There appear to be some small departures from
a simple fractionation behavior for sample EK-1-4-1, but there is no high-
precision, reliable Si data to justify a firm conclusion regarding the presence
of a peculiar nuclear component in Si.

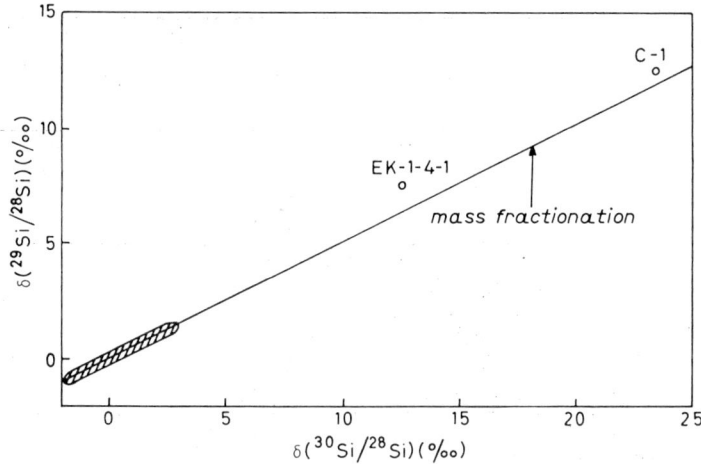

Fig. 10. – Silicon composition for FUN inclusions EK-1-4-1 and C-1. The hatchured area shows the extent of variations in Si isotope composition along the calculated mass fractionation line for meteorites and non-FUN Allende inclusions. Large displacement of the FUN samples along the mass fractionation line yields Si fractionation effects which are comparable to those observed for Mg in these samples. Deviations off the Si fractionation line are not well resolved analytically but suggest the presence of nuclear effects (relative deficiency of ^{30}Si or excess of ^{29}Si). (After [36, 37].)

Calcium. The chart of the nuclides in the neighborhood of Ca is shown in fig. 11. The abundances given in the chart are not the percent abundance of each isotope as is usually given but the « solar » abundance of each nuclear species (see [38]). This tabulation permits a direct comparison of observed isotopic effects with the number of nuclei which must be altered, relative to the « solar » values with Si $\equiv 10^6$. All variations in isotopic abundances are given relative to terrestrial standards and are expressed in fractional deviations either in parts in 10^3 (permil) called δ units or in parts in 10^4 called ε units. Note that the relative abundances of the Ca isotopes range over a factor of $\sim 3 \cdot 10^4$. The nucleosynthesis of the Ca isotopes is not well understood, but they are produced by distinctive processes (explosive oxygen and silicon burning, neutron capture on seed nuclei [39, 40]). As a result of the wide range in relative abundances and nucleosynthetic processes, Ca was considered as a key element for study. From a technical point of view the measurement of relative Ca isotopic abundances is made difficult by both the wide range in these abundances and the large mass differences which may result in large instrumental mass fractionation. In the case of large fractionation the particular « law » used to reduce the data can affect both the quantitative and qualitative nature of « nonlinear » isotopic effects. Most samples of Ca in Nature show only small fractionation effects. A thorough discussion of both procedures and data is given by RUSSELL, PAPANASTASSIOU and TOMBRELLO [41].

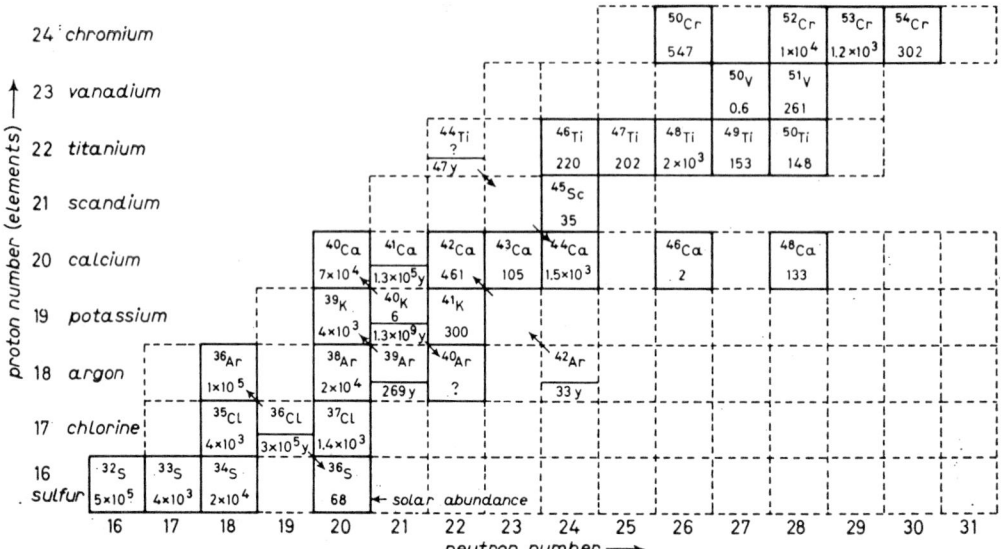

Fig. 11. – Chart of the nuclides in the vicinity of Ca. Dashed lines indicate unstable nuclides. The entries in each square include the solar abundance of each nuclide from [38] using the ^{28}Si $\equiv 10^6$ normalization. Half-lives are given for the longer-lived radioactive nuclides. Note the existence of very neutron-rich Ca isotopes as well as the large abundance difference of the Ca isotopes ($7 \cdot 10^4$ to 2).

The results for C-1 and EK-1-4-1 are shown in fig. 12 [42]. We have also included results from a third FUN inclusion which was recently discovered [43]. The experimental errors for ^{46}Ca are $\pm \sim (10 \div 20)\%_0$ and have been omitted from the figure for clarity. If we assume that ^{40}Ca and ^{44}Ca are unaffected by nuclear effects, neither of these samples appears to be seriously fractionated as determined by measurements using a double spike [42]. Sample C-1 shows all isotopes to be normal within error except for a depletion in ^{48}Ca by about $3\%_0$. It is surprising that ^{46}Ca shows no large anomalies, since it is only 1/60th as abundant as ^{48}Ca. Shortage of the same number of nuclei at ^{46}Ca as is observed at ^{48}Ca would cause a $180\%_0$ effect. As all other isotopes appear normal in C-1, it is plausibile to attribute the pattern to a deficiency in ^{48}Ca. This indicates that the ^{48}Ca which was last injected to produce the average solar-system value must have been generated by a rather special mechanism which favored this nucleus or that ^{48}Ca was preferentially destroyed in a parcel of terrestrial-type Ca.

The data on EK-1-4-1 are shown on the same figure and indicate very large relative enhancements at ^{48}Ca of $14\%_0$ and at ^{42}Ca of $1.6\%_0$. The EK-1-4-1 results appear to indicate addition of ^{48}Ca and ^{42}Ca to average solar material with the fractional effects being ten times larger at 48 relative to 42. The Ca data on C-1 and EK-1-4-1 taken together suggest that solar-system Ca is the

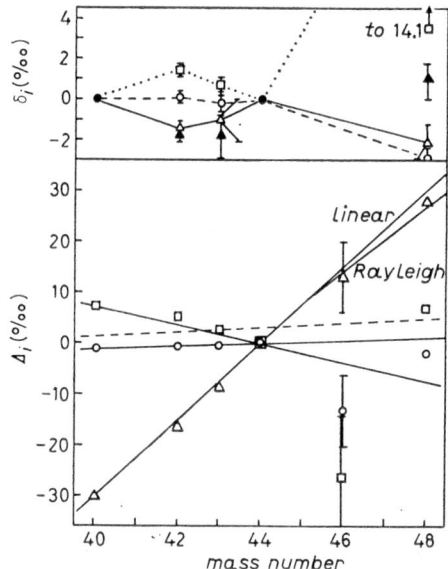

Fig. 12. – Fractional differences of Ca isotopes relative to normal ($\delta = 0$ line) normalized to ^{40}Ca and ^{44}Ca: △ HAL, ○ C-1, □ EK-1-4-1. The upper graph shows nonlinear effects after correction for mass fractionation. Sample EK-1-4-1 shows excesses in ^{42}Ca and ^{48}Ca (off scale), while sample C-1 shows a ^{48}Ca deficiency. The HAL sample shows only small nonlinear effects at ^{42}Ca, but effects at ^{48}Ca are not clearly resolved due to uncertainties as to which fractionation law is applicable. The lower graph shows ratios relative to ^{44}Ca only which demonstrate the large fractionation observed for Ca isotopes in the HAL, hibonite-rich inclusion. Linear and Rayleigh fractionation laws (for HAL) are shown. (After [42, 43].)

result of the incomplete mixing between the uncontaminated bulk material of the solar nebula which was substantially deficient in ^{48}Ca and slightly deficient in ^{42}Ca relative to normal and a special nuclear component consisting of ^{48}Ca and ^{42}Ca in the ratio of 2.5 to 1 by number. In this interpretation normal Ca is also a mixture and the deficiency in ^{48}Ca implies that the special nuclear component accounts for at least $3 \cdot 10^{-3}$ of the ^{48}Ca and $5 \cdot 10^{-6}$ of the total Ca presently in the solar system.

Strontium. The Sr isotopic variations in C-1 and EK-1-4-1 are shown in fig. 13 with isotopes 86 and 88 used for normalization. Both samples show a deficit of the rare isotope ^{84}Sr which is produced solely by the p process [44]. It would be simplest to assign the effect to a shortage in ^{84}Sr (p) or an enhancement in $^{86, 87, 88}$Sr which are all dominantly s products. However, this is not in accord with the evidence in Ba, Nd and Sm in these samples. In particular, EK-1-4-1 shows excesses in the r-process isotopes for Ba, Nd and Sm, and a p-process excess in Sm but null p isotope effects in Ba; C-1

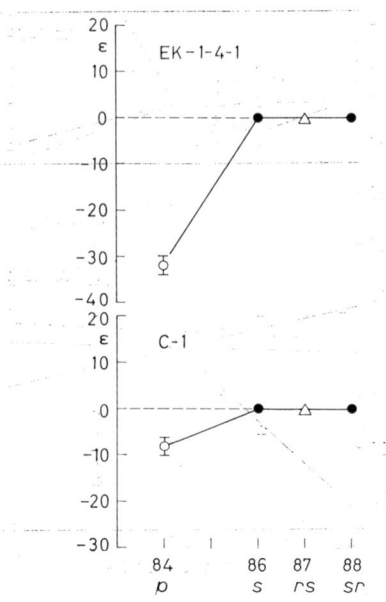

Fig. 13. – Sr isotope composition for FUN inclusions. In this and subsequent graphs the pair of isotopes assumed to have normal composition are shown as full circles; the rest of the isotopes are shown as open circles. The units are in parts in 10^4 and denoted by ε. The canonical processes contributing to the isotopes are listed under the isotope mass. The patterns for Sr, for the normalization to ^{86,88}Sr, correspond to relative deficiencies in the p-process ^{84}Sr (or uniform excesses in ^{86,87,88}Sr). ^{87}Sr (triangle) has been corrected for decay of ^{87}Rb for 4.6 AE. (After [44, 45].)

shows no effects for Ba and Nd (except for a small ^{135}Ba deficiency) and a p-excess for Sm.

The assignment of the Sr isotopic anomaly to ^{84}Sr is not unique. A check can be made using ^{87}Sr, the only other isotope not used for normalization. The isotope ^{87}Sr is produced dominantly by the s process and partly by the decay of ^{87}Rb, so that it is not possible to assign a very precise normal value to its abundance and to establish whether further nuclear effects are present in this isotope. ^{87}Sr increases by $\sim 0.005\,\%$ per 10^6 y due to radioactive decay of ^{87}Rb in a mix of solar proportions. The observed values of ^{87}Sr/^{86}Sr when corrected for ^{87}Rb decay for 4.6 AE lie close to the best estimates of the early solar-system value using the ^{86}Sr/^{88}Sr normalization. This appears to favor the case for a ^{84}Sr deficiency, however it cannot be considered a strong argument. Alternate interpretations consistent with the Sr data include a) a deficiency in the r-process component in ^{88}Sr, or b) excesses in ^{86}Sr and in ^{87}Sr which would not coincide with those excesses expected from the canonical s-process. If we arbitrarily attribute the observed effects to a shortage in ^{84}Sr, this would require the addition of at least $\sim 2\cdot10^{-5}$ of the total Sr nuclei as exotic ^{84}Sr to the nebula to make up the average normal Sr composition. A shortage attributed to ^{88}Sr would correspond to a factor of 150 larger contribution of exotic Sr nuclei.

Barium. The data for Ba in C-1 and EK-1-4-1 [46] are shown in figure 14, normalized to ^{134}Ba and ^{138}Ba. Sample C-1 appears to be normal for all isotopes within errors except for a small but significant deficit of ε135 $= -1.8$ at ^{135}Ba.

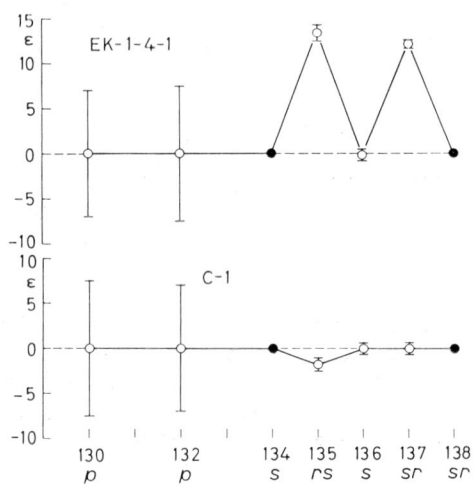

Fig. 14. – Ba isotopic composition for the two FUN inclusions (cf. fig. 13 for conventions). EK-1-4-1 shows excesses in ^{135}Ba and ^{137}Ba, while C-1 shows a small deficiency in ^{135}Ba. The less abundant p-process isotopes ^{130,132}Ba show no effects but with significantly worse precision. C-1 shows no well-defined deficiency in ^{137}Ba mirroring the excess in EK-1-4-1. (After [46].)

Sample EK-1-4-1 shows a very different pattern with marked excesses ε135 and ε137 of 13 and 12 (parts in 10^4), respectively. All other isotopes for these samples have terrestrial abundances. Comparison of the data for these objects indicates that EK-1-4-1 has some nuclei added to the terrestrial value at masses 135 and 137, while C-1 appears to be the same as terrestrial except for a deficiency at mass 135. Inspection of the chart of the nuclides in the region of Ba (see fig. 15) shows that the isotopes ^{130,132}Ba are attributed to the p process. Neither of the FUN inclusions shows discernible effects at these masses within the larger uncertainties due to their low abundances. The other shielded isotopes ^{134,136}Ba are pure s process and ^{135,137,138}Ba are dominantly s process, but ^{135}Ba and ^{137}Ba have significant r-process components. The pattern exhibited by Ba in EK-1-4-1 strongly indicates the addition of r-process material with a ratio $(^{137}\text{Ba}/^{135}\text{Ba})_r \sim 1.6$. This value is close to the observed terrestrial value. This should be compared with average r-process production

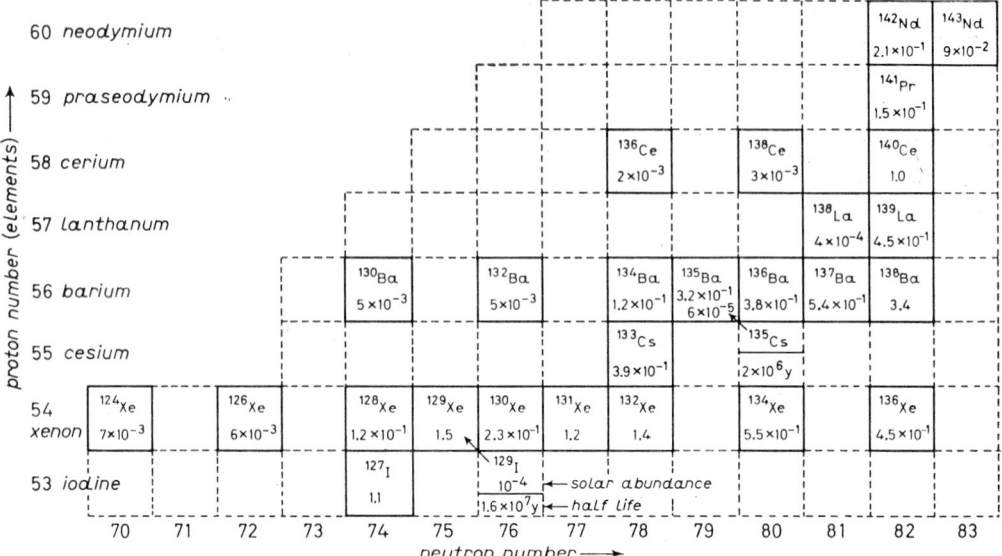

Fig. 15. – Chart of the nuclides in the vicinity of I and Ba (cf. fig. 11 for conventions). Note the low abundance of p-process isotopes ^{130,132}Ba, shielded isotopes ^{134,136}Ba and contribution of $6.5 \cdot 10^{-5}$ at ^{135}Ba. This appears to represent the minimum amount of ^{135}Ba added to the solar nebula in the last event, and is comparable to the abundance of ^{129}I. The possibility of ^{135}Cs decay is hinted at from the data on inclusion C-1.

estimates [47]. The addition of a comparable number of nuclei to ^{138}Ba as appear to be added at 135 and 137 would not cause a measurable effect, since ^{138}Ba is ~ 10 times more abundant.

The general discussion of the significance of deficiencies has been presented earlier. The simplest explanation is that the solar-system Ba had an original isotopic composition which was deficient in ^{135}Ba and ^{137}Ba and that these isotopes were then added in fixed relative proportions to produce the average solar value and other mixtures with both residual deficiencies or excesses. In this case we would expect the observed deviations to follow the rule $\varepsilon135 = (^{137}\text{Ba}/^{135}\text{Ba})_\odot(^{135}\text{Ba*}/^{137}\text{Ba*})\ \varepsilon137$. From the EK-1-4-1 data and terrestrial values we would expect for C-1 deficiencies in 137 about equal to 135 within $\sim 10\%$. The Ba spectrum of C-1 shows a deficiency in 135, but does not appear to include deficiencies at both 135 and 137 which are complementary to the excesses at these mass numbers as observed in EK-1-4-1. It is possible to attribute a deficiency to ^{137}Ba in C-1 just within the limit of errors and with the same ^{137}Ba*/^{135}Ba* ratio for the injected matter as for EK-1-4-1. An explanation of the ^{135}Ba deficiency in C-1 might be due to the late addition of r-process material with a holdup at ^{135}Cs ($\tau_{\frac{1}{2}} = 2.3 \cdot 10^6$ y). The presence of ^{135}Cs would be compatible with the time scale for late injection defined by ^{26}Al. The ^{135}Ba deficiency is of great importance, as it provided the first lower limit for the fraction of intermediate-atomic-weight nuclei which had to be added to the presolar-system material to bring it up to the average value. The amount of ^{135}Ba required is $\sim 0.5 \cdot 10^{-4}$ atoms relative to a cosmic abundance of Si $\equiv 10^6$. This is commensurate in magnitude with the observed abundance of ^{129}I which is $\sim 10^{-4}$ atoms on the same scale (see fig. 15).

Neodymium. In general, isotopic anomalies in the rare-earth elements are of special importance because of the fact that these elements are not readily subject to chemical fractionation. As a result, the number of exotic nuclei in most of the rare-earth elements should be a reliable measure of the composition of injected material. The data for Nd [46] in samples C-1 and EK-1-4-1 normalized to ^{142}Nd and ^{144}Nd are shown in fig. 16. It can be seen that the

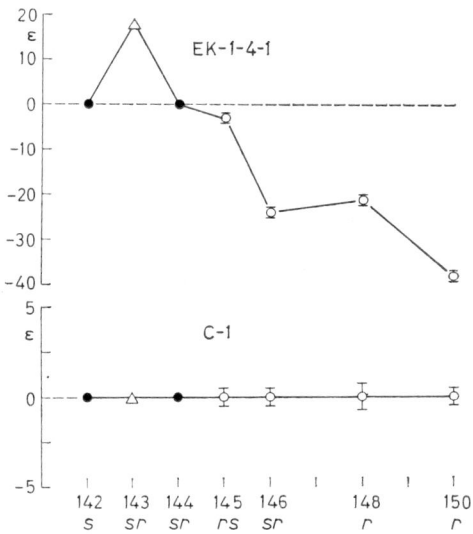

Fig. 16. – Nd isotopic composition in FUN inclusions using a ^{142,144}Nd normalization (cf. fig. 13 for conventions). The pattern for Nd shows no regularities corresponding to the *s* and *r* processes indicated below mass numbers and indicates the need for alternate normalizations (cf. fig. 18). Nd composition for C-1 is normal. Note the absence of *p*-process isotopes in Nd. Data for ^{143}Nd (triangle) have been corrected for ^{147}Sm decay. (After [46].)

Nd in C-1 has normal composition to within analytical errors. In contrast, the results for EK-1-4-1 show large deviations from the normal values. The magnitudes of the effects are very large for several isotopes. However, in this representation there are no regularities which would indicate simple addition of exotic nuclei to normal Nd. As discussed in the introduction, the choice of two isotopes for normalization is arbitrary and a selection which is more readily interpretable from nuclear systematics is preferable. From the data on Ba in EK-1-4-1 there is strong indication of addition to the unshielded isotopes for this sample. Inspection of the chart of the nuclides in the region of Nd (see fig. 17) shows that only ^{142}Nd is shielded (by ^{142}Ce). If we assume that addition to all the unshielded isotopes also occurs for Nd, we may obtain a revised form of the deviations with an additional assumption of the magnitude of the yield at another isotope. Taking a contribution to ^{144}Nd of $\varepsilon 144 = 18$ (18 parts in 10^4), we obtain the relative spectrum with additions in all the unshielded isotopes as shown in fig. 18. This representation is in full agreement with the model of r-process excess (or s deficiency) inferred from the Ba results on the same sample.

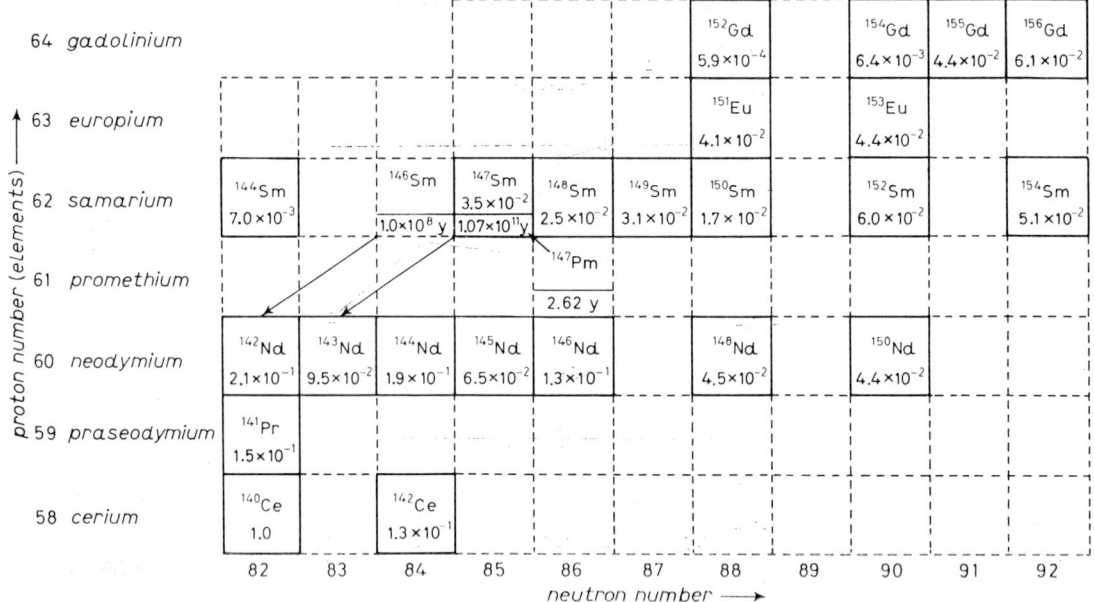

Fig. 17. – Chart of the nuclides for Nd and Sm (cf. fig. 11 for conventions). Note that for Nd only ^{142}Nd is shielded from the r process, and that no p-process Nd isotopes exist. For Sm, ^{144}Sm is a relatively abundant p-process isotope, while ^{148}Sm and ^{150}Sm are shielded by their Nd isobars.

The absence of any effect in Nd from C-1 is somewhat surprising, however it should be noted that the deficiency in ^{135}Ba is quite small in magnitude. To explain both the C-1 and EK-1-4-1 data by addition of exotic material to a batch of « standard » material which is slightly deficient in some isotopes, and assuming no chemical fractionation between Ba and the rare-earth elements (REE), would require that the contribution of exotic nuclei at Nd ($Z = 60$) is somewhat less than at Ba ($Z = 56$).

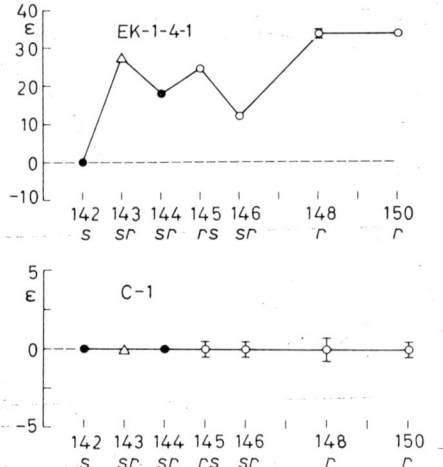

Fig. 18. – Nd isotope composition in FUN samples using ^{142}Nd and ^{144}Nd for normalization after adding 18 parts in 10^4 to ^{144}Nd for EK-1-4-1 to reflect expected r-process contributions to this isotope also. Resulting pattern clearly reflects roughly commensurate relative excesses in the unshielded Nd isotopes (cf. fig. 16). Inclusion C-1 retains its normal composition (including ^{144}Nd). (After [46].)

Samarium. Observations on Sm are particularly important as this element has three shielded isotopes, one of which (^{144}Sm) is attributed to the p process and the other two (^{148}Sm, ^{150}Sm) are pure s-process. Sm is the only rare-earth element with two pure s isotopes and can thus be used to test models of r-process addition or enhancement relative to s-process nuclei as first indicated by the observations on Ba. The chart of the nuclides in the neighborhood of Sm is shown in fig. 17. The Sm isotopic compositions for the samples C-1 and EK-1-4-1 are shown in fig. 19 normalized to 144 and 148. This choice of normalization is the preferred one and most clearly exhibits the nature of the anomalies [48]. Data for Sm on EK-1-4-1 have also been reported by LUGMAIR, MARTI and SCHEININ [49]; these workers have also confirmed the Nd measurements on EK-1-4-1 of McCulloch and Wasserburg [46]. The data for C-1 indicate that all Sm isotopes are terrestrial in composition with the exception of the p-process ^{144}Sm, which has an enrichment of $\varepsilon 144 = 15 \pm 5$. In contrast, EK-1-4-1 shows large excesses for the four unshielded isotopes and also an excess for the p-process ^{144}Sm. These results on Sm provide additional strong evidence for r-process enrichment relative to the s isotopes.

Summary of Mg, Si, Ca, Sr, Ba, Nd *and* Sm *in two* FUN *samples.* The isotopic analyses for Ba, Nd and Sm on C-1 and EK-1-4-1 which we have discussed here were performed on bulk samples of these Ca-Al-rich inclusions and on separated mineral phases. Each of the samples analyzed had different ratios

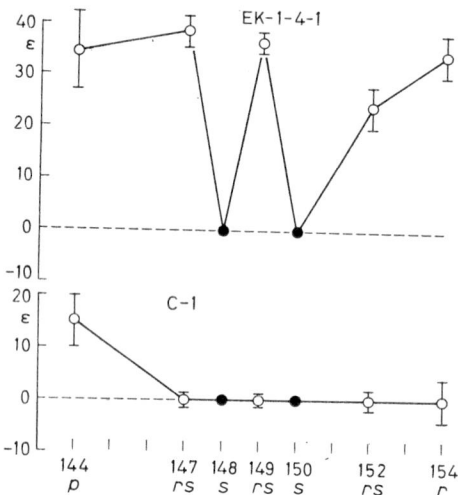

Fig. 19. – Sm isotopic composition in the two FUN inclusions normalized to the two s-process only isotopes ^{148}Sm and ^{150}Sm (cf. fig. 11 for conventions). Sample EK-1-4-1 shows excesses of relatively equal magnitude in the unshielded isotopes and in the p-process isotope 144. Sample C-1 shows only an excess in the p-process isotope, indicating that r and p process effects can be decoupled. (C-1 and EK-1-4-1 data after [48] and EK-1-4-1 data after [49].)

of Ba/Nd/Sm. The different phases within each inclusion showed the same peculiar isotopic composition for these elements. The isotopic compositions of the two inclusions were distinct. This observation is in support of similar results for Mg, Ca and Sr and distinct from that observed for oxygen. We, therefore, conclude that each inclusion represents a sample of isotopically peculiar material which was made isotopically homogeneous for refractory elements at the time of crystallization. Presolar dust grains, which may have supplied material making up these inclusions, have been destroyed as distinguishable objects for the major oxide phases.

The elements Ba, Nd and Sm in EK-1-4-1 have isotope shifts which can all be explained by the addition of a generic r-type component to normal material or the enhancement of r-process relative to s-process material. The relative abundances of the added nucleosynthetic component is rather close to that for average r-process nuclei (see table I). The Sm data, however, also require a concomitant enhancement in the p-process nuclide in Sm which is not observed in Ba. Inspection of fig. 15 and 17 and table I shows that the solar abundances of ^{144}Sm, ^{130}Ba and ^{132}Ba are comparable, so the addition of the same number of ^{130,132}Ba* nuclei as ^{144}Sm* nuclei to solar material should produce similar effects. While the uncertainties in these rare Ba isotopes are relatively large, there is no hint of any effect. It is not clear why ^{144}Sm should be preferentially enhanced considering the abundances of possible pro-

TABLE I. – *Nucleosynthetic components in* Ba, Nd *and* Sm *for* EK-1-4-1 (after [48]).

Nuclide	$N(A)$ [a]	$\sigma(A)$ [b]	$N_s(A)$ [c]	$N_r(A)$ [d]	$\varepsilon(A)$ [e]	$\varepsilon(A) \cdot N(A) \cdot 10^{-4}$ [f]	$\dfrac{\varepsilon(A)N(A)}{\varepsilon(154)N_r(A)}$ [g]
^{130}Ba	0.005 [m]	2000 [k]	0.0	0.0	0.0	0.0	
^{132}Ba	0.005 [m]	650 [k]	0.0	0.0	0.0	0.0	
^{134}Ba	0.116	225 [h]	0.116	0.0	0.0	0.0	
^{135}Ba	0.316	470 [i]	0.055	0.261	13.4	$4.23 \cdot 10^{-4}$	0.47
^{136}Ba	0.375	70 [h]	0.375	0.0	0.0	0.0	
^{137}Ba	0.543	72.6 [i]	0.363	0.180	12.3	$6.68 \cdot 10^{-4}$	1.08
^{138}Ba	3.44	5.7 [h]	3.30	0.14	0.0	0.0	
^{142}Nd	0.211	45 [j]	0.211	0.0	0.0	0.0	
^{143}Nd	0.095	332 [i]	0.025	0.07	27.8 [l]	$2.64 \cdot 10^{-4}$	1.1
^{144}Nd	0.186	67 [i]	0.116	0.07	18.0	$3.35 \cdot 10^{-4}$	1.39
^{145}Nd	0.065	485 [i]	0.016	0.049	24.9	$1.61 \cdot 10^{-4}$	0.96
^{146}Nd	0.134	105 [i]	0.069	0.065	12.0	$1.61 \cdot 10^{-4}$	0.72
^{148}Nd	0.045	210 [k]	0.0	0.045	33.6	$1.50 \cdot 10^{-4}$	0.98
^{150}Nd	0.044	240 [k]	0.0	0.044	33.6	$1.47 \cdot 10^{-4}$	0.98
^{144}Sm	0.007 [m]	120 [k]	0.0	0.0	34.4	$0.24 \cdot 10^{-4}$	
^{147}Sm	0.035	1150 [k]	0.006	0.029	38.5	$1.34 \cdot 10^{-4}$	1.35
^{148}Sm	0.025	260 [k]	0.025	0.0	0.0	0.0	
^{149}Sm	0.031	1620 [k]	0.004	0.027	36.5	$1.14 \cdot 10^{-4}$	1.23
^{150}Sm	0.017	370 [k]	0.017	0.0	0.0	0.0	
^{152}Sm	0.060	450 [k]	0.014	0.046	24.1	$1.46 \cdot 10^{-4}$	0.92
^{154}Sm	0.051	380 [k]	0.0	0.051	34.3	$1.76 \cdot 10^{-4}$	1.0

(a) Cosmic abundance (per Si $\equiv 10^6$) [38].
(b) 30 keV cross-sections (millibarn).
(c) *s*-process abundance.
(d) *r*-process abundance given by $N_r(A) = N(A) - N_s(A)$.
(e) Deviations from normal, in parts in 10^4, from [46, 48].
(f) Excess atoms.
(g) Ratio of excess atoms to cosmic *r*-process atoms, normalized to ^{154}Sm.
(h) Ref. [50].
(i) Ref. [51].
(j) Ref. [52].
(k) Ref. [53].
(l) Corrected for ^{147}Sm decay for an age of $4.56 \cdot 10^9$ years.
(m) *p*-process only nuclide.

genitors for the *p* isotopes of Ba and Sm, whether by successive p, γ or γ, n reactions. This may suggest that the late *p*-type process was very specific or it may reflect the fact that ^{144}Sm is a very abundant *p* isotope and hence is always favored. More detailed theoretical analysis of the p process is clearly needed. An investigation of p-process addition to heavier rare-earth elements in C-1 and EK-1-4-1 should aid in clarifying this problem of process yields. The presence of ^{144}Sm excesses in C-1, which is otherwise normal, and in EK-1-4-1, which also has excesses in the unshielded isotopes, indicates that the p and r components are not coupled, but are independent. This rule, if valid, should place strong constraints on possible astrophysical sites and processes for p and r nuclide production.

The isotopic patterns for Ba, Nd and Sm appear to be most readily recognizable ones. Insofar as the patterns are dominated by excesses of unshielded isotopes relative to shielded isotopes, it is reasonable to associate them with processes with very high neutron fluxes. This may be some type of r process. It is difficult to conceive of proton bombardment mechanisms which would yield these effects, unless there were some special mechanism for abundant neutron production. In consideration of all the isotopic anomalies reported here, the results on the heavy elements seem to point most strongly toward a supernova.

Some of the results could be explained by the spontaneous or induced fission of an unknown progenitor. As the isotopic effects are independent of the relative chemical abundances within the individual phases in both inclusions, the fission would have had to take place prior to crystallization. Using the deficiency of ^{135}Ba in C-1 as an estimate, this would require an abundance of $\sim 3 \cdot 10^{-5}$ fissionable parent nuclei (100 % fission) in units of 10^6 Si atoms. The fission hypothesis does not explain the ^{144}Sm excess and could not be the source of the anomalies found at low Z.

In considering the very short time scale as determined both by ^{26}Al and ^{107}Pd, there must have been late production and injection of nuclei with a wide range in mass. It is likely that some very heavy fissionable nuclei (uranium and transuranics) were produced which contributed to the material added to the protosolar cloud. Such nuclei could alter the isotopic effects to be expected from a « simple r-process » addition, and should be taken into account in any detailed theory. However, if the yield curve for the injected nuclei is in any way similar to the solar abundances, we would expect such contributions to be comparatively small.

We note a clear distinction in interpretation if we extend our considerations to the observations on the lower-atomic-number elements Mg, Si, Ca and Sr for the two FUN inclusions. For example, while there is a clear indication of p excess, s deficiency and r excess for Sm in EK-1-4-1, the Sr shows a p deficiency or s excess or r deficiency. The Ca effects are not attributable to any clearly defined process. The Mg shows a deficiency in ^{24}Mg, an excess in ^{25}Mg, or a deficiency in ^{26}Mg. Again there is no clearly defined process from purely nuclear considerations. This apparent lack of correlation between isotopes of different elements in the same samples, both in the types of processes involved as well as in the sign of the effects, indicates that a mixture of specific nucleo-synthetic processes, astrophysical production sites and models of extraction of materials from a star are required to interpret the effects. Alternatively the isotopic effects depend in detail on the specific synthesis processes operating possibly in a restricted fashion or on pre-existing seed nuclei, so that the effects cannot be directly associated with the canonical, average r, s and p processes. For some nuclei there will have to be a better definition of the production scheme.

There is a further distinction in the presence of isotopic fractionation for some of the low-Z elements (O, Mg, Ca and Si) which is not apparent in the heavier elements. This is undoubtedly a reflection of the much greater atomic weight, but may also depend on the complex nature of the chemical (and ion) kinetics and the distribution of these elements in the local parcels of gas and dust from which the inclusions formed. So far it has not been possible to predict which elements in the FUN inclusions should show marked fractionation. The basic reason for the presence of nuclear anomalies in matter which shows strong fractionation in some elements is not obvious. More poking around in Pandora's box of the nuclides will be required to establish true regularities and firm rules.

6. – Time scales.

One of the basic problems concerning the observed isotopic heterogeneities is the time to be associated with the production of the various nuclides relative to the formation of the solar system. The identification of extinct radioactive nuclides in early solar-system material is the principal means of determining the times of production. It is plausible that the time scale for chemical and isotopic mixing between gas and dust is very long under cool-interstellar-cloud conditions. In addition, dust grains and possibly macroscopic fragments could have been formed in different stellar sites and preserved in the aggregation of larger bodies. These considerations may complicate interpretation of some observations. If it can be shown that relatively short-lived nuclides were present in macroscopic objects which formed as isotopically homogeneous bodies within the solar system, then it is possible to establish a time scale that links the time of nucleosynthetic processes with the solar-system time scale. As emphasized by CLAYTON [54], it is possible that a short-lived nuclide decayed in an object, say a dust grain, prior to being incorporated into the solar system. In this case only a lower limit to the time scale can be obtained and there is no direct connection between a nucleosynthetic event and the solar-system chronology.

The pertinent radioactive nuclei used in cosmochronologies are listed in table II. The ratio of two nuclei just after the termination of continuous

TABLE II.

Nuclide	Mean life (y)	N_i/N_j at termination		Observed	Δ(y)
		YONI (*)			
^{232}Th	$2 \cdot 10^{10}$	^{238}U/^{232}Th ~ 0.38		0.41	—
^{238}U	$6.5 \cdot 10^{9}$				
^{235}U	$1.0 \cdot 10^{9}$	^{235}U/^{238}U $\sim (0.31)$		0.31	—
^{244}Pu	$1.2 \cdot 10^{8}$	^{244}Pu/^{232}Th $\sim 6.6 \cdot 10^{-3}$		$5 \cdot 10^{-3}$	—
^{129}I	$2.3 \cdot 10^{7}$	^{129}I/^{127}I $\sim 7.7 \cdot 10^{-3}$		$1 \cdot 10^{-4}$	10^{8}
^{107}Pd	$9.4 \cdot 10^{6}$	^{107}Pd/^{110}Pd $\sim 10^{-3}$		$4 \cdot 10^{-5}$	$\leqslant 2 \cdot 10^{7}$
^{26}Al	$1.1 \cdot 10^{6}$	^{26}Al/^{27}Al $\sim 10^{-7}$		$5 \cdot 10^{-5}$	$\leqslant 3 \cdot 10^{6}$
??	1?				

(*) $N_j/N_i = p_j \tau_j (1 - \exp[-\lambda_j T])/p_i \tau_i (1 - \exp[-\lambda_i T])$, $T \sim 8.7$ AE.

uniform nucleosynthesis for 8.7 AE are also shown. The ^{235}U/^{238}U ratio was chosen to be in accord with the value at the time of formation of the solar system and is thus shown in parenthesis. Pairs of nuclei were chosen to minimize errors due to chemical fractionation in the solar system. With due regard for uncertainties in the production rates, the values are in reasonable agreement for the isotopes with mean lives over 10^8 years. The column labeled Δ is the time interval between termination of galactic nucleosynthesis and the formation of planetary objects in order to account for the difference between the YONI and observed values. The discrepancies in the last three entries with different mean lives (^{129}I, ^{107}Pd, ^{26}Al) show that it is not possible to account for these differences by free decay and show that the differences must result from a nonuniform rate of production over a time scale short compared to 10^8 years.

Until recently the only well-identified short-lived nuclides were extinct ^{129}I ($\tau_{\frac{1}{2}} = 16 \cdot 10^6$ y) and the not quite extinct ^{244}Pu ($\tau_{\frac{1}{2}} = 83 \cdot 10^6$ y). The evidence for the existence of ^{129}I is the presence of excesses of ^{129}Xe relative to other Xe isotopes and the correlation of this excess with the presence of ^{127}I [55-57]. The evidence for ^{244}Pu is from observations of the unshielded neutron-rich isotopes of Xe with a unique abundance pattern for ^{131}Xe, ^{132}Xe, ^{134}Xe, ^{136}Xe [58]. This unique Xe spectrum is correlated with the presence of fission tracks in minerals rich in U, Th and REE [59, 60]. Plutonium is known to have chemical characteristics similar to Nd and Sm and should thus be associated with these elements. The Xe spectrum of ^{244}Pu spontaneous fission was found to correspond precisely with the unique Xe spectrum found earlier in some meteorites [61]. Some of the meteorites which contain the ^{244}Pu SF products are obviously the result of crystallization from a silicate melt and the product of planetary differentiation [62, 63]. There must, of course, be some ^{244}Pu remaining today and a hint exists in one preliminary experiment.

The existence of ^{244}Pu in the early solar system and in planetary objects does not by itself provide a strict time scale because of the long half-life. The estimated abundance of ^{244}Pu relative to ^{232}Th at the time of formation of the solar system is ^{244}Pu/^{232}Th $\sim 5 \cdot 10^{-3}$ [64]. This value in conjunction with a plausible relative production rate for these two nuclides yields a time scale of $\sim (3 \div 4) \cdot 10^8$ y. ^{129}Xe excesses have been found in a variety of meteorites including basaltic achondrites and in stony and sulphide inclusions in iron meteorites which are planetary differentiates [63, 65, 66]. This ^{129}Xe is strongly correlated with ^{127}I and it must, therefore, be considered as resulting from the *in situ* decay of ^{129}I in the solar system. The ^{129}I must have been included not only in aggregates of more primitive material (including Allende inclusions) but in differentiated planetary objects (iron meteorites and a eucrite) after their formation. The abundance of ^{129}I relative to ^{127}I at the time of solar-system formation is ^{129}I/^{127}I $\sim 1 \cdot 10^{-4}$. This implies a time interval of $(1 \div 2) \cdot 10^8$ y from the last period of ^{129}I production. Until very recently the ^{129}I/^{127}I and ^{244}Pu/^{232}Th abundances were used in conjunction with the ^{235}U/^{238}U and ^{238}U/^{232}Th abundances to construct self-consistent time-dependent models of galactic nucleosynthesis [67]. The rate of *r*-type nucleosynthesis with time as determined by this approach depends on the ^{244}Pu/^{232}Th ratio. In particular the question of whether there was a terminal spike is sensitive to the ^{244}Pu/^{232}Th, ^{235}U/^{238}U and ^{129}I/^{127}I ratios. However, in all cases the last time of production is governed by the ^{129}I/^{127}I ratio, since ^{129}I has the shortest half-life of these nuclides.

For many years there have been intermittent searches for evidence of shorter-lived nuclides. Searches were carried out for ^{26}Al ($\tau_{\frac{1}{2}} = 0.7 \cdot 10^6$ y) with the daughter product ^{26}Mg and ^{107}Pd ($\tau_{\frac{1}{2}} = 6.5 \cdot 10^6$ y) with the daughter product ^{107}Ag. Until recently no positive results had been brought forward which withstood careful tests of verification. If the observed ratio of ^{129}I/^{127}I $\sim 10^{-4}$ is the result of sudden production (with a ratio of production rates of order unity) followed by free decay without any further contributing nucleosynthetic events, then a nuclide with a half-life one-half of that of ^{129}I would only have an abundance relative to a stable isotopic cogener of 10^{-8}. Such a disparity in abundances is so enormous that a free-decay interval of $\sim 10^8$ y would absolutely prohibit the persistence of shorter-lived nuclides. It should be noted that, in the case of ^{26}Al, the solar value is $(^{27}$Al/^{26}Mg$)_{\odot} \sim 0.7$. If ^{26}Al/^{27}Al is as high as 10^{-3}, then, for the solar system, the total increase in ^{26}Mg due to ^{26}Al decay would be 0.07%. This means that to observe any effects it is necessary to have the most ancient materials and they must also be highly enriched in Al relative to Mg.

For almost two decades the astrophysical and cosmochemical communities were quite satisfied with a scenario in which the interstellar cloud from which the solar system formed was isolated from either local or galactic nucleosynthesis for $\sim 1 \cdot 10^8$ y. Attempts by FOWLER, GREENSTEIN and HOYLE [68] to account for D, Li, Be and B by local bombardment did not appear acceptable both because of the observational data on isotopic homogeneity and the specialized geometry assumed for a bombardment by energetic solar protons. Some workers, particularly CAMERON, had argued for a supernova trigger for the formation of the solar system [69]. The fall of the Allende meteorite in 1969 brought large quantities of material containing Ca and Al rich inclusions that approximated in chemical composition what had been calculated theoretically to be the early condensates from a hot medium of solar composition [12, 13]. Similar Ca-Al-rich inclusions had first been recognized by CHRISTOPHE-MICHEL-LÉVY [70] in the Vigarano meteorite. The large mass of the Allende meteorite made the Ca-Al-rich inclusions more readily available for scientific research. This meant that more « primitive » samples of early solar-system material could be studied in contrast to the more evolved or more thoroughly mixed objects which had previously been available.

The existence of abundant samples, in conjunction with the development of more advanced methods used in the study of lunar samples, led to a new generation of high-precision, high-sensitivity meteorite studies. An extensive investigation by GRAY, PAPANASTASSIOU and WASSERBURG [71] showed that some of the Ca-Al-rich inclusions in Allende were the oldest known objects in the solar system as manifested by the very low $^{87}Sr/^{86}Sr$ ratios. The work by GRAY et al. [71] showed that one Ca-Al-rich chondrule (D7) contained initial $^{87}Sr/^{86}Sr$ which was 3 ε units below BABI (= 0.698 98). This result was obtained in a sample with an extremely low $^{87}Rb/^{86}Sr$ ratio and was reproducible (see fig. 20). It follows that the Sr in this sample separated from the solar nebula at about $10 \cdot 10^6$ y prior to the time corresponding to BABI if the shift was due to the decay of ^{87}Rb in the solar nebula $((Rb/Sr)_\odot = 0.5)$. Data for other inclusions did not lie on an isochron (see fig. 20 and 21) and the system has clearly been subjected to disturbances by elemental migration some aeons after the solar system formed [72, 73]. It is evident from these observations that the Allende inclusions are vestigial cosmic virgins containing definitive evidence of ancient age and of recent tampering.

GRAY et al. [71] indicated that a new search for ^{26}Al was necessary. CAMERON [74] proposed that the primitive character of Sr found in one of the Allende inclusions by these workers might not be due to a shorter time for ^{87}Rb decay in the solar system but rather incomplete mixing of s and r process material from a late injection. While this interpretation of the Sr results could not be substantiated, the possibility of a shortened time scale was again brought into consideration. This was greatly enhanced by the discovery by CLAYTON et al. [18] of isotopic anomalies in oxygen. A search of isotopic anomalies in Mg led to the discovery by GRAY and COMPSTON [75] and LEE and PAPANASTASSIOU [30] of distinct shifts in the isotopic abundance of Mg in some Allende inclusions which could not be simply explained by mass fractionation. Subsequent work by LEE, PAPANASTASSIOU and WASSERBURG [32, 33, 76-78] showed that in two inclusions there were isotopic anomalies in Mg which were due to

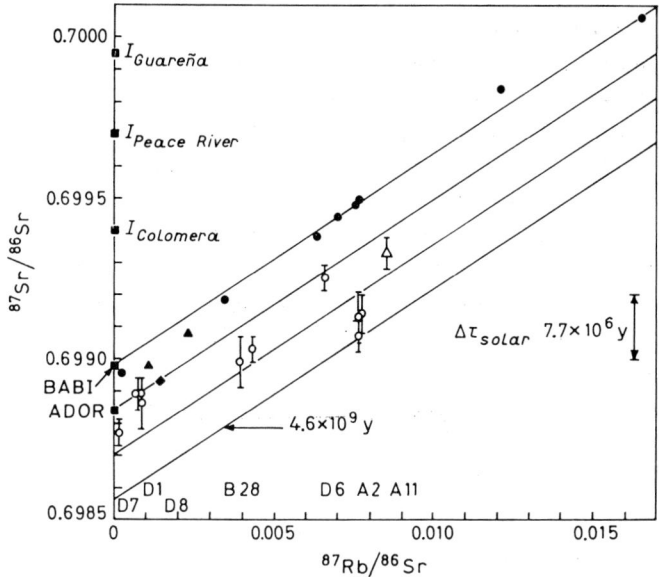

Fig. 20. – Rb-Sr evolution diagram for those Ca-Al-rich inclusions from Allende which have extremely low ^{87}Rb/^{86}Sr. The Allende data points carry labels and error bars. Inclusion D7 contains the lowest ^{87}Sr/^{86}Sr ever measured and establishes the most primitive value for ^{87}Sr/^{86}Sr; the initial ^{87}Sr/^{86}Sr intercept with y-axis calculated from the data on D7 is essentially equal to the measured value and distinctly lower than the initial ^{87}Sr/^{86}Sr in Angra dos Reis (ADOR) and that obtained for the basaltic achondrite suite (BABI). The difference in initial ^{87}Sr/^{86}Sr corresponds to about a $10 \cdot 10^6$ y interval of growth in the solar system (Rb/Sr ~ 0.5). Note that all the Allende data points do not lie on a single line and indicate later element redistribution. (After [71]; see also fig. 21.)

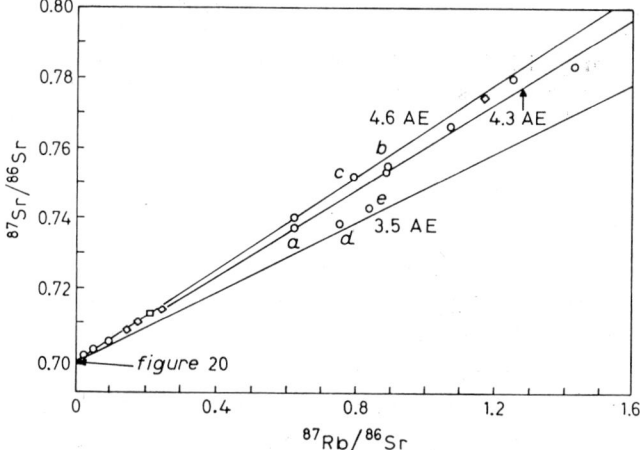

Fig. 21. – Rb-Sr evolution diagram for Ca-Al-rich inclusions from Allende which are alkali rich. The data points do not fall on a single line, but indicate a late disturbance at a time younger than 3.5 AE. (After [71].)

excesses in ^{26}Mg (^{26}Mg*) (relative to terrestrial Mg) and that these excesses were strongly correlated with the Al/Mg ratio of the coexisting mineral phases Evidence of the existence of an extinct nuclide is provided by demonstrating that excesses of the daughter nuclide are present in several different objects when the other isotopes of the daughter element are unchanged. Demonstration that the now extinct parent nuclide was present in an object at the time of its formation requires that the observed distribution of the daughter nuclide follows the chemical distribution of the parent nuclide in the different phases of that object (cf. fig. 22). This requires that objects be well preserved since the time of formation. Figure 23 shows the relationship of Mg isotopic composition

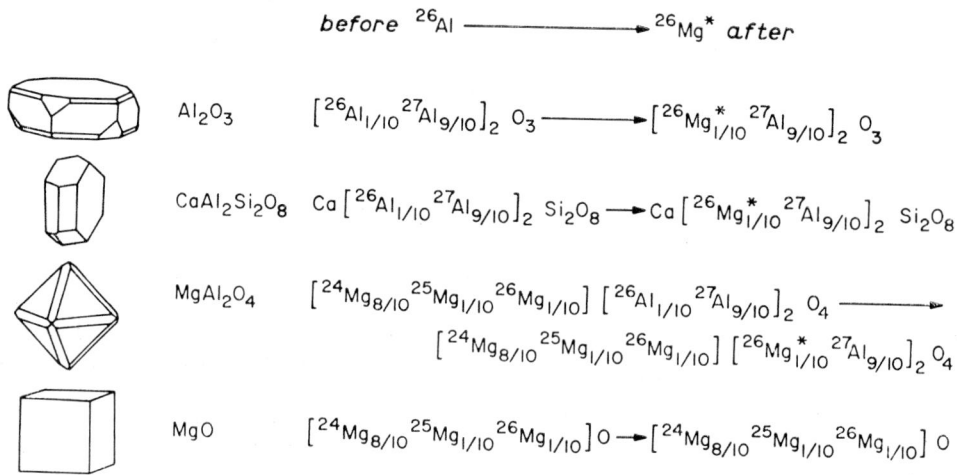

Fig. 22. – Means of identification of ^{26}Mg produced by *in situ* decay of ^{26}Al. Exaggerated abundances are shown for purposes of illustration. Different crystals are seen to contain different stoichiometric ratios of Al/Mg, but to have formed with a uniform Al isotopic composition consisting of 1/10 ^{26}Al and 9/10 ^{27}Al. The formulae are grouped so as to show that, subsequent to decay, the radiogenic ^{26}Mg* replaces ^{26}Al atoms in the crystal lattices. If some of these mineral phases (and crystals) coexisted in one inclusion and were extracted, their ^{26}Mg/^{24}Mg ratio would correlate precisely with ^{27}Al/^{24}Mg.

and the Al/Mg ratio for coexisting phases formed at a single time from a piece of matter which was initially isotopically uniform. The time « zero » refers to the time at which the matter was isotopically uniform. If the isotopically uniform state (of, say, Mg and Al) refers to the whole solar nebula, then the lines correspond to times relative to « zero » and may, therefore, be used as a chronometer. As samples may be chemically fractionated at different times, the intercepts (^{26}Mg/^{24}Mg)$_I$ may differ for isochronous objects, but the slopes will be equal. If the solar nebula is heterogeneous (with regard to the isotopic composition of Mg and Al), then the demonstration of *in situ* decay is still

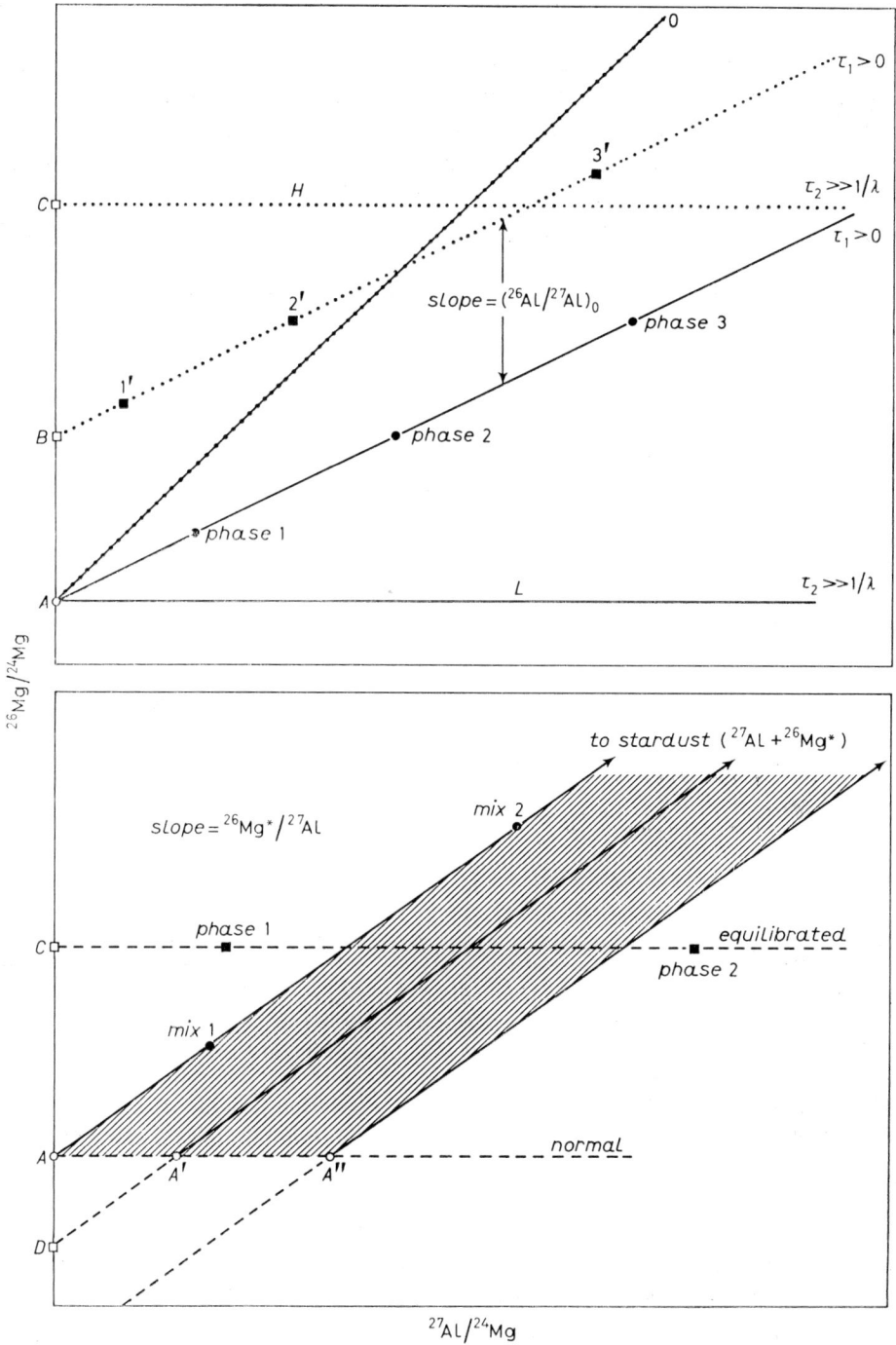

Fig. 23. – (top) Al-Mg evolution diagram for two isochronous systems which were initially isotopically homogeneous. Phases in each system are colinear. Slopes correspond to $^{26}Al/^{27}Al$ in the systems. Differences in initial $^{26}Mg/^{24}Mg$ reflect differences in Al/Mg (high, low) in the parent reservoirs. (bottom) Systematics of mixing of normal material with stardust containing ^{27}Al and fossil $^{26}Mg^*$. Mixtures of stardust and of solar-system materials containing Al and Mg will fall in the stippled area and will not in general define a straight line. (After [76].)

possible, but the decay scheme does not provide a chronometer for different samples of the solar nebula. However, it does provide an approximate chronometer for the time between production of the parent nuclide and the object, if production rates and dilution factors can be estimated and if the half-life is short. A striking example of this correlation was found in inclusion WA and is shown in fig. 24. The plagioclase ($CaAl_2Si_2O_8$) has a high ratio of Al/Mg

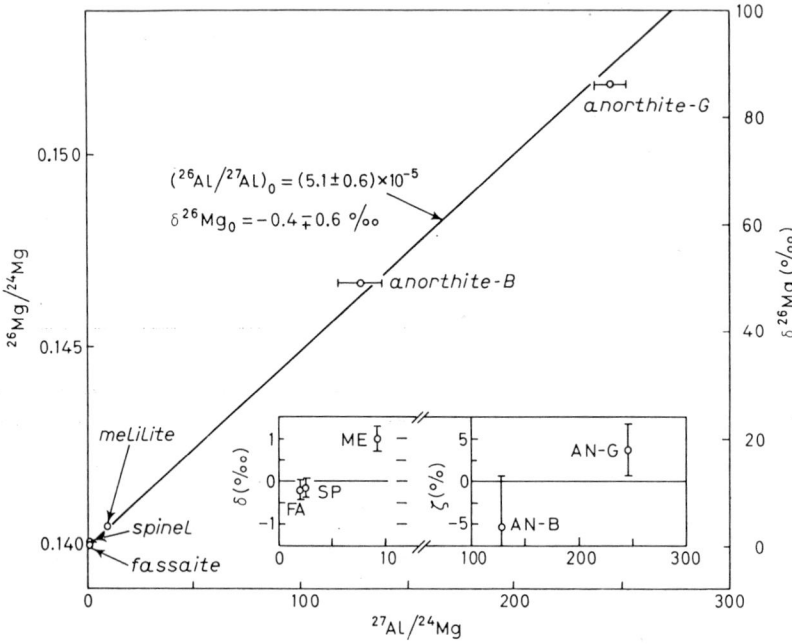

Fig. 24. – ^{26}Mg-^{26}Al evolution diagram for coexisting mineral phases in Ca-Al-rich inclusion WA from Allende. Anorthite samples show very large excesses of ^{26}Mg (right-hand axis) up to 85 permil correlating with ^{27}Al/^{24}Mg. Melilite also shows significant ^{26}Mg excess. Insert shows deviations of the data from the best-fit line; note that the melilite (ME) analysis is not on the best-fit line. The slope of the line corresponds to an initial ^{26}Al/^{27}Al $= 5 \cdot 10^{-5}$ at the time the inclusion formed. The initial Mg in the inclusion had normal ^{26}Mg/^{24}Mg composition. This correlation provides the basic evidence demonstrating the *in situ* decay of ^{26}Al in the early solar system. (After [76].)

and yields an excess of $\sim 8.5\%$ of ^{26}Mg relative to terrestrial Mg. This correlation was shown to hold for individual crystals (~ 100 μm) as well as for macroscopic samples of these crystals. By assuming that the ratio ^{25}Mg/^{24}Mg was the same as terrestrial, it was possible to show that there was an apparent excess of ^{26}Mg in a phase (melilite) with an intermediate Al/Mg ratio in the same inclusion. The data point for the melilite lies close to, but is not on, the correlation line. This correspondence in ^{26}Mg* with Al content was also found in an inclusion BG2-6 where the high Al phase is grossular ($Ca_3Al_2Si_3O_{12}$). This correlation has been substantiated in other Allende inclusions by independent workers using different techniques [79-82]. Some of the data are, however, of much lower precision, so that strict tests of a linear correlation are not possible. In the case of inclusion WA, excesses of up to 50 % in ^{26}Mg have been found [79]. There are some clear violations of the correlation shown in fig. 24 [35, 77]. These are explainable by late-stage melting of the inclusions, chemical alteration of the inclusions or incomplete mixing of ^{26}Al in the proto-solar nebula. Existence of ^{26}Mg excesses in meteorites other than Allende have been reported by LORIN et al. [81] and LORIN and CHRISTOPHE [82], so that the evidence for ^{26}Al includes at least two meteorites. It appears that there is strong evidence for the existence of ^{26}Al in the early solar system at the time that centimeter-size objects were formed. There will have to be more extensive investigations to adequately document the case for in situ ^{26}Al decay (particularly for phases which intrinsically contain both Al and Mg in their crystal structure) and to obtain a better estimate of $^{26}Al/^{27}Al$ for the solar system. There is as yet no clear evidence for large $^{26}Al/^{27}Al$ ratios in samples which show general isotopic anomalies for many elements (such as C-1 and EK-1-4-1) and there is no correlation with the oxygen isotopic effects. Whether the ^{26}Al production is connected in time with the general isotopic anomalies (O, Mg, Ca, Kr, Sr, Xe, Ba, Nd, Sm) remains to be demonstrated. At present it appears that ^{26}Al was present in the solar system with an abundance of $^{26}Al/^{27}Al \sim 5 \cdot 10^{-5}$ in some objects, but that ^{26}Al was not uniformly distributed. In some cases there has been serious alteration of the ^{26}Mg-^{26}Al parent-daughter relation by element redistribution due to chemical reactions between the early condensates and a later lower-temperature gaseous regime.

The abundance ratio $(^{26}Al/^{27}Al)_\odot$ is an extremely sensitive indicator of the number of exotic nuclei added in a late stage. If the ratio $(^{26}Al/^{27}Al)_{SN}$ is the value in the source with due regard for dilution in the envelope, then this implies that at least a fraction $(^{26}Al/^{27}Al)_\odot/(^{26}Al/^{27}Al)_{SN}$ of the total Al in the solar system was injected by this late event. For an effective production ratio of $(^{26}Al/^{27}Al)_{SN} \sim 10^{-3}$ [83, 84] and a solar-system value of $5 \cdot 10^{-5}$, this corresponds to a contribution of at least 5 % of all the Al in the Sun.

The time scale determined by the existence of ^{26}Al in the solar system is at most $10 \cdot 10^6$ y and more reasonably $\sim 3 \cdot 10^6$ y. It follows that the time for the last injection of freshly synthesized nuclear material is at least an order of magnitude less than that obtained from ^{129}I. The time scale of $3 \cdot 10^6$ y is short compared to a galactic year and suggests the formation of the solar system in a region where many stars are being formed and which contains massive, rapidly evolving stars which can contribute fresh nuclear debris over the time the solar system is accumulating. An alternative possibility is that the early Sun went through a phase of intense activity with a large flux (10^{23} p/cm^2) of energetic protons which produced ^{26}Al in a surrounding dust cloud [23]. It is not apparent how this mechanism could produce the observed excesses in unshielded isotopes of Ba, Nd and Sm as discussed elsewhere in this paper. From the observed anomalies in both neutron-rich and neutron-poor isotopes of widely different Z it has become difficult to accept local modification in the solar system as the cause.

The ^{26}Al time scale is extremely short; it is commensurate with the time scale for gravitational free fall and much shorter than the Kelvin-Helmholtz time. The free-fall time $\left(\tau_{\text{free fall}} \sim 4 \cdot 10^7/\sqrt{n_H} \text{ y}\right)$ is a rough estimate of the time scale for collapse from a uniform initial state with n_H protons/cm³. It follows that, if the free-fall time is commensurate with the mean life of ^{26}Al, this requires an initial density of $n_H \sim 2 \cdot 10^3/\text{cm}^3$. This is a typical value for dense interstellar clouds. If we assume a supernova source of $20 M_\odot$, then a time scale of between 10^5 and 10^6 y would permit a reasonable dilution factor with a mean velocity of expansion of 100 km/s (1 parsec in 10^4 y). However, to understand the evolution from a state of highly dispersed material to collapse and the formation of the Sun and planets in a time of less than 10^7 years will demand imaginative and careful analysis. This condensation and formation time scale is required if the ^{26}Al is due to injection from external sources such as a supernova. The sequence of formation of the smaller planetary objects, the terrestrial and Jovian planets and the Sun is an open issue. There appears to be a major time difference between the formation of small objects (marbles to meteorite parent bodies) and the Moon and the Earth. If the young ages of the Earth and Moon ($\Delta T \sim 0.1$ AE) are not times of internal differentiation, then the terrestrial planets were accumulated from « meteoritic » debris long after the decay of ^{26}Al. This means that accumulation of silicate planetary bodies up to ~ 100 km was very rapid and the subsequent accumulation of larger terrestrial bodies was late, most plausibly after the formation of the Sun and the Jovian planets. The absence of ^{26}Al in the Earth, Moon, Mars, Venus and Mercury would again leave the initial heat sources in these bodies as a problem.

An abundance of ^{26}Al/^{27}Al$\sim 5 \cdot 10^{-5}$ provides an intense heat source for planets which form promptly. This could explain the rapid melting of small asteroidal-size objects (down to a few kilometers) as well as for the larger terrestrial planets [85]. This abundance in the gas and dust of the solar nebula would also be a major source of ionizing radiation [86]. The applicability of ^{26}Al as a general planetary heat source and as a radiation source in the solar nebula must depend on the large-scale abundance of this nuclide and the rate of planet formation. This can only be addressed by more extensive sampling of early solar-system debris.

The issue regarding time scales is summarized in fig. 25. In the past few years we have gone from $\sim 10^8$ y to $\sim 10^6$ y for the last time of injection. It is clear that some of the intermediate-lived nuclei (*e.g.* ^{244}Pu) must come from somewhat earlier periods of nucleosynthesis $((1 \div 4) \cdot 10^8 \text{ y})$. Earlier contributions for ^{244}Pu are required in order to explain the solar ^{244}Pu/^{232}Th

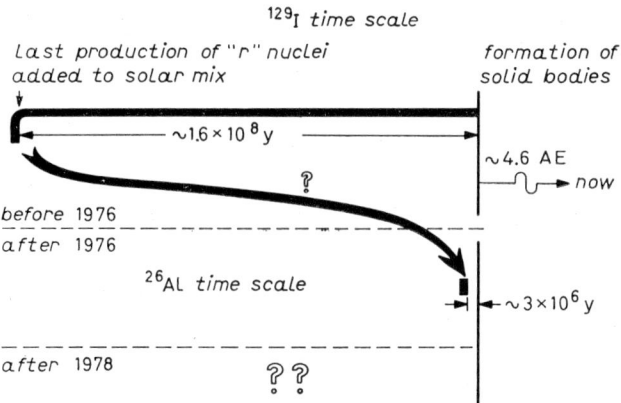

Fig. 25. – Cartoon showing the required change in the length of the time interval between isolation of the solar system and condensation and aggregation of meteorites. Most plausible interpretation is the addition of ^{26}Al, ^{107}Pd, some ^{129}I, (^{135}Cs?) within $\sim 3 \cdot 10^6$ y of last production (after [87]).

ratio, since the production of ^{244}Pu (and ^{235}U) on a 10^6 y time scale is small unless the production pattern for high-mass ($A > 200$) nuclei is enriched by a factor of 100 relative to those at intermediate mass for the late-stage addition. The longer-lived and stable nuclei must dominantly come from much more ancient events. The rate of nucleosynthesis as estimated from the most recent events which we sample in the solar system gives a clear hint of the granularity of the nucleosynthetic processes, but does not yet permit a firm estimate of the rates of nucleosynthesis over the history of the Galaxy. It is possible that extrapolations backward using the short-lived products as an estimate of production rates is an inadequate means of constructing the long-time-scale rates of nucleosynthesis. However, it is certain that we are now looking through and resolving the granular structure of nucleosynthesis.

7. – Common and uncommon anomalies—The unmade connection.

The FUN inclusions show isotopic anomalies in almost every refractory element so far measured. These FUN inclusions are relatively rare, as they have been found to comprise only a few percent of the class of Ca-Al chondrules and aggregates. While the searches are not extensive, the isotopic anomalies found in the refractory elements in the two FUN inclusions are not observed in a variety of other meteorites. In contrast to this behavior, the isotopic anomalies observed for oxygen are common to almost all of the Ca-Al-rich chondrules and aggregates. The relative fractional magnitudes of the oxygen effects are one to two orders of magnitude greater than found in the FUN samples for the refractory elements. Small but distinct differences in oxygen composition are found in a variety of meteorites. In general, samples which are found to have large oxygen effects are found to have no observable effects in other elements. It, therefore, appears that the isotopic anomalies in the FUN inclusions are not correlated with the widespread oxygen anomaly. The connection with the oxygen effects found in the FUN samples is seen in a departure from the typical mixing line between O_E and O_N and is most plausibly due to mass fractionation as is found for several elements. This observation is at present the only link between what otherwise appear to be distinctive phenomena. The other major element (in the solar system) that shows major and widespread isotopic anomalies is Ne. The Ne effects appear to be related to differences in isotopic composition of the solar gas and that impregnated into dust (or solids). Because of the chemically inert nature of Ne, clear correlations with condensable material are not apparent. Some of the observations in oxygen hint at differences between gas and dust. Certainly the degree of chemical reactivity between the different reservoirs or chemical compounds must be important in isolating them. From the magnitude of the oxygen effects (and possibly the Ne) it seems that the nuclear processes responsible for the O_E component must be extremely specific with only minor contributions to most other elements (both low and high Z). The exotic nuclides responsible for the UN effects in C-1 and EK-1-4-1 thus appear to come from a source distinct from that which either made or isolated O_E: At best it would appear that a connection between the common anomaly in oxygen and the uncommon anomalies in the refractory elements has not been made.

The fundamental difference in chemical composition between gas and dust is one obvious and reliable astrophysical fact that requires more careful consideration. As most of the heavy elements are stored in interstellar dust, we may presume that the oxygen which resides in the dust is comprised of oxygen nuclei which were produced during heavy-element synthesis and which were precipitated out with the heavy elements. This will naturally produce a difference between the isotopic composition of the oxygen in the dust as compared with the oxygen in the gas phase insofar as the rate of exchange between gas and dust is small. This mechanism must play a significant role in the problem of isotopic anomalies; however, as implied earlier, it does not explain either the anomalies in Ne, Kr and Xe (unless there is some plausibile entrapment mechanism) or the variations in composition of the refractory « heavy » elements. The « special » late addition appears to be required ([26]Al, [107]Pd) and so it is reasonable that the observed effects must be due to isotopic heterogeneity between gas and dust in the ambient interstellar medium coupled with some late addition. There is also the possibility of isotopic heterogeneity within the old « dust phase » itself on a small astronomical scale.

8. – Conclusions.

There is good reason to believe that continuing study of the late nucleosynthetic processes may give us better understanding of solar-system formation and direct insight in element processing-reprocessing in massive rapidly evolving stars.

The difficulty with any canonical astrophysical explanation (*e.g.* a supernova) of the observations reviewed here is that no one has come forward with a plausible scenario that explains the observed isotopic anomalies or which clearly predicts new discoveries of isotopic anomalies. It is hoped that this report will provoke serious scholarly inquiry into the basic nuclear astrophysical processes that must be the cause of the isotopic effects. More intensive searches for samples with correlated isotope anomalies will be necessary. The demonstration of isotopic effects in a chain of elements which are directly related to nuclear reactions should prove a key to identifying the actual processes causing the anomalies. There are also major problems in our understanding of the condensation process and the chemical and ion reactions which occur between early and late formed parcels of matter in the early solar system. We will, no doubt, now begin to focus on the chemical « alteration » processes of « primitive » material—an issue which has been mostly ignored. Study of the « alteration » process will provide a fundamental advance in early-solar-system chemistry and physics and may well aid in identifying the basic reservoirs of unmixed interstellar materials.

One of the gravest problems which we will face over the next few years will be the shortage of reliable high-quality isotopic data and of sound predictive theories. Now that the extraordinary is known to exist, it may be too readily found and explained where it is in fact absent. Every effort must be made to test and confirm both isotopic effects and nucleosynthetic models.

* * *

This work was supported by grants from the National Science Foundation and the National Aeronautics and Space Administration. Without this ongoing support the precise data summarized in this work could not have been obtained.

The contributions of M. McCulloch to part of the data summarized here merit special mention. Discussion with and stimulation from our colleagues in the Kellogg Radiation Laboratory is gratefully acknowledged. In this report we have omitted a general discussion of the rare gases and, in particular, xenon and krypton anomalies. These observations and ideas are of great importance and have been the subject of extensive investigation and discussion over the past two decades. It is hoped that our noble colleagues with affection for atomic numbers 36-10-54 will recognize the limited nature of this report. This work very closely follows the review paper submitted to the 22nd Liège International

Astrophysical Symposium, June 20-22, 1978. The original report of the correlated isotopic anomalies in C-1 and EK-1-4-1 was made at the Welch Foundation Conference on Cosmochemistry [87].

ADDENDUM

A variety of other intermediate- to short-lived radioactive nuclear species may have been produced in the terminal event depending on the details of the particular nuclear processes involved. Searches for other extinct radio-activities are underway (cf. the so far negative results for ^{41}Ca ($\tau_{\frac{1}{2}} = 1.3 \cdot$ $\cdot 10^5$ y) [88]). Subsequent to the presentation at the Liège meeting and the Varenna summer school, KELLY and WASSERBURG [89] reported the discovery of 4% excess of ^{107}Ag in an iron meteorite with a ratio Pd/Ag $\sim 10^4$. They consider this as most plausibly due to the decay of ^{107}Pd and infer

$$(^{107}\mathrm{Pd}/^{110}\mathrm{Pd})_\odot \geqslant 2 \cdot 10^{-5} .$$

This is compatible with the ^{26}Al time scale, the $(^{129}\mathrm{I}/^{127}\mathrm{I})_\odot$ value and the magnitude of the amount of addition ($\sim 10^{-4}$) of intermediate-Z nuclei to the solar system. Certainly some late and nonuniform addition of ^{129}I may have occurred. This implies also the formation of differentiated planetary bodies within $\sim 20 \cdot 10^6$ y of the last stage of nucleosynthesis. More recent results by KELLY and WASSERBURG [90] show a correlation of ^{107}Ag/^{109}Ag with ^{110}Pd/^{109}Ag in the Santa Clara meteorite and the presence of ^{107}Ag* in a second meteorite. A third meteorite with a high Pd/Ag ratio showed no effect. It is obvious that confirmation of this excess of ^{107}Ag and an adequate demonstration of the *in situ* decay of ^{107}Pd are needed. The existence of ^{107}Pd directly connects the time of last nucleosynthesis and planetary differentiation, while the ^{26}Al connects the last nucleosynthesis with formation of dust and marbles. It now appears to be certain that the time scale between late nucleosynthesis and *small* planet formation is less than $2 \cdot 10^7$ years.

REFERENCES

[1] ANGRA DOS REIS CONSORTIUM: *Earth Planet. Sci. Lett.*, **35**, 271 (1977).
[2] P. GOLDREICH and W. R. WARD: *Astrophys. J.*, **183**, 1051 (1973).
[3] V. S. SAFRONOV: *Evolution of the Protoplanetary Cloud and the Formation of the Earth and Planets* (in Russian) (Moscow, 1969). English translation as NASA TT F-677, NTIS (Springfield, Va., 1972).
[4] F. TERA, D. A. PAPANASTASSIOU and G. J. WASSERBURG: *Earth Planet. Sci. Lett.*, **22**, 1 (1974).
[5] E. M. BURBIDGE, G. R. BURBIDGE, W. A. FOWLER and F. HOYLE: *Rev. Mod. Phys.*, **29**, 547 (1957).
[6] A. G. W. CAMERON: *Publ. Astron. Soc. Pac.*, **69**, 201 (1957).

[7] J. W. TRURAN, J. J. COWAN and A. G. W. CAMERON: *Astrophys. J. Lett.*, **222**, 63 (1978).

[8] S. E. WOOSLEY and W. M. HOWARD: *Astrophys. J. Suppl. Ser.*, **36**, 285 (1978).

[9] F. A. PODOSEK: *Annu. Rev. Astron. Astrophys.*, **16**, 293 (1978).

[10] F. BEGEMANN, J. GEISS and D. C. HESS: *Phys. Rev.*, **107**, 540 (1957).

[11] H. VOSHAGE: *Earth Planet. Sci. Lett.*, **40**, 83 (1978).

[12] H. C. LORD III: *Icarus*, **4**, 279 (1965).

[13] L. GROSSMAN: *Geochim. Cosmochim. Acta*, **36**, 597 (1972).

[14] D. A. WARK and J. F. LOVERING: *Proceedings of the VIII Lunar Science Conference* (Houston, Tex., 1977) p. 95.

[15] M. BLANDER and L. H. FUCHS: *Geochim. Cosmochim. Acta*, **39**, 1605 (1975).

[16] U. B. MARVIN, J. A. WOOD and J. S. DICKEY jr.: *Earth Planet. Sci. Lett.*, **7**, 346 (1970).

[17] H. P. TAYLOR, M. B. DUKE, L. T. SILVER and S. EPSTEIN: *Geochim. Cosmochim. Acta*, **29**, 489 (1965).

[18] R. N. CLAYTON, L. GROSSMAN and T. K. MAYEDA: *Science*, **182**, 485 (1973).

[19] R. N. CLAYTON, N. ONUMA, L. GROSSMAN and T. K. MAYEDA: *Earth Planet. Sci. Lett.*, **34**, 209 (1977).

[20] R. N. CLAYTON, N. ONUMA and T. K. MAYEDA: *Earth Planet. Sci. Lett.*, **30**, 10 (1976).

[21] R. N. CLAYTON and T. K. MAYEDA: *Earth Planet. Sci. Lett.*, **40**, 168 (1978).

[22] A. O. NIER, M. B. McELROY and Y. L. YUNG: *Science*, **194**, 68 (1976).

[23] T. LEE: *Astrophys. J.*, **224**, 217 (1978).

[24] P. G. WANNIER, R. LUCAS, R. A. LINKE, P. J. ENCRENAZ, A. A. PENZIAS and R. W. WILSON: *Astrophys. J. Lett.*, **205**, 169 (1976).

[25] D. S. P. DEARBORN, B. M. TINSLEY and D. N. SCHRAMM: *Astrophys. J.*, **223**, 557 (1978).

[26] D. C. BLACK: *Geochim. Cosmochim. Acta*, **36**, 377 (1972).

[27] P. EBERHARDT: *Earth Planet. Sci. Lett.*, **24**, 182 (1974).

[28] P. EBERHARDT: *Proceedings of the IX Lunar Science Conference* (Houston, Tex., 1978), p. 1027.

[29] J. GEISS, F. BUEHLER, H. CERUTTI, P. EBERHARDT and CH. FILLEUX: *Solar wind composition experiment, Apollo 16 Preliminary Science Report*, NASA SP-315, sect. **14**-1 (1972).

[30] T. LEE and D. A. PAPANASTASSIOU: *Geophys. Res. Lett.*, **1**, 225 (1974).

[31] R. N. CLAYTON and T. K. MAYEDA: *Geophys. Res. Lett.*, **4**, 295 (1977).

[32] T. LEE, D. A. PAPANASTASSIOU and G. J. WASSERBURG: *Geophys. Res. Lett.*, **3**, 109 (1976).

[33] G. J. WASSERBURG, T. LEE and D. A. PAPANASTASSIOU: *Geophys. Res. Lett.*, **4**, 299 (1977).

[34] L. GROSSMAN: *Geochim. Cosmochim. Acta*, **39**, 433 (1975).

[35] T. M. ESAT, T. LEE, D. A. PAPANASTASSIOU and G. J. WASSERBURG: *Geophys. Res. Lett.*, **5**, 807 (1978).

[36] H.-W. YEH and S. EPSTEIN: in *Lunar and Planetary Science IX*, Lunar and Planetary Institute (Houston, Tex., 1978), p. 1289.

[37] R. N. CLAYTON, T. K. MAYEDA and S. EPSTEIN: *Proceedings of the IX Lunar Science Conference* (Houston, Tex., 1978), p. 1267.

[38] A. G. W. CAMERON: *Space Sci. Rev.*, **15**, 121 (1973).

[39] S. E. WOOSLEY, W. D. ARNETT and D. D. CLAYTON: *Astrophys. J. Suppl. Ser.*, **26**, 231 (1973).

[40] W. M. HOWARD, W. D. ARNETT, D. D. CLAYTON and S. E. WOOSLEY: *Astrophys. J.*, **175**, 201 (1972).

[41] W. A. RUSSELL, D. A. PAPANASTASSIOU and T. A. TOMBRELLO: *Geochim. Cosmochim. Acta*, **42**, 1075 (1978).

[42] T. LEE, D. A. PAPANASTASSIOU and G. J. WASSERBURG: *Astrophys. J. Lett.*, **220**, 21 (1978).

[43] T. LEE, W. A. RUSSELL and G. J. WASSERBURG: *Astrophys. J. Lett.*, **228**, L93 (1979).

[44] D. A. PAPANASTASSIOU and G. J. WASSERBURG: *Geophys. Res. Lett.*, **5**, 595 (1978).

[45] D. A. PAPANASTASSIOU, J. C. HUNEKE, T. M. ESAT and G. J. WASSERBURG: in *Lunar and Planetary Science IX* (Houston, Tex., 1978), p. 859.

[46] M. T. McCULLOCH and G. J. WASSERBURG: *Astrophys. J. Lett.*, **220**, 15 (1978).

[47] R. A. WARD, M. J. NEWMAN and D. D. CLAYTON: *Astrophys. J. Suppl. Ser.*, **31**, 33 (1976).

[48] M. T. McCULLOCH and G. J. WASSERBURG: *Geophys. Res. Lett.*, **5**, 599 (1978).

[49] G. W. LUGMAIR, K. MARTI and N. B. SCHEININ: in *Lunar and Planetary Science IX*, Lunar and Planetary Institute (Houston, Tex., 1978), p. 672.

[50] A. R. DE L. MUSGROVE, J. W. BODLMAN and R. L. MACKLIN: *Nucl. Phys. A*, **256**, 173 (1976).

[51] J. A. HOLMES, S. E. WOOSLEY, W. A. FOWLER and B. A. ZIMMERMAN: *At. Data Nucl. Data Tables*, **18**, 305 (1976).

[52] J. CONRAD: *Analysis of the s-process and the nuclear synthesis of the elements*, Ph. D. Thesis (Heidelberg, 1976).

[53] B. J. ALLEN, H. H. GIBBONS and R. L. MACKLIN: *Adv. Nucl. Phys.*, **4**, 205 (1971).

[54] D. D. CLAYTON: *Astrophys. J.*, **199**, 765 (1975).

[55] J. H. REYNOLDS: *Phys. Rev. Lett.*, **4**, 8 (1960).

[56] J. H. REYNOLDS: *J. Geophys. Res.*, **68**, 2939 (1963).

[57] P. M. JEFFERY and J. H. REYNOLDS: *J. Geophys. Res.*, **66**, 3582 (1961).

[58] M. W. ROWE and P. K. KURODA: *J. Geophys. Res.*, **70**, 709 (1965).

[59] G. J. WASSERBURG, J. C. HUNEKE and D. S. BURNETT: *J. Geophys. Res.*, **74**, 4221 (1969).

[60] R. S. LEWIS: *Geochim. Cosmochim. Acta*, **39**, 433 (1975).

[61] E. C. ALEXANDER jr., R. S. LEWIS, J. H. REYNOLDS and M. C. MICHEL: *Science*, **172**, 837 (1971).

[62] C. M. HOHENBERG: *Geochim. Cosmochim. Acta*, **34**, 185 (1970).

[63] G. J. WASSERBURG, F. TERA, D. A. PAPANASTASSIOU and J. C. HUNEKE: *Earth Planet. Sci. Lett.*, **35**, 294 (1977).

[64] F. A. PODOSEK: *Earth Planet. Sci. Lett.*, **8**, 183 (1970).

[65] F. A. PODOSEK: *Geochim. Cosmochim. Acta*, **36**, 755 (1972).

[66] S. NIEMEYER: in *Lunar and Planetary Science IX* (Houston, Tex., 1978), p. 808.

[67] D. N. SCHRAMM and G. J. WASSERBURG: *Astrophys. J.*, **162**, 57 (1970).

[68] W. A. FOWLER, J. L. GREENSTEIN and F. HOYLE: *Geophys. J. R. Astron. Soc.*, **6**, 148 (1962).

[69] A. G. W. CAMERON: *Icarus*, **1**, 339 (1963).

[70] M. CHRISTOPHE-MICHEL-LÉVY: *Bull. Soc. Fr. Minéral. Cristallogr.*, **91**, 212 (1968).

[71] C. M. GRAY, D. A. PAPANASTASSIOU and G. J. WASSERBURG: *Icarus*, **20**, 213 (1973).

[72] J. H. CHEN and G. R. TILTON: *Geochim. Cosmochim. Acta*, **40**, 635 (1976).

[73] M. TATSUMOTO, D. M. UNRUH and G. A. DESBOROUGH: *Geochim. Cosmochim. Acta*, **40**, 617 (1976).

[74] A. G. W. CAMERON: *Nature (London)*, **246**, 30 (1973).

[75] C. M. GRAY and W. COMPSTON: *Nature (London)*, **251**, 495 (1974).

[76] T. LEE, D. A. PAPANASTASSIOU and G. J. WASSERBURG: *Astrophys. J. Lett.*, **211**, 107 (1977).

[77] D. A. PAPANASTASSIOU, T. LEE and G. J. WASSERBURG: in *Comets, Asteroids, Meteorites*, edited by A. H. DELSEMME (Toledo, O., 1977), p. 343.

[78] G. J. WASSERBURG, T. LEE and D. A. PAPANASTASSIOU: *Meteoritics*, **11**, 384 (1976).

[79] J. G. BRADLEY, J. C. HUNEKE and G. J. WASSERBURG: *J. Geophys. Res.*, **83**, 244 (1978).

[80] I. D. HUTCHEON, I. M. STEELE, J. V. SMITH and R. N. CLAYTON: *Proceedings of the IX Lunar and Planetary Science Conference* (1978), p. 1345.

[81] J.-C. LORIN, N. SHIMIZU, M. CHRISTOPHE-MICHEL-LÉVY and C. J. ALLÈGRE: *Meteoritics*, **12**, 299 (1977).

[82] J.-C. LORIN and M. CHRISTOPHE-MICHEL-LÉVY: *Short Papers of the IV ICGCIG* (Denver, Colo., 1978), p. 257.

[83] W. D. ARNETT and J. P. WEFEL: *Astrophys. J. Lett.*, **224**, 139 (1978).

[84] J. W. TRURAN and A. G. W. CAMERON: *Astrophys. J.*, **219**, 226 (1978).

[85] H. C. UREY: *Proc. Nat. Acad. Sci. USA*, **41**, 127 (1955).

[86] G. J. CONSOLMAGNO: *Meteoritics*, **12**, 200 (1977).

[87] G. J. WASSERBURG: *Proceedings of the R. A. Welch Foundation Conference on Chemical Research XXI Cosmochemistry*, Nov. 7-9, 1977, edited by W. O. MILLIGAN (Houston, Tex., 1978), p. 95.

[88] F. BEGEMANN and W. STEGMANN: *Nature (London)*, **259**, 549 (1976).

[89] W. R. KELLY and G. J. WASSERBURG: *Geophys. Res. Lett.*, **5**, 1079 (1978).

[90] W. R. KELLY and G. J. WASSERBURG: in *Lunar and Planetary Science X* (Houston, Tex., 1979) (abstract), p. 652.

Nuclear Bombardment in Space.

D. LAL

Physical Research Laboratory - Ahmedabad 380 009, *India*

Any matter exposed at any time in the past in the interplanetary space has been subjected to diverse bombardment by radiation and particles: stellar electromagnetic radiation, nuclear particles of the cosmic radiation, solar-wind plasma and solid objects of various sizes. This bombardment induces both physical and chemical changes in matter. The cosmic-ray bombardment leads to discernible changes in terrestrial and extraterrestrial materials by *in situ* production of isotopes and changes in the lattice structure of crystalline solids [1-3]. Likewise, appreciable compositional as well as physical changes occur due to trapping of solar-wind ions in materials exposed in free space or on the surface of bodies devoid of a sensible atmosphere or a global magnetic field [4-5]. Besides these chemical and isotopic changes, physical and chemical changes occur due to impact of solid bodies in the interplanetary space. Impacts produce erosion and craters, mix materials and may lead to changes in the orbital elements. The impact of solid bodies is one of the most important processes that has taken place on the planets and smaller objects in the solar system [6]. This process has been important in the early history of the solar system during accumulation of planetesimals and subsequently in the evolutionary history of minor and major bodies of the solar system. The cosmic-ray and solar-wind bombardment, therefore, generally proceeds under changing conditions of geometry of irradiation whether the object is exposed in free space or on a planetary surface—as a result of alterations in the physical size and shape of the sample due to erosion/fragmentation and changing amounts of overlying materials due to impact-induced transport and mixing. Clearly, therefore, if the nature of chemical and isotopic changes induced by the bombardment is understood, specifically the magnitude of the various effects expected for different irradiation geometries, one may expect to delineate the evolutionary history of different components of a given sample. Implicit in this statement is an *a priori* knowledge of the particle fluxes in the past. But, since the extraterrestrial samples constitute the only record of ambient particle fluxes in the past, one has to make a judicious choice of samples to obtain information on temporal variations in particle fluxes in the past. Fortunately, because of the availability of large amounts of extraterrestrial samples of diverse exposure history, namely the meteorites and the lunar rock/soil samples, this task is feasible in certain cases. Based on such studies, one can made plausible assumptions about long-term averaged fluxes of particles in the interplanetary space.

Studies of chemical, isotopic and solid-state changes can indeed provide valuable clues about the early and late evolutionary history of extraterrestrial samples. Production of stable and radioactive isotopes by cosmic rays in a meteorite subsequent to its ejection from the parent body and prior to capture by the Earth allows one to deduce the duration of its exposure in space—conventionally, this interval is called the cosmic-ray exposure age. Implantation of solar-wind gas and solid-state damage induced by solar-flare ions are examples of processes which have occurred in the early history of the meteorite—prior to its compaction. Some of these processes occurred soon after the formation of the solar nebula, some at a later stage [7-10]. In favourable cases it becomes possible to determine meaningful limits on the time of solar-wind and solar-flare irradiation, provided it happened very early in the history of the solar system, within a time span of a few hundred million years [11].

At this Summer Course, we are primarily concerned with the early-solar-system processes. We will, therefore, be concerned both with nuclear reactions leading to production of long-lived or stable isotopes and solid-state damage of materials in meteorites/lunar samples. Any attempt to decipher the evolutionary history of objects will involve an assessment of the expected rates of production of a given effect due to different incident particles, for different exposure geometries. In this lecture, I will briefly discuss the types of particles of interest and the nature of their interactions. The effective ranges of interaction will be considered in each case. I will discuss our present-day knowledge of the composition and flux of cosmic-ray and solar-wind particles and also the information about their long-term fluxes based on the study of extraterrestrial samples.

It has to be borne in mind that impacts due to interplanetary particles can wipe out or diffuse some of the cosmogenic records discussed above. Further, since the cosmogenic irradiation proceeds in the presence of changing geometry due to erosion, fragmentation and transport, it will, therefore, be necessary to consider the depth and time scales of impact cratering in relation to nuclear effects. In the next lecture, I will consider this problem and discuss how the nuclear labelling of materials allows one to delineate the evolutionary history of planetary regoliths with particular reference to the course of events leading to the formation of gas-rich meteorites.

1. – Particles in space and their interactions.

Amongst the nuclear particles, the most energetic are the cosmic-ray particles, the corpuscular radiation consisting of particles accelerated by our

Sun and by stars in our Galaxy. Acceleration and propagation from the Sun occurs during solar-flare activity and hence solar contributions to the cosmic-ray flux in the solar system are sporadic. The continuous flux of cosmic-ray particles that is incident on the solar system is due to acceleration of particles in the galactic supernovae and other stellar environments; the steady cosmic-ray flux is a manifestation of the fairly long storage time of cosmic-ray particles in the Galaxy, $\sim 10^6$ y [12]. The term cosmic radiation is still used today since the early days when we knew nothing about their origin; the particles of galactic origin are called the *galactic cosmic radiation* and those produced by our Sun as the *solar cosmic radiation*.

Besides cosmic-ray particles, the corpuscular radiation in the solar system is of solar origin. Suprathermal particles and solar-wind ions are emitted by the Sun.

We will now briefly discuss the energies and fluxes of the corpuscular radiation in the solar system and later consider their interactions.

a) Galactic cosmic radiation. For reviews of the composition of the galactic cosmic radiation, reference is made to [13-18]. In fig. 1 the measured differential energy spectra of the relatively abundant cosmic-ray nuclei are shown for the kinetic-energy region extending to 10^3 GeV/nucleon; 1 GeV $= 10^9$ eV. Below a few GeV/nucleon, the particle fluxes depend on the general level of solar activity. The solar modulation effect is anticorrelated, *i.e.* the cosmic-ray particle fluxes are depressed during periods of high solar activity. During solar minimum, the flux can be conveniently described by the expression [13]

$$(1) \qquad\qquad dN/dE = K_i(1 + E)^{-2.5} ,$$

where K_i is a constant for the nuclear component i; this expression is valid for the region $(0.5 \div 10^3)$ GeV/n. E is the kinetic energy in GeV/n. The cumulative flux of protons above 0.5 GeV kinetic energy is

$$3100 \text{ particles/m}^2 \cdot \text{s} \cdot \text{sr} ,$$

or

$$4100 \text{ nucleons/m}^2 \cdot \text{s} \cdot \text{sr} .$$

The average kinetic energy of cosmic-ray nucleons in the vicinity of the Earth is 3.6 GeV, corresponding to an energy flux of $4.4 \cdot 10^9$ eV/cm$^2 \cdot$s. The cosmic-ray energy density is 0.6 eV/cm^3, which is approximately equal to the energy density of starlight in our neighbourhood.

Recent measurements of cosmic-ray intensity gradients in the solar system, based on the Pioneer 10 and 11 measurements, show that over the heliocentric radial range $(1 \div 5)$ AU the gradients are positive but less than or of the order

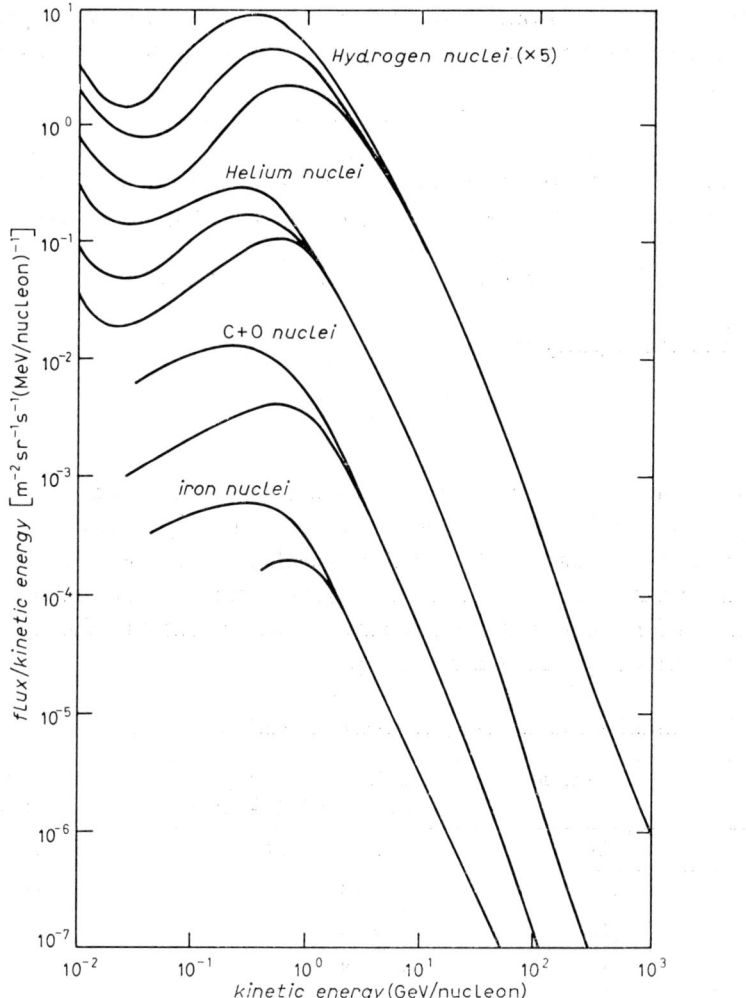

Fig. 1. – Differential energy spectra of the more abundant nuclear species in cosmic radiation as measured near the Earth during different periods of solar activity. After [17].

of a few tens of percent per astronomical unit [19] for protons and helium nuclei of $(10 \div 100)$ MeV/n. Thus the fluxes given in fig. 1 are valid up to 5 AU in the ecliptic.

In fig. 1, the spectra turn up below about $(20 \div 30)$ MeV/n. The origin of these low-energy particles is probably now understood as being due to local acceleration of nucleons near the boundary regions where shocks are observed, characterized by jumps in the magnetic field [20]. In any case, at energies below 100 MeV/n, the galactic cosmic-ray flux is of little consequence compared to the solar cosmic-ray particles, on a time-averaged basis, as we shall see later.

b) *Solar cosmic radiation.* The energy spectra of solar cosmic-ray nuclei can be expressed over a narrow energy interval as a power law in kinetic energy:

$$(2) \qquad\qquad dN = \text{const} \cdot E^{-\gamma} dE \,.$$

The value of γ, however, changes rapidly with energy in the interval $(1 \div \div 100)$ MeV/n. A better fit is the exponential rigidity spectrum:

$$(3) \qquad\qquad dN(R) = \text{const} \cdot \exp\left[- R/R_0\right] dR \,,$$

where R is the rigidity in volt

$$R = \frac{pc}{Ze} \,.$$

The value of the slope γ in eq. (2) lies within $1 \div 4$ and the characteristic rigidity R_0 within $(50 \div 200)$ MV over the kinetic-energy interval $(1 \div 100)$ MeV/n. For a detailed discussion of solar-flare spectra, reference is made to [2], where the relationship between γ and R_0 is given and also typical data on energy spectra of solar protons and alpha-particles are discussed. For a general review of solar cosmic-ray phenomena, see [21].

TABLE I. – *Relative abundance of elements normalized to a base of 1.0 for oxygen* [22].

Z	Element	Solar cosmic rays of $E > 20$ MeV/amu	Solar photosphere
2	He	103 $\pm$ 10	110 $\pm$ 25
3	Li	< 0.02	—
4	Be	< 0.02	—
5	B	< 0.02	—
6	C	0.56 $\pm$ 0.06	0.60 $\pm$ 0.10
7	N	0.19 $\pm$ 0.03	0.15 $\pm$ 0.05
8	O	1.0	1.0
10	Ne	0.16 $\pm$ 0.03	0.13 $\pm$ 0.03
12	Mg	0.056 $\pm$ 0.014	0.051 $\pm$ 0.004
14	Si	0.028 $\pm$ 0.010	0.060 $\pm$ 0.007
16	S	0.008 $\pm$ 0.006	0.028 $\pm$ 0.004
18	A	< 0.017	—
20	Ca	< 0.010	0.004 $\pm$ 0.0005
22	Ti	—	—
24	Cr	—	—
26	Fe	0.050 $\pm$ 0.020	0.045 $\pm$ 0.015

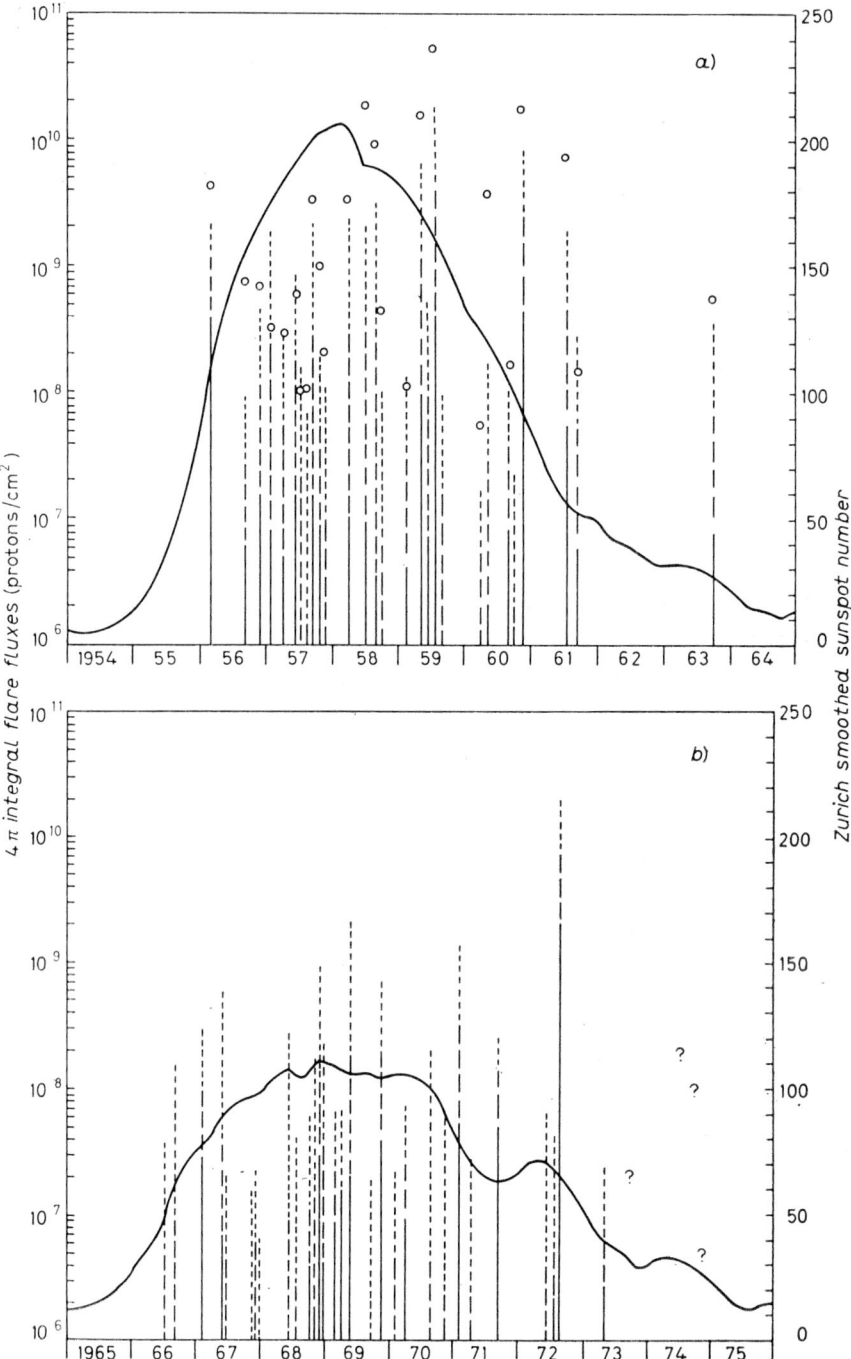

Fig. 2. – Omnidirectional integral fluxes of protons above 10, 30 and 60 MeV during solar cycle 19 (1954-64) and solar cycle 20 (1965-75) are shown in *a*) and *b*), respectively: $---$ $E > 10$ MeV, $-- --$ $E > 30$ MeV, $\overline{\quad\quad}$ $E > 100$ MeV. After [23].

The relative abundances of elements in solar cosmic rays, normalized to oxygen $= 1$, are given in table I [22, 24]; the comparative values for solar photosphere are also given in table I.

Estimated integral fluxes of solar protons since 1956 have been summarized by REEDY [23]. In fig. $2a)$, $b)$, the 4π integral solar proton fluxes have been given for solar cycles 19 and 20, respectively, for kinetic energies above 10, 30 and 100 MeV. The great variability between individual events is clearly seen. One or two large events generally contribute to more than 90% of the cumulative flux during the whole cycle. The cumulative integrated 4π omni-directional fluxes of protons above 10 MeV kinetic energy are deduced to be 90 and $380/\text{cm}^2 \cdot \text{s}$ for solar cycles 20 and 19, respectively. The corresponding time-averaged cumulative proton fluxes above 10 MeV are $7 \cdot 10^4$ and $3 \cdot 10^5/\text{m}^2 \cdot \text{s} \cdot \text{sr}$. This has to be compared with the cumulative flux of $3 \cdot 10^3/\text{m}^2 \cdot \text{s} \cdot \text{sr}$ cited earlier for the galactic cosmic-ray protons, down to the lowest energies.

Although there is a general agreement between the abundances in the solar photosphere and solar cosmic radiation for nuclei of $E > 20$ MeV/n, at lower energies $\left(E = (5 \div 10)\ \text{MeV/n}\right)$ the solar cosmic-ray beam is progressively enriched in heavier nuclei. The abundance enhancement factors have been discussed extensively in the literature [22, 25].

$c)$ *Solar wind.* The observed typical values for the solar-wind flux and densities at 1 AU are summarized in table II, based on published values [26].

Solar-wind abundances based on direct measurements for elements from H to Fe agree within a factor of two with those in the solar surface. The elemental and isotopic abundances in solar wind, based on Apollo 11, 12, 14, 15, 16 and Zond 8 foil collection technique are given in table III [27].

The radial dependence of the solar-wind flux for $(1 \div 4)$ AU follows an approximate inverse square law [26].

In fig. 3 we show the spectra of suprathermal particles emitted from the Sun, from low-energy plasmas to relativistic particles. The integral spectra

TABLE II. – *Typically observed solar-wind properties at* 1 AU.

1)	flow speed (radial) (*)	$400\ \text{km} \cdot \text{s}^{-1}$
2)	proton and electron density	$7\ \text{cm}^{-3}$
3)	proton temperature	$8 \cdot 10^4\ \text{K}$
4)	electron temperature	$(1 \div 1.5) \cdot 10^5\ \text{K}$
5)	proton flux	$3 \cdot 10^8\ \text{cm}^{-2}\ \text{s}^{-1}$
6)	kinetic-energy flux	$2.5 \cdot 10^{11}\ \text{eV cm}^{-2}\ \text{s}^{-1}$
7)	kinetic-energy density	$6 \cdot 10^3\ \text{eV cm}^{-3}$
8)	mass flux	$4 \cdot 10^{-16}\ \text{g cm}^{-2}\ \text{s}^{-1}$

(*) The range in speeds is $(275 \div 800)\ \text{km} \cdot \text{s}^{-1}$.

TABLE III. – *Elemental and isotopic abundances in solar wind, solar surface and Earth's atmosphere* [27].

	Earth's atmosphere	Average solar wind	Outer convective zone of the Sun (estimates)
^{4}He/^{20}Ne	0.3	570 $\pm$ 70	650
^{20}Ne/^{36}Ar	0.52	28 $\pm$ 9	20
^{36}Ar/^{84}Kr	49	—	5000
^{84}Kr/^{132}Xe	28	—	20
^{4}He/^{3}He	$7 \cdot 10^5$	2350 $\pm$ 120	2600
^{20}Ne/^{22}Ne	9.80 $\pm$ 0.08	13.7 $\pm$ 0.3	13.4
^{22}Ne/^{21}Ne	34.5 $\pm$ 1.0	30 $\pm$ 4	30
^{36}Ar/^{38}Ar	5.32 $\pm$ 0.01	$5.33^{+0.08}_{-0.03}$	5.3
^{40}Ar/^{36}Ar	296	< 0.15	$\sim 10^{-4}$

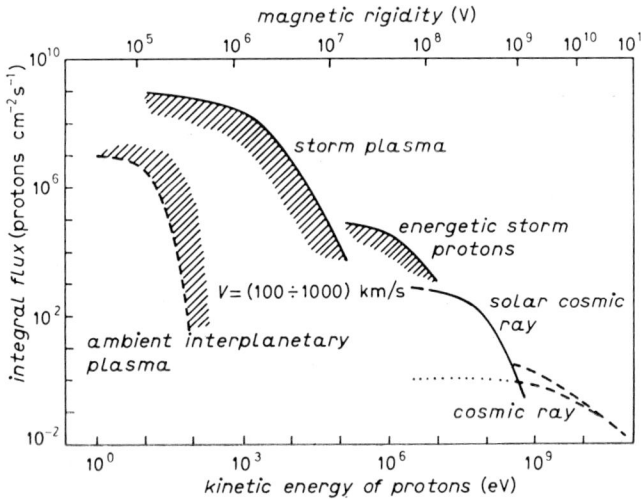

Fig. 3. – The energy spectrum of solar particles near the orbit of the Earth, representing a typical moderate subrelativistic solar cosmic-ray flare event. After [28].

shown are characteristic of moderate subrelativistic solar cosmic-ray flare events [28]. The characteristic rigidity values for different solar particles are summarized below:

	R_0 (MV)
solar cosmic rays	200
energetic storm protons	50
solar-flare plasma	1

It is interesting to see from fig. 3 how the fluxes of energetic particles originating from different mechanisms monotonically rise at lower energies.

2. – Processes associated with particle bombardment in space

We have already discussed that any material exposed in the interplanetary space will be subjected to a diverse particle bombardment. Besides the destructive changes (*i.e.* erosion, comminution and melting) due to impacts of micrometeorites and larger objects, solar-wind ions can get trapped in the outer layers of the matter exposed. For a discussion of solar-wind gases in extraterrestrial samples reference is made to [4, 29-31].

The solar-wind bombardment produces changes in the lattice structure—a solid-state damage. Likewise, appreciable solid-state damage occurs in insulating materials during the transit of multiply charged ions. The damage due to cosmic-ray nuclei of atomic number exceeding 20 in common rock minerals is sufficient to allow enlarging the damaged regions by a chemical treatment. FLEISCHER, PRICE and WALKER showed more than a decade ago that chemically etched tracks can be revealed in a great variety of insulating materials due to fission fragments and cosmic-ray nuclei ($Z > 20$). Reference is made to [32] for a comprehensive account of these developments.

The other processes induced by cosmic-ray particles, solar or galactic, are all nuclear in character—nuclear interactions of high-energy nucleons/nuclei, or capture of thermal neutrons. In these interactions, spallation products are produced—both stable and radioactive nuclei. Thus the types of processes produced in the matter exposed in the interplanetary space can be summarized as belonging of one of the three categories:

1) physical and chemical changes due to impacts—erosion, comminution and heating/melting;

2) trapping of suprathermal solar-wind ions in the lattice structure of the target matrix;

3) changes in the solid-state structure of the target matrix as a result of ionization losses suffered by the heavy cosmic-ray nuclei and fission fragments, leading to chemically etchable tracks;

4) production of stable and radioactive isotopes due to solar and galactic cosmic-ray nuclei and the secondary particles produced as a result of their interactions.

The above processes occur in the outer surfaces of materials as well as *in situ* up to considerable depths. The depth dependence in the rate of a given process is variable and depends on the nature of the process and the radiation causing it. For example, isotopes are produced within the target both by low-energy solar-flare particles and also by the secondary particles produced in nucleonic cascades of galactic cosmic-ray particles. The effects due to relatively low-energy solar cosmic-ray particles will be confined to the upper regions of the target, whereas galactic cosmic-ray effects will be seen up to greater depths. The same holds for etch tracks due to the solar and galactic cosmic-ray particles.

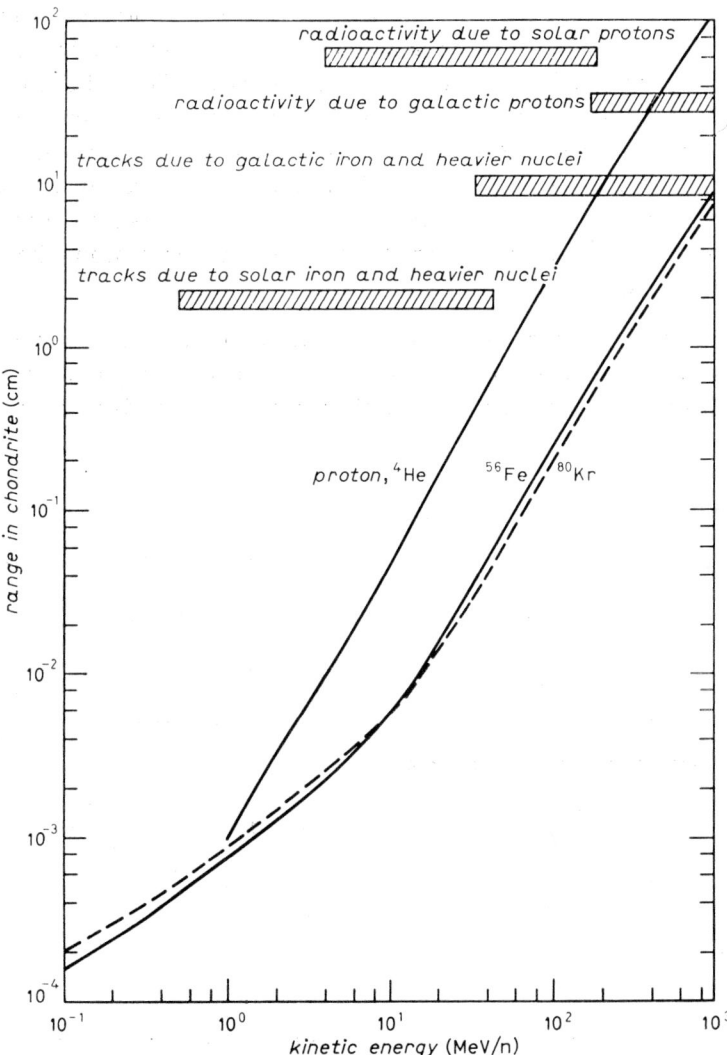

Fig. 4. – Residual ranges in chondrite, as a function of kinetic energy, for ^{1}H, ^{4}He, ^{56}Fe and ^{80}Kr. The approximate energy intervals responsible for the cosmic-ray record found in lunar/meteoritic samples are indicated for the solar and galactic cosmic radiation. After [33].

It is the variety of processes occurring in a depth-dependent manner in matter exposed in interplanetary space which affords the possibility of deducing the evolutionary history of materials. In order to appreciate these effects, we have to consider the depth dependence of the various processes. When these processes are considered in relation to the erosionary, fragmentation and transport processes induced by impacts of micrometeorites/meteorites, the evolutionary history of materials can then be delineated.

The solar-wind ions are implanted within the top layers, < 500 Å in the case of common rock silicates, in view of the low velocity of the solar plasma. Presence of any magnetic field or a gaseous envelope around the target can eliminate the solar-wind ions completely [34]. Nuclear effects due to solar and galactic cosmic rays are more extensive and we will now consider the expected depth dependence in the case of formation of tracks, *i.e.* the latent solid-state damage which can be enlarged by chemical etching, and in the production of isotopes. For doing this, it is instructive to examine the range-energy relation for protons and heavy nuclei. In fig. 4 we show this for protons, helium, iron and krypton nuclei.

    a) Track production rates. Track formation occurs in a narrow energy interval towards the end of range of a particle. In a given solid, the etchable track length depends on the atomic number (fig. 5). Considering fig. 4 and 5,

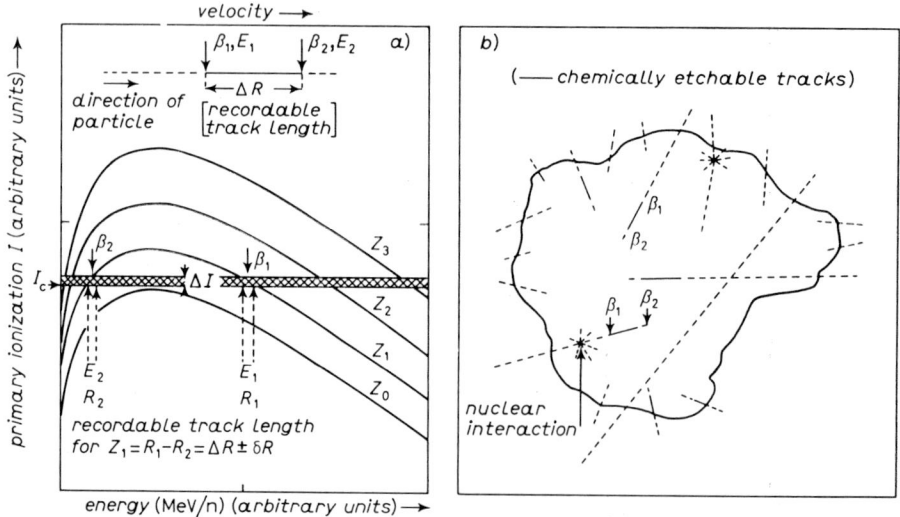

Fig. 5. – In an insulating solid, chemically etchable tracks form along the trail of a charged particle of atomic number Z, if its rate of primary ionization exceeds the critical value I_c (*a*)). The maximum chemically etchable range of a particle is R, corresponding to the distance travelled between E_1 and E_2. In a solid object exposed to cosmic radiation in space, tracks revealed within correspond to particles of different primary kinetic energies. After [2].

it becomes clear that solar-flare tracks will be confined to shallow depths within
the target, since the flux of heavy nuclei in solar cosmic radiations drops quickly
with energy (eqs. (2) and (3)); it is essentially confined to particles of energy
below 50 MeV/n. However, in the galactic radiation the flux of heavy nuclei
is appreciable down to energies of few BeV/n (fig. 1) and tracks form to greater
depths in the target near the end of their range (fig. 4). Within the target,

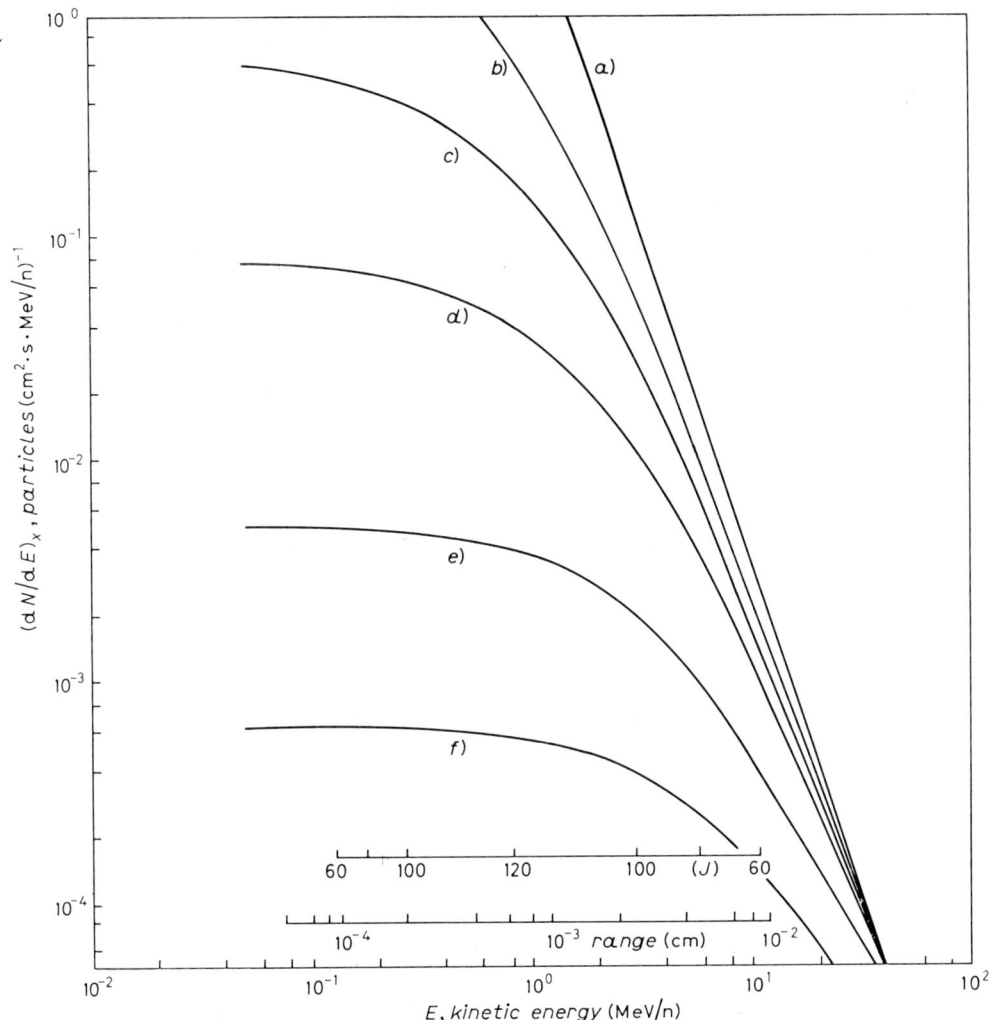

Fig. 6. – Calculated values of « secondary » differential kinetic-energy spectra (ionization modified) at different depths inside a lunar rock for VH group nuclei with an initial
power law kinetic-energy spectrum ($dN = KE^{-3} dE$): *a)* $x = 0$, *b)* $x = 5$ μm, *c)* $x =$
$= 10$ μm, *d)* $x = 20$ μm, *e)* $x = 50$ μm, *f)* $x = 100$ μm. The two scales parallel to
the energy axis give the residual range and primary ionization values as a function
of energy. After [25].

the energy spectrum of heavy nuclei (^{56}Fe, ^{80}Kr, etc.), which produce latent tracks, is modified due to ionization losses and nuclear fragmentation. In the case of the solar cosmic rays, the former effect dominates; the spectrum gets quickly modified due to ionization losses [2, 25]. As an illustration, we show in fig. 6 the expected changes in the energy spectrum for the primary iron group nuclei (nuclei of $Z = 20 \div 28$; this group is designated as VH, the very

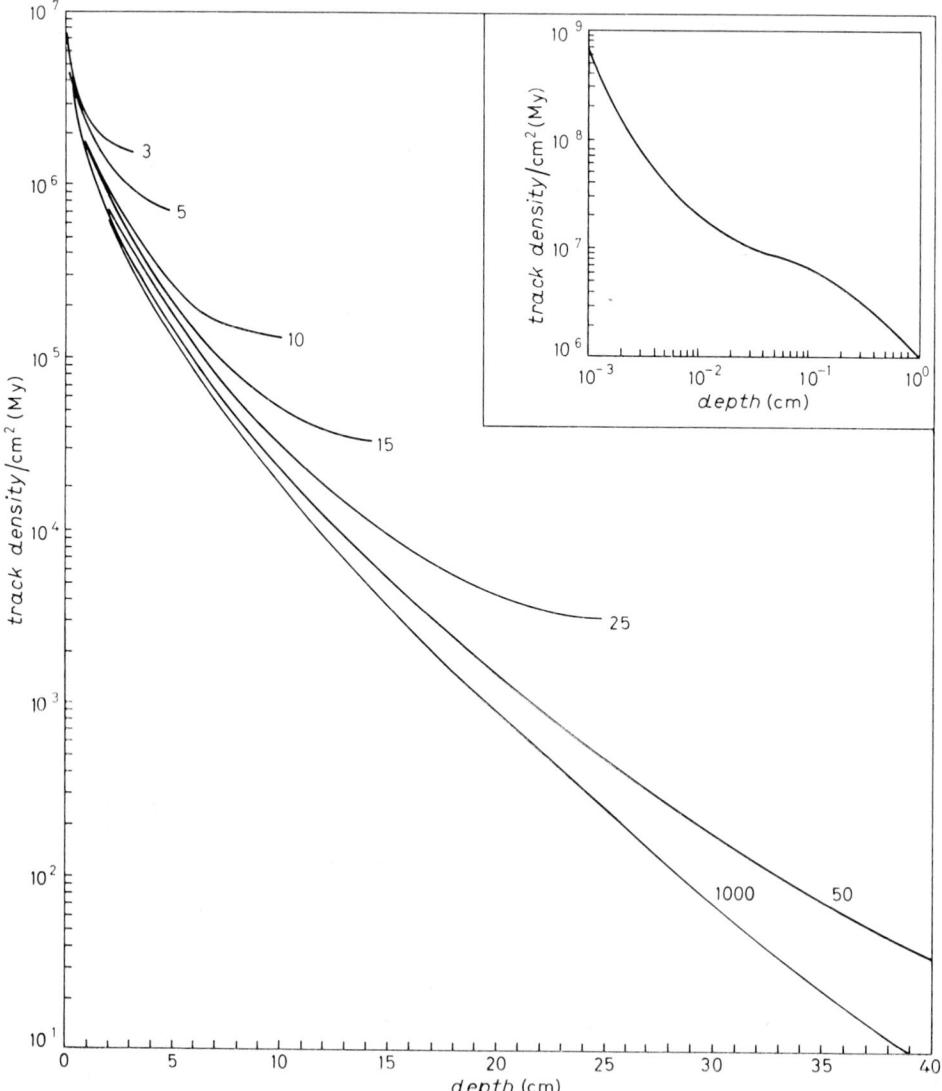

Fig. 7. – Calculated rate of track formation in chondritic meteorites of different radii based on long-term average energy spectrum of VH nuclei [35]. The inset shows production rates at shallow depths in the case of erosion equilibrium up to a depth of $5 \cdot 10^{-2}$. After [2].

heavy nuclei; those above $Z = 30$ are termed VVH nuclei, the very very heavy nuclei) due only to ionization losses. In the case of the more energetic galactic heavy nuclei one would have to consider their losses due to nuclear interaction as well. The mean free paths for interaction in chondrites for ^{56}Fe, ^{84}Kr and ^{132}Xe nuclei in a chondrite are of the order of 6, 5 and 4 cm, respectively [36].

The track production rates have been estimated for solar and galactic cosmic-ray nuclei [2, 36-38] in meteorites and in the Moon. In fig. 7, we show the estimated track production rates due to long-term averaged solar and galactic cosmic-ray nuclei of $Z \geqslant 20$—based on contemporary and prehistoric flux estimates [2, 38].

b) Isotope production rates. The primary cosmic-ray beam gets modified quickly within the target due to ionization losses and nuclear cascade. In the case of the solar cosmic radiation, where the beam is mostly composed of particles having low energies, with ranges small compared to the mean free path for interaction, the changes in energy spectrum within the target can be considered to be caused primarily due to ionization losses. The method for deriving the ionization-modified energy spectrum has been discussed [2]. As an illustration, we show in fig. 8 how the energy spectrum of solar-flare particles will change with depth in the Moon, for an initial spectrum of the exponential rigidity type (eq. (3)) with the characteristic rigidity $R_0 = 100$ MV.

The nucleonic-cascade effects have to be considered explicitly in the case of galactic particles, since the primary spectrum is a modulated one, being deficient in low-energy particles ($E < 0.5$ GeV; see fig. 1), which are produced profusely in nuclear interactions of high-energy particles ($E > 1$ GeV). The basic nuclear processes are similar to those discussed for the case of the atmosphere by LAL and PETERS [1], except for the behaviour of π-mesons. With a good approximation, the collisions of cosmic-ray nucleons with target nuclei can be described in terms of collisions between free nucleons, provided their energy is well above the nuclear binding energy. The large majority of collisions is peripheral, so that the incident nucleon retains a large fraction η ($\simeq 0.5$) of its kinetic energy and receives a small transverse momentum, $\leqslant 500$ MV/c. This behaviour holds for particles from a few hundred MeV to at least several thousand GeV [39]. Thus, the primary nucleons, during their traversal through the target, produce recoil nucleons of few hundred MeV energy during successive interactions in which they emerge with an energy η times the original energy. Before being reduced to an energy of ~ 500 MeV, comparable to the energy of recoil nucleons, a primary nucleon of energy W_0 will make a large number of interactions n':

$$n' = \log (W_0/W)/\log (1/\eta).$$

The value of n' lies between zero and nine for primary energies between 0.5 and ~ 100 GeV. In each interaction a recoil nucleon of few hundred MeV

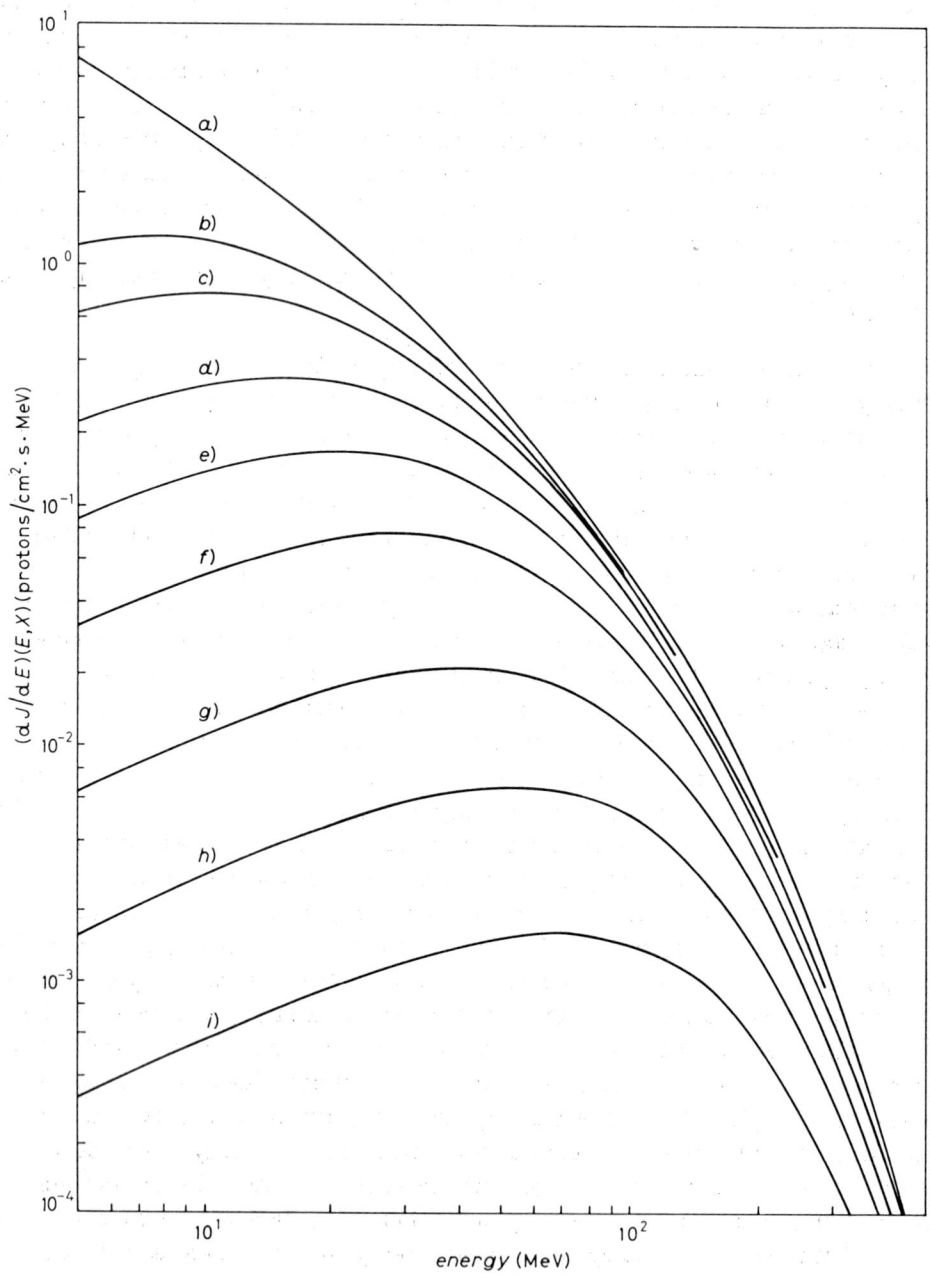

Fig. 8. – Calculated values of differential flux of solar protons at different depths in the Moon, for the case of omnidirectional flux of 100 protons cm⁻² s⁻¹ for kinetic energies above 10 MeV: *a*) 0 g/cm², *b*) 0.1 g/cm², *c*) 0.2 g/cm², *d*) 0.5 g/cm², *e*) 1.0 g/cm², *f*) 2.0 g/cm², *g*) 5.0 g/cm², *h*) 10 g/cm², *i*) 20 g/cm². The primary spectrum was assumed to have an exponential rigidity spectrum with $R_0 = 100$ MV. After [40].

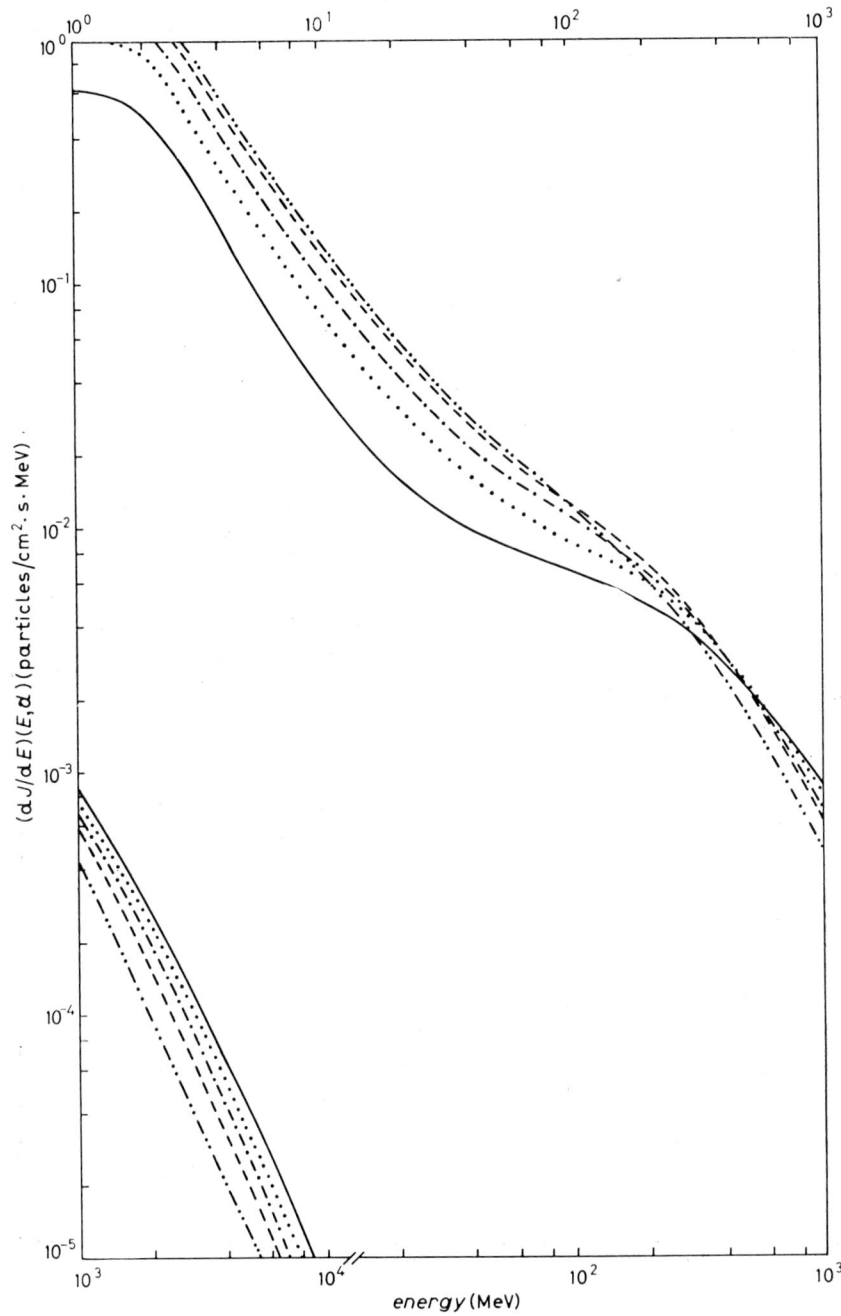

Fig. 9. – Calculated differential-flux values for all nuclear active galactic cosmic-ray particles at various depths in the Moon, up to 70 g cm^{-2}: ——— 0 g/cm^2, ⋯ 10 g/cm^2, —·—·— 20 g/cm^2, — — — 40 g/cm^2, —··—··— 70 g/cm^2. After [40].

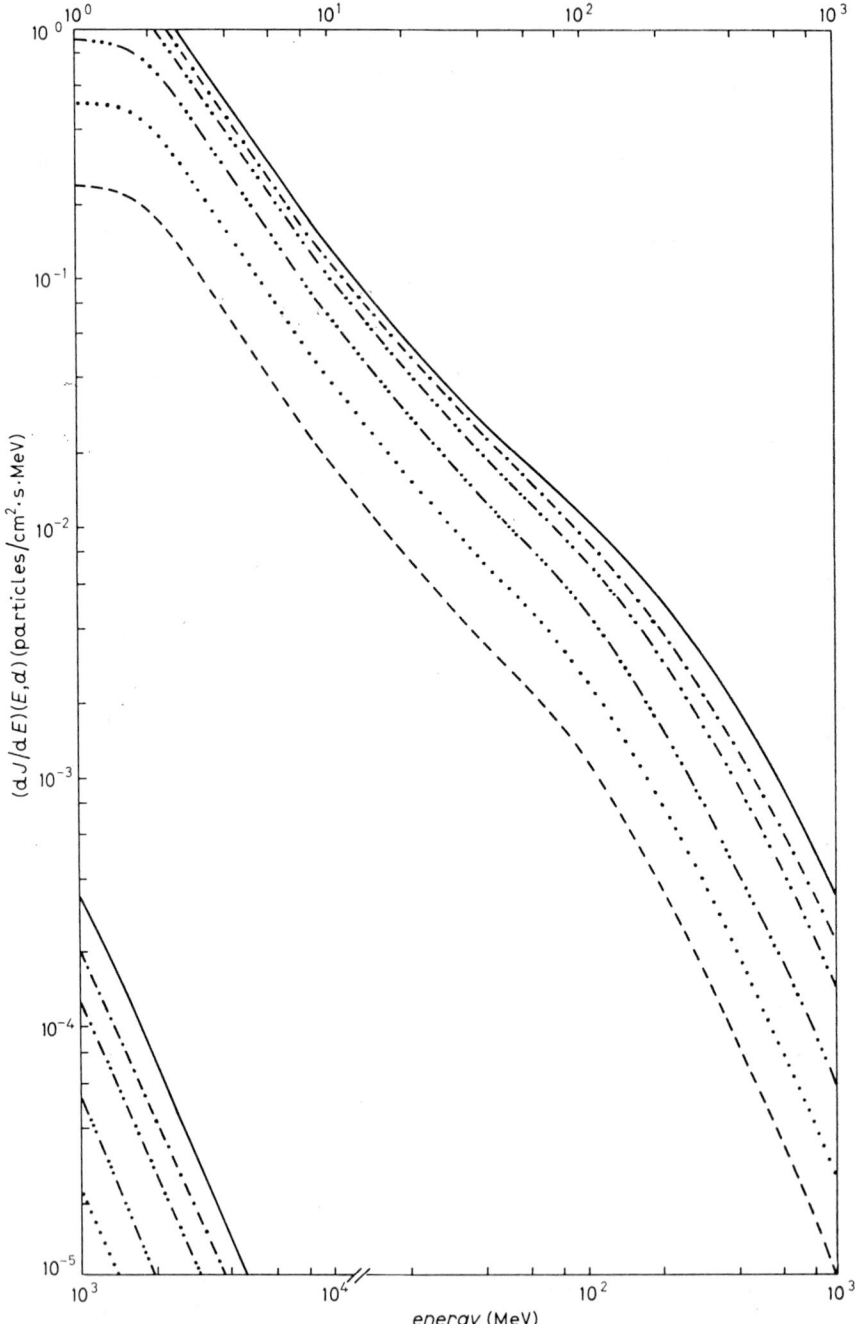

Fig. 10. – Calculated differential-flux values for nuclear active particles in the Moon, as in fig. 9, but for depths up to 500 g cm^{-2}: ——— 100 g/cm^2, —·—·— 150 g/cm^2, —··—··— 200 g/cm^2, —···—···— 300 g/cm^2, ··· 400 g/cm^2, — — — 500 g/cm. After [40].

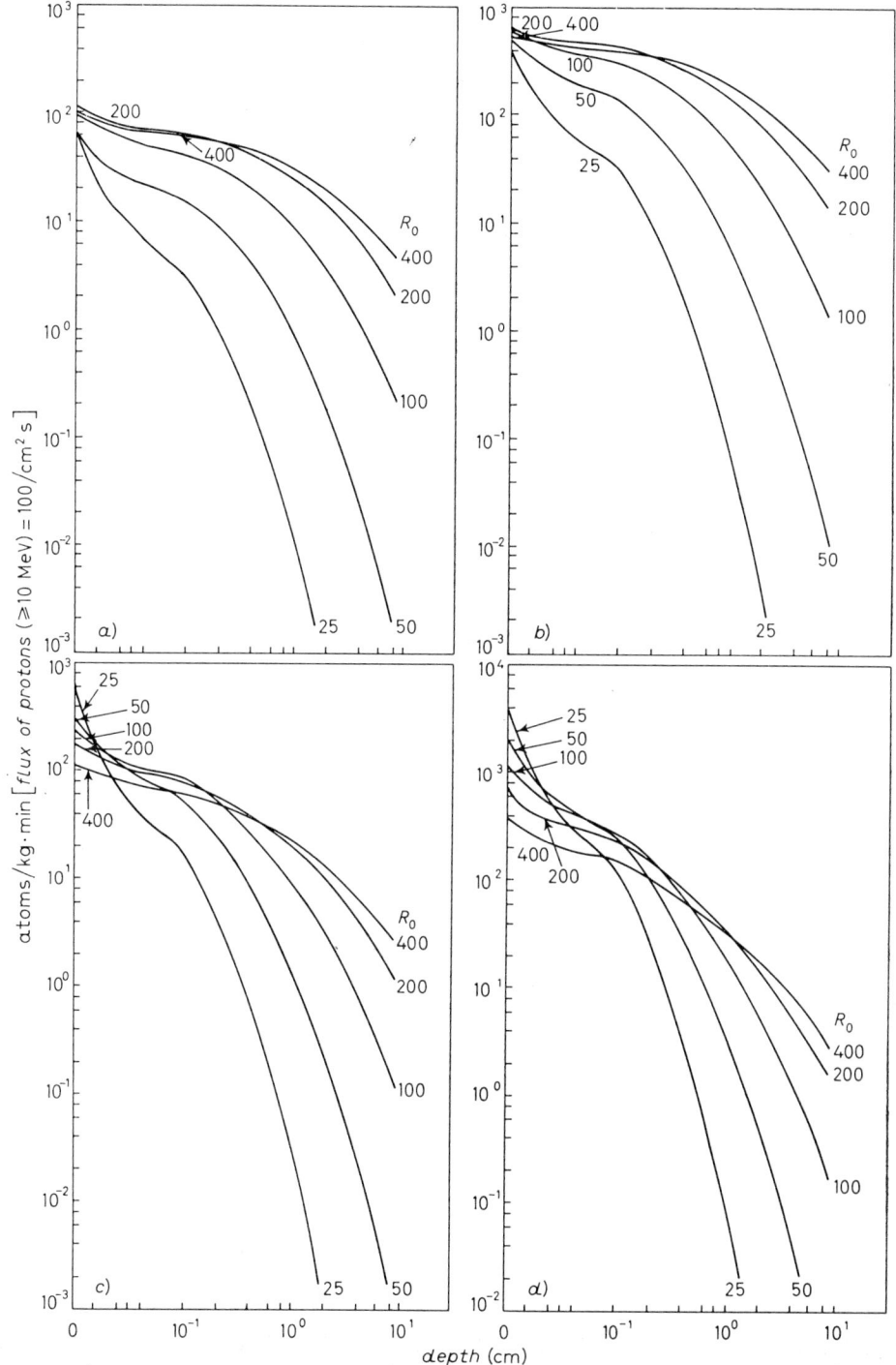

Fig. 11. – Calculated production rates of radioisotopes, a) ^{22}Na, b) ^{26}Al, c) ^{53}Mn and d) ^{56}Co in the Moon due to solar protons. An exponential rigidity spectrum was assumed with an integrated proton flux above 10 MeV $= 100$ cm^{-2} s^{-1}. After [2].

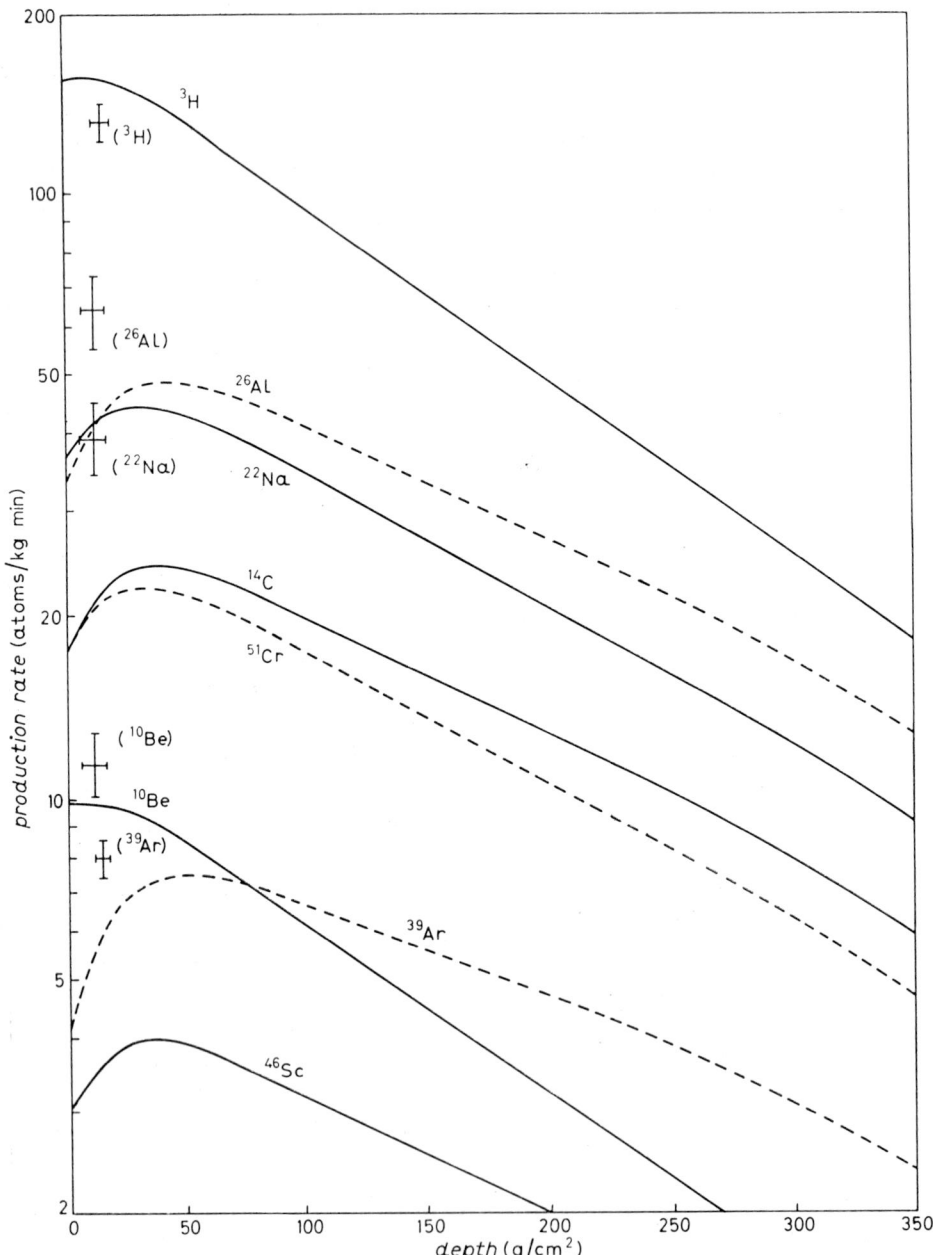

Fig. 12. – Calculated production rates for several radionuclides due to galactic cosmic-ray primaries and secondaries as a function of depth in the Moon, based on the chemical composition of Apollo 12 rock 12002. (From [40].) Experimental data points for Moon rock 12002 are shown.

is produced. Recoil protons being subject to ionization losses are less effective than recoil neutrons for producing further interactions. The neutrons produce further interactions $(2 \div 4)$ before being slowed down. Besides protons and neutrons, the cascade also consists of unstable particles which can induce further nuclear reactions.

In the Earth's atmosphere, the π-mesons mostly decay, except those produced at very high energies, before they can interact [1]. However, in condensed matter, such as in a target of chondritic composition, the majority will interact rather than decay.

The development of the nucleonic cascade in meteorites has been considered in detail by ARNOLD *et al.* [41]. They concluded that the spectrum of protons below 500 MeV kinetic energy is quickly generated by a nucleonic cascade. The total flux of nuclear active particles increases down to a depth of $\sim (100 \div 200)$ g·cm^{-2}, depending on the geometry of the target, this increase being due to the low-energy nucleons, of $E < 500$ MeV, produced in the nucleonic cascade. At greater depths, the total flux of nuclear active particles decreases. As the cascade develops, the spectral shape of nuclear interacting particles above $\simeq 2$ BeV kinetic energy remains nearly the same as in the primary spectrum; the spectrum below 100 MeV continues to become softer until depths of ~ 250 g·cm^{-2}, beyond which spectral-shape changes are not appreciable.

It would be instructive to examine the energy spectra of nuclear active particles in the Moon. The calculated differential-flux values [40] for the kinetic-energy interval $(1 \div 10^4)$ MeV are shown in fig. 9 and 10. Note that, even at the lunar surface, there exists a considerable flux of nucleons below 500 MeV kinetic energy. This arises due to the upward propagation of secondary nucleons produced in interactions at different depths below the surface.

Solar and galactic isotope production rates have been calculated in meteorites and in the Moon, based on the energy spectrum estimates discussed above and the available data on excitation functions for the production of isotopes from different target elements. For a detailed discussion, reference is made to [2, 36, 40-43].

As illustrations, we show the calculated isotope production rates for ^{22}Na ^{26}Al, ^{53}Mn and ^{56}Co in the Moon due to solar protons and for ^{3}H, ^{26}Al, ^{22}Na, ^{14}C, ^{51}Cr, ^{39}Ar and ^{46}Sc in the Moon due to galactic cosmic radiation in fig. 11 and 12, respectively.

3. – Interpretation of observed nuclear effects in extraterrestrial samples.

Even though the estimates of the expected track and isotope production rates are available, the interpretation of the observed nuclear effects in extraterrestrial samples is not a straightforward matter. Uncertainties arise due to a number of factors: the secular variations in the intensity of the cosmic-ray components and changes in the exposure geometry. The flux of nuclear active particles within the target matrix could be a continuous function of time during irradiation, due to physical changes in the target (*e.g.* erosion and fragmentation) as well as due to changes in the shielding depth arising from mixing and transport processes during accretion on a planetary or asteroidal surface, for instance. Information on the place of irradiation is also not known generally. Rather, except for lunar samples, the information is not known in the case of other extraterrestrial samples. For meteorites whose orbits are photographed, their recent orbital parameters are known. But, even during this recent exposure, uncertainties must remain on account of any changes in the exposure geometry due to collisions. Changes in exposure geometry during irradiation are clearly a manifestation of the fact that irradiation in space is a dynamic one. The tasks of a nuclear physicist or a chemist are made challenging by the dynamic variability in the parameters characterizing the exposure history of the extraterrestrial samples. As a matter of fact, it is one of the important purposes of these studies to explicitly evaluate the changes in the irradiation geometry—in other words the evolutionary histories of the samples. However, in order that one may carry out any such investigations, one should have an *a priori* knowledge of the temporal and spatial changes in the fluxes of particles emitted by Sun, and in the intensity of the galactic cosmic radiation. This information can be obtained from the study of suitable samples where the changes in the irradiation geometry are minimal.

Information on the exposure geometry can be delineated by examining the gradients in different types of cosmogenic effects. As can be seen from fig. 13, the production of tracks and isotopes is a sensitive function of depth. The presence of large gradients in cosmogenic effects, for example, will indicate an exposure at shallow depths, *i.e.* small shielding. For a sample exposed in a complex manner, the effective exposure age T_i, based on the observed magnitude of cosmogenic effect i, will differ from that based on another effect [2, 33]. It is rather easy to see that, if the geometry of irradiation is changing during exposure, *i.e.* in a dynamically evolving sample, the effective time of irradiation will differ for the different cosmogenic effects studied. Conversely, if the estimated irradiation times based on different cosmogenic effects are the same, the sample must have received its irradiation for most part during a certain exposure geometry.

Excavation of a deep-seated rock from the lunar surface, where it lay shielded from cosmogenic effects, and exposure on the lunar surface for a period T before being picked up by an astronaut is an example of simple exposure history. Cases of simple exposure histories may be expected in the case of meteorites—excavation from a shielded region in the parent body and subsequent exposure in space as a small body until its capture by the Earth.

Most lunar rocks, however, have had a complicated exposure history based

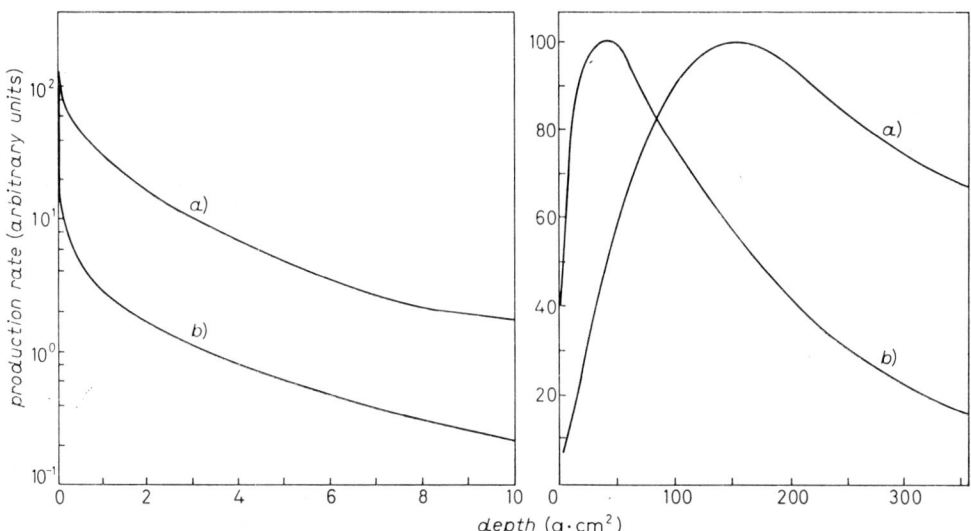

Fig. 13. – Profiles at production for radioisotopes due to galactic cosmic radiation are given on the r.h.s. as a function of depth: *a*) thermal-neutron capture rate, *b*) typical production rate profile for low-threshold isotopes. Figure on the l.h.s. gives the depth dependence in *a*) the production of ^{26}Al due to solar cosmic radiation ($R_0 = 100$ MV) and *b*) fossil tracks (based on observations in Apollo rocks). The ordinate is arbitrarily fixed in each case. After [2].

on an intercomparison of cosmogenic effects. The inequalities in effective irradiation times follow the relationship given below, as expected on theoretical grounds [2, 33], considering their depth dependence at production:

$$(4) \qquad T_{\text{SCR}}(\text{rad}) < T_{\text{FT}} < T_{\text{GCR}}(\text{rad}) < T_{\text{GCR}}(\text{stable}) < T_{\text{GCR}}(\text{n capture}),$$

where T denotes the effective irradiation period based on the various cosmogenic effects; the subscripts SCR and GCR refer to isotopes produced by solar and galactic cosmic rays, respectively. The subscript FT refers to fossil tracks produced by low-energy cosmic rays. Stable isotopes, of course, would provide longer time integrals compared to those based on radioactive (rad) isotopes, for cases of complex irradiation history. Dramatic cases of inequalities have indeed been observed in the case of lunar rocks; the different ages in principle allow one to write down a plausible history of the evolution of the rock in the lunar regolith [33, 44-46].

From considerations as above, it becomes possible to delineate samples of relatively simple exposure history and thereby estimate long-term average fluxes of cosmic-ray particles, both solar and galactic. The information available to date has been summarized [2, 33, 35, 47]. Figure 14 summarizes the deduced long-term averaged spectra for solar protons and heavy nuclei. No

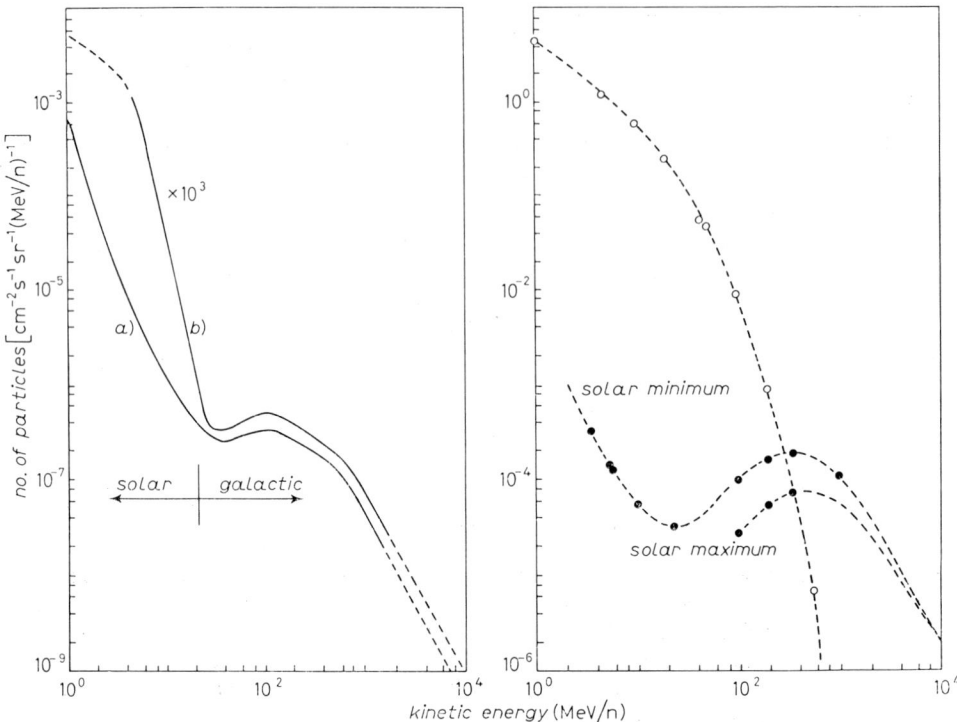

Fig. 14. – Long-term averaged differential kinetic-energy spectra for a) VH ($Z \geqslant 20$) and b) VVH ($Z \geqslant 30$) nuclei are shown in the l.h.s. diagram for both the solar and the galactic radiation. The long-term averaged solar-proton spectra based on studies of lunar rocks (—o—o—) are shown in the r.h.s. diagram along with the galactic cosmic-ray proton data (—•—•—) observed during solar maximum and minimum periods. After [33].

energy spectrum estimates are possible for the galactic protons. It can, however, be concluded that the time-averaged galactic cosmic-ray fluxes have remained essentially the same as today, during the past few million years, similar to the case for solar cosmic radiation. If one goes farther back in time, the resolution in time decreases. Although the time-averaged fluxes seem to have remained the same over periods of the order of 10^9 y, one cannot exclude existence of epochs of duration of $\sim 10^8$ y with very low or high cosmic-ray fluxes [48].

4. – Summary.

We have discussed the various atomic and nuclear processes which occur in materials exposed in the interplanetary space. These interactions label matter, their surfaces and their interior regions in a very conspicuous manner during time periods as encountered in the evolution of planetary objects in the solar system. If a piece of rock is exposed in space for periods of the order of $(10^2 \div 10^6)$ y, a number of prominent physical and chemical changes will occur in the sample. (We assume that solar-wind ions can reach unimpeded the region of space considered.) The sample would be subjected to bombardment by solid particles, whereby craters of submicrometre to centimetre size will be formed. In the near-surface regions of the exposed rock, solar-wind ions will be trapped. These ions will penetrate to depths $\leqslant 1000$ Å. Solar-flare heavy nuclei of atomic number exceeding 20 with kinetic energies of $E < 50$ MeV/n will cause appreciable solid-state damage in the structure of matter which, unless annealed, can be quantitatively studied. These effects will primarily be confined to depths of ~ 500 μm. At greater depths the effects produced due to galactic cosmic radiation can be seen—both tracks and radioactivity. Figure 15 shows qualitatively the approximate terminal depths for the various cosmogenic effects considered. The present discussion is based on the fluxes and energy spectra as observed for the contemporary radiation in space, and also deduced for the long-term averages for the $(1 \div 3)$ AU space.

The magnitude of the various effects discussed above will depend sensitively on the erosionary/fragmentation history of the rock. Macrodestruction processes would have a stochastic character and hence marked spatial variations will be observed in the concentrations of solar-wind ions, solar-flare-induced tracks and radioactivity. Some surfaces may attain an equilibrium with production, others not.

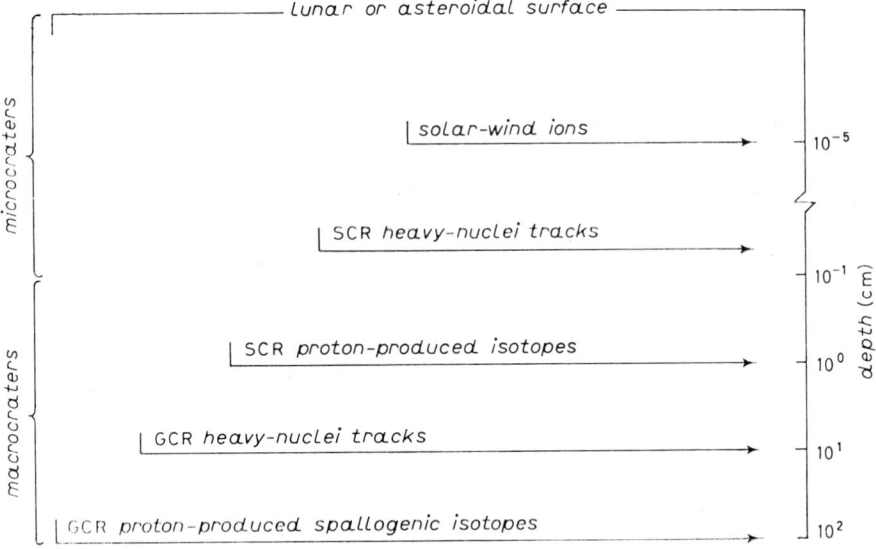

Fig. 15. – Approximate terminal depths for formation of craters, implantation of solar-wind ions and production of tracks and spallogenic products in matter of chondritic composition. SCR = solar cosmic radiation, GCR = galactic cosmic radiation.

The particle bombardment processes are extremely damaging, so that the outer surfaces of materials are continuously renewed by the deeper layers [5, 49-51]. Materials which have been heavily irradiated with solar-wind ions and solar-flare radiation can be preserved only if they are prevented from further irradiation as a result of fragmentation/burial caused by impacts. It must, therefore, be realized immediately that the extraterrestrial samples which have registered the solar particles are amongst those which have not been over-irradiated—they are indeed only those which have the vestigial records.

The above information is based on high-resolution physical and chemical examinations of meteorites and lunar samples. The lunar regolith samples contain a rich and diverse record of the near-surface processes and have in fact provided a great deal of insight about processes which occur to materials exposed in space. Characteristic near-surface effects such as formation of micro-craters and agglutinates, implantation of solar-wind ions and formation of solar-flare tracks are indeed found in individual grains in chondrites. These observations are indicative of processes which have occurred prior to their compaction in the meteorite parent bodies. These studies are progressively becoming more useful towards understanding the history of formation of gas-rich meteorites, as we shall see in the next lecture.

In closing, it may be mentioned that we have not considered at least two special cases of irradiation:

a) early irradiation of matter in solar system during the *T-Tauri* phase of the Sun (cf. [52]), possibly leading to anomalies in isotopes like Mg, O; and

b) direct implantation of stable and radioactive isotopes, *e.g.* ^{3}He, ^{14}C, ^{3}H, contained in the suprathermal solar particles including the solar-flare cosmic-ray particles (cf. [53-55]).

The above may be of particular relevance in delineating the early irradiation history of materials, but here our present krowledge is fairly limited to present a discussion of the expected features in a quantitative fashion.

REFERENCES

[1]   D. LAL and B. PETERS: *Handbuch der Physik*, **46/2** (Berlin, 1967), p. 551.
[2]   D. LAL: *Space Sci. Rev.*, **14**, 3 (1972).
[3]   R. M. WALKER: *Annu. Rev. Earth Planet. Sci.*, **3**, 99 (1975).
[4]   H. E. SUESS, H. WANKE and F. WLOTZKA: *Geochim. Cosmochim. Acta*, **28**, 595 (1964).
[5]   M. MAURETTE and P. B. PRICE: *Science*, **187**, 121 (1975).
[6]   E. M. SHOEMAKER: in *Impact and Explosion Cratering* (New York, N. Y., 1977).
[7]   D. LAL: *Proceedings of the XXI Nobel Symposium, From Plasma to Planet (held at Saltsjobaden, Sweden, September 6-10, 1971)*, edited by A. ELVIUS (Stockholm, 1972), p. 49.

[8] P. PELLAS: *Proceedings of the XXI Nobel Symposium, From Plasma to Planet* (*held at Saltsjobaden, Sweden, September 6-10, 1971*), edited by A. ELVIUS (Stockholm, 1972), p. 65.

[9] J. C. HUNEKE, S. P. SMITH, R. S. RAJAN, D. A. PAPANASTASSIOU and G. J. WASSERBURG: *Abstracts, Lunar Science VIII* (*held at Houston, March 14-18, 1977*), part 1 (1977), p. 484.

[10] L. SCHULTZ and P. SIGNER: *Earth Planet. Sci. Lett.*, **36**, 363 (1977).

[11] J. D. MACDOUGALL and B. K. KOTHARI: *Earth Planet. Sci. Lett.*, **33**, 36 (1976).

[12] R. R. DANIEL and S. A. STEPHENS: *Space Sci. Rev.*, **17**, 45 (1975).

[13] B. PETERS: *J. Geophys. Res.*, **64**, 155 (1959).

[14] M. M. SHAPIRO and R. SILBERBERG: *Philos. Trans. R. Soc. London Ser. A*, **277**, 319 (1974).

[15] V. L. GINZBURG: *Philos. Trans. R. Soc. London Ser. A*, **277**, 463 (1974).

[16] P. MEYER: *Philos. Trans. R. Soc. London Ser. A*, **277**, 349 (1974).

[17] P. MEYER, R. RAMATY and W. R. WEBBER: *Phys. Today*, **27**, No. 10, 23 (1974).

[18] C. J. WADDINGTON: in *Fundamentals of Cosmic Physics*, Vol. **3** (New York, N. Y., 1977), p. 1.

[19] R. B. MCKIBBEN: *Rev. Geophys. Space Phys.*, **13**, 1088 (1975).

[20] C. W. BARNES and J. A. SIMPSON: *Astrophys. J.*, **210**, 191 (1976).

[21] M. A. POMERANTZ and S. P. DUGGAL: *Rev. Geophys. Space Phys.*, **12**, 343 (1974).

[22] S. BISWAS: *Bull. Astron. Soc. India*, **3**, 68 (1975).

[23] R. C. REEDY: *Proceedings of the VIII Lunar Science Conference, Geochim. Cosmochim. Acta, Suppl.* 8, **1**, 825 (1977).

[24] S. BISWAS and C. E. FICHTEL: *Space Sci. Rev.*, **4**, 709 (1965).

[25] J. N. GOSWAMI and D. LAL: *Proceedings of the VI Lunar Science Conference, Geochim. Cosmochim. Acta, Suppl.* 6, **3**, 3541 (1975).

[26] M. NEUGEBAUER: *Space Sci. Rev.*, **17**, 221 (1975).

[27] J. GEISS: in *On the Origin of the Solar System*, edited by H. REEVES (1972), p. 217; *Proceedings of the XIII International Cosmic Ray Conference*, Vol. **5** (1973), p. 3375.

[28] T. OBAYASHI: *Space Sci. Rev.*, **3**, 79 (1964).

[29] H. DUCATI, S. KALBITZER, J. KIKO, T. KIRSTEN and H. W. MULLER: *Moon*, **8**, 210 (1973).

[30] P. EBERHARDT: *Proceedings of the III Solar Wind Conference* (Asilomar, 1974).

[31] T. KIRSTEN: *Philos. Trans. R. Soc. London Ser. A*, **285**, 391 (1977).

[32] R. L. FLEISCHER, P. B. PRICE and R. M. WALKER: *Nuclear Tracks: Principles and Applications* (Berkeley, Cal., 1975).

[33] D. LAL: *Philos. Trans. R. Soc. London Ser. A*, **285**, 69 (1977).

[34] J. N. GOSWAMI: *Nature (London)*, **261**, 675 (1976).

[35] D. LAL: *Philos. Trans. R. Soc. London Ser. A*, **277**, 395 (1974).

[36] S. K. BHATTACHARYA, J. N. GOSWAMI, S. K. GUPTA and L. LAL: *Moon*, **8**, 253 (1973).

[37] R. L. FLEISCHER, P. B. PRICE, R. M. WALKER and M. MAURETTE: *J. Geophys. Res.*, **72**, 331 (1967).

[38] S. K. BHATTACHARYA, J. N. GOSWAMI and D. LAL: *J. Geophys. Res.*, **78**, 8356 (1973).

[39] Y. PAL and B. PETERS: *K. Dan. Vidensk. Selsk. Mat.-Fys. Medd.*, **33**, No. 15 (1964).

[40] R. C. REEDY and J. R. ARNOLD: *J. Geophys. Res.*, **77**, 537 (1972).

[41] J. R. ARNOLD, M. HONDA and D. LAL: *J. Geophys. Res.*, **66**, 3519 (1961).

[42] T. P. KOHMAN and M. L. BENDER: *High-Energy Nuclear Reactions in Astrophysics*, edited by B. P. SHEN (New York, N. Y., 1967), p. 169.

[43] T. A. KIRSTEN and O. A. SCHAEFFER: *Elementary Particles* (New York, N. Y., and London, 1971), p. 75.

[44] D. LAL, D. MACDOUGALL, L. WILKENING and G. ARRHENIUS, *Proceedings of the Apollo 11 Lunar Science Conference, Geochim. Cosmochim. Acta, Suppl.* 1, **3**, 2295 (1970).

[45] N. BHANDARI, S. G. BHAT, D. LAL, G. RAJAGOPALAN, A. S. TAMHANE and V. S. VENKATAVARADAN: *Proceedings of the II Lunar Science Conference, Geochim. Cosmochim. Acta, Suppl.* 2, **3**, 2611 (1971).

[46] G. CROZAZ, R. DROZD, C. HOHENBERG, C. MORGAN, C. RALSTON, R. WALKER and D. YUHAS: *Proceedings of the V Lunar Science Conference, Geochim. Cosmochim. Acta, Suppl.* 5, **3**, 2475 (1974).

[47] O. A. SCHAEFFER: *Proceedings of the XIV International Cosmic Ray Conference* Vol. **11** (Munich, 1975), p. 3508.

[48] J. GEISS, H. OESCHGER and M. SCHWARZ: *Space Sci. Rev.*, **1**, 197 (1962).

[49] D. E. GAULT, F. HORZ, D. E. BROWNLEE and J. B. HARTUNG: *Proceeding of the V Lunar Science Conference, Geochim. Cosmochim. Acta, Suppl.* 5, **3**, 2365 (1974).

[50] J. P. BIBRING, Y. LANGEVIN, M. MAURETTE, R. MEUNIER, B. JOUFFREY and C. JOURET: *Earth Planet. Sci. Lett.*, **22**, 205 (1974).

[51] J. AUDOUZE, J. P. BIBRING, J. C. DRAN, M. MAURETTE and P. M. WALKER: *Astrophys. J.*, **206**, L185 (1976).

[52] D. HEYMANN and M. DZICZKANIEC: *Science*, **191**, 79 (1976).

[53] J. D. ANGLIN: *Astrophys. J.*, **198**, 733 (1975).

[54] F. BEGEMANN, W. BORN, H. PALME and H. WANKE: *Abstract, Lunar Science III*, edited by C. WATKINS (revised abstracts of papers presented at the *Third Lunar Science Conference, Houston, January 10-13, 1972*) (1972), p. 53.

[55] J. E. LUPTON: *J. Geophys. Res.*, **78**, 8330 (1973).

Surface Evolution Records in Chondrites and Lunar Regolith: Implications to Early Accretion in the Solar System.

D. LAL

Physical Research Laboratory - Ahmedabad 380 009, India

A variety of atomic and nuclear processes label matter exposed in the interplanetary space. A great deal of this labelling survives as a vestigial record of the evolution of small and large planetary objects in the solar system. In the previous lecture, the basic processes were outlined; the fluxes of solar wind, solar and galactic cosmic-ray particles as well as the magnitude of the cosmogenic effects were discussed. Lunar regolith is a rich storehouse for studying the nature and magnitude of these processes—as it provides samples of diverse exposure history. Already before the acquisition of the lunar samples, meteorites had provided a clear-cut evidence for the implantation of solar-wind ions [1, 2] and formation of tracks due to solar-flare iron group and heavier nuclei [3, 4] in gas-rich meteorites prior to their compaction in parent bodies. Lunar samples corroborated these deductions and additionally provided the impetus to look for microcraters and glassy splashes in gas-rich meteorites. These effects were indeed found [5, 6]. The use of surfaces labelled due to particle bombardment (impact of solid particles, solar-wind and solar-flare irradiation) as a tool for understanding the early history of gas-rich meteorites, specifically the processes which occurred prior to their compaction in a parent body, has thus clearly been established. Another very useful information in this context was provided by the technique developed by MACDOUGALL and KOTHARI [7]. They exploited the fact that fission fragments from the primordial ^{244}Pu would be expected to lead to an appreciable track contribution on the surfaces of mineral grains, over and above that due to ^{244}Pu and ^{238}U contained within, if the grains were compacted with the matrix during the early history of the solar system. The method works well for determining compaction ages greater than $4.2 \cdot 10^9$ y.

Besides providing clues about regolith-type processes experienced by gas-rich meteorites, the lunar studies have now proven invaluable for the study of optical maturation of surfaces exposed without an atmosphere, providing baseline information for the interpretation of asteroid spectral reflectance data.

An intercomparison shows very large differences in the evolutionary paths of regoliths of asteroids, Moon and Mercury, due to differing micrometeoroid impact velocities, gravity and composition [8].

ANDERS has discussed that the place of origin of stony meteorites can be determined from their trapped solar-wind gases [9, 10]. He has considered a number of irradiation features observed and came to the interesting conclusion that most stony meteorites came from the asteroidal belt, and not from the long- or short-period comets, or the Trojan asteroids, or the stray bodies in thinly populated parts of the inner solar system. These conclusions are based on the thesis for which there seems to be a general consensus that gas-rich meteorites formed in a regolith, by analogy with the lunar resolith [6, 11-14].

In a recent study of the details of the solar wind and flare irradiation of the carbonaceous chondrites, GOSWAMI and LAL [15] have provided strong evidence that the irradiation of individual components of carbonaceous chondrites, however, did not take place in a regolith. An irradiation in a regolith would have given to the olivine grains in $C2$ chondrites a very different solar-flare irradiation pattern. Besides, the observed track densities cannot result from an irradiation in a regolith within time periods of the order of $\leqslant 10^6$ y as constrained by the contents of spallogenic neon isotopes and radiogenic ^{26}Al in these meteorites [16-18].

We will attempt in this paper to discuss the types of irradiation patterns found in the lunar regolith and in gas-rich meteorites and discuss briefly the implications thereof to the accretionary history of the parent bodies of the gas-rich meteorites. It becomes obvious that sufficient information is available in the form of vestigial records of precompaction events. Future studies should hopefully provide definitive information on the history of formation of carbonaceous chondrites and the ordinary gas-rich chondrites.

1. – Irradiation pattern observed in lunar regolith.

Different lunar soils show large variations in their concentrations of solar-wind gases, particle tracks, cosmogenic radioactive and stable isotopes and agglutinates. Although one can define the average solar-wind concentration of a given soil sample, for example, this is at best only an operational definition because of the large variations in the solar-wind contents of different grains. The lunar regolith is indeed very complex, because the mixing and transport processes on the lunar surface are caused by meteorite impact, which is a random process. No two spots at a given site on the lunar surface have identical surface exposure and evolution histories. Even in regions characterized by stratigraphic layers, the material within a layer is quite unique. Variability is the key word in lunar soils and it becomes clear that, to understand the evolutionary history of a given soil sample, the irradiation history of individual grains has to be understood.

The stochastic nature of mixing of the lunar regolith has been discussed in detail by GAULT *et al.* [19]. Particle track studies provide a means of examining differences in the exposure history of different grains [20-24]. Subsequently GOSWAMI *et al.* [25] studied microstratigraphy in epoxy-impregnated drill cores from Apollo 15 and 16 missions. The highly stochastic nature of evolutionary history of individual grains is quite clearly seen from the data on track density distributions plotted in fig. 1*a* and *b*. As illustrations, in fig. 2

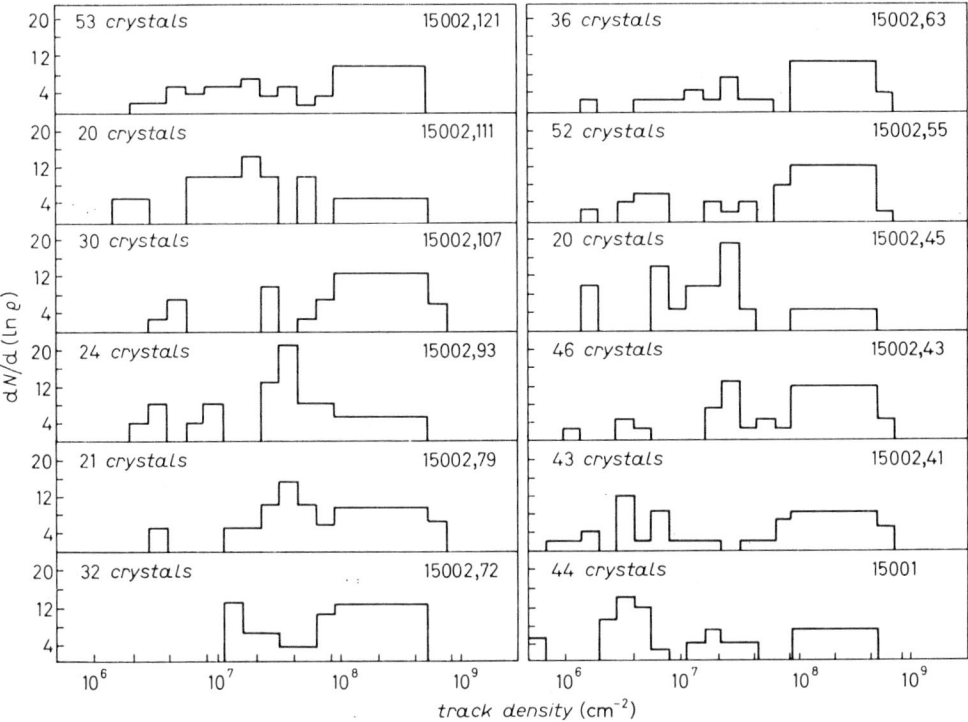

Fig. 1*a*. – Histograms showing track density frequency distributions in grains from several soil samples collected from the lunar regolith at the Apollo 15 site. After [26].

are shown photomicrographs of heavy-nuclei tracks seen in individual grains from the lunar regolith and carbonaceous chondrites.

High track densities in grains, $> 10^8$ cm^{-2}, can arise only from irradiation of grains in near-surface regions. This becomes clear from an examination of the track production rates given in fig. 3. The production rates are based on the published values estimated by BHANDARI *et al.* [26], BLANFORD *et al.* [27], HUTCHEON *et al.* [28] and WALKER and YUHAS [29]. For depths exceeding

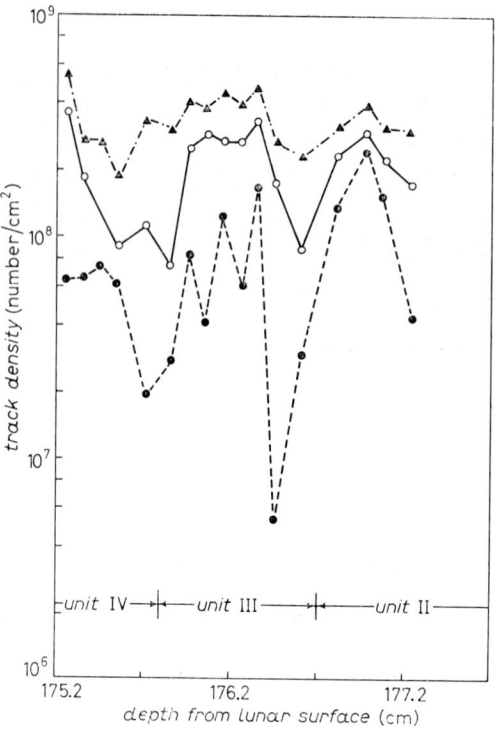

Fig. 1*b*. – Observed variations in the track parameters ϱ_{med} (median track density) (▲—·—▲—·—▲), ϱ_q (quartile track density) (o———o———o) and ϱ_{min} (minimum track density) (●— — —●— — —●) in an impregnated drill core section sampled at the Apollo 16 site. After [25].

0.1 cm, the estimates agree fairly well, but are in discord for shallower depths [30]. However, for the purpose of the present discussion, we have adopted a production profile intermediate between the extremes. The uncertainty in production rates is, however, of no important consequence to our present deductions.

We will now consider the expected track density distribution in soil samples on the basis of track production rates (fig. 1) for different cases of mixing.

For the simplest case we consider a *static* irradiation, starting from a regolith with no previous irradiation history. We consider a column of depth L and ask what will be track density distribution of grains in this column after an exposure of T y. For the present calculations, we make the simplifying assumptions:

1) Track production rate, as a function of depth X, is given by

(1)
$$\dot{\varrho}(X) = \frac{d\varrho}{dt} = K(A + X)^{-\alpha}.$$

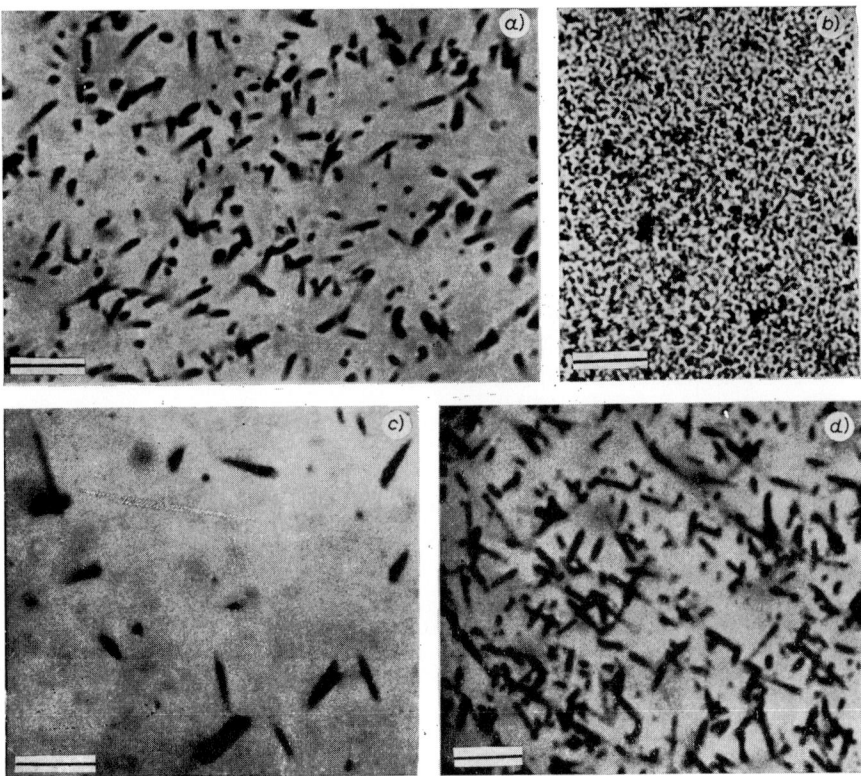

Fig. 2. – Photomicrographs of heavy cosmic-ray nuclei tracks as seen in grains taken from the lunar regolith (a) and b)) and from the carbonaceous chondrite Murchison (c) and d)). Scale bar is 10 μm.

2) All grains in the regolith are identical cubes of side d; the number density is n per cm³.

The normalized number track density distribution is then expected to be given by

$$(2) \qquad \frac{\mathrm{d}N}{\mathrm{d}\varrho} = -\frac{1}{\alpha L}(KT)^{1/\alpha}\varrho^{-(1+1/\alpha)}$$

for the interval in X corresponding to $T\varrho(X)$ at $X = 0$ and $X = L$. The expected distributions for $L = 3$ cm and different values of T are shown in fig. 4.

However, if the layer of depth L was continuously kept in a rapid agitation, so that the grains spent equal time at all depths, all grains will have a uniform track density given by

$$(3) \qquad \varrho_{\mathrm{mix}}(0-L) = \frac{T}{L}\int_0^L K(A+X)^{-\alpha}\,\mathrm{d}X\,.$$

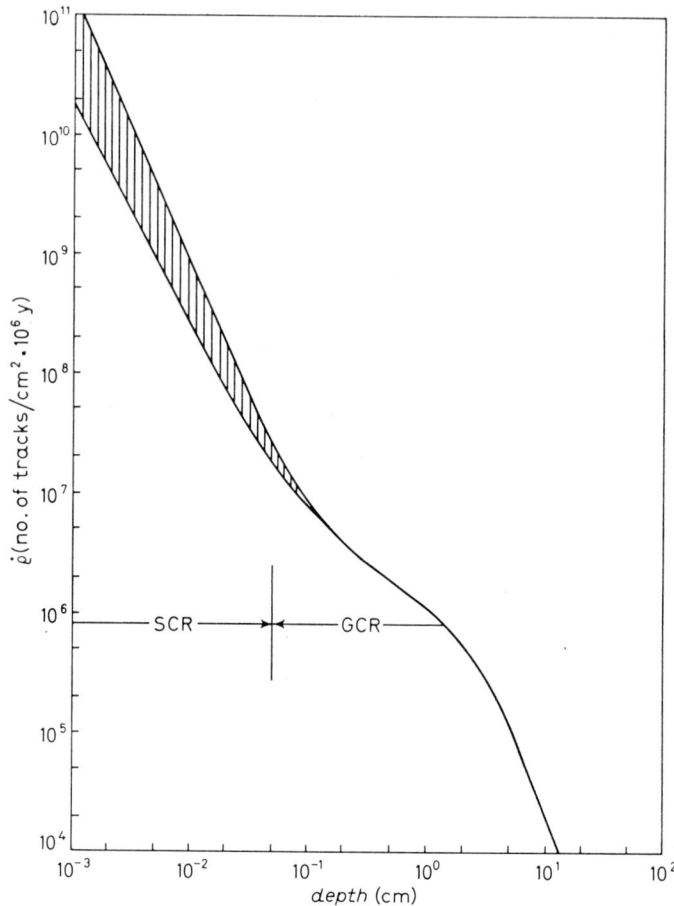

Fig. 3. – Track production rate in the Moon for material of density = 3.4 g·cm⁻³, based on studies of fossil track records in lunar rocks and meteorites, given as a function of depth for VH ($Z > 20$) nuclei [30].

The expected number track density distribution corresponding to $L = 3$ cm and $T = 10 \cdot 10^6$ y is shown in fig. 5a) for the static case (eq. (2)) and in fig. 5c) for rapid mixing. If, however, a slow mixing occurs, some of the low-track-density grains will attain higher track density values, with a corresponding depletion in the population of low grains. Qualitatively, the expected situation for a hypothetical case of slow mixing is also shown in fig. 5b).

The essence of the exercise above is that, with the steep track production function, high track densities are attained easily in the case of mixing during the short time spent by grains in the upper layers. And, in fact, the only way of obtaining high ϱ values, $> 10^8$ cm², is to expose the grains within the top few millimetres of the regolith. In the lunar cores track density patterns similar

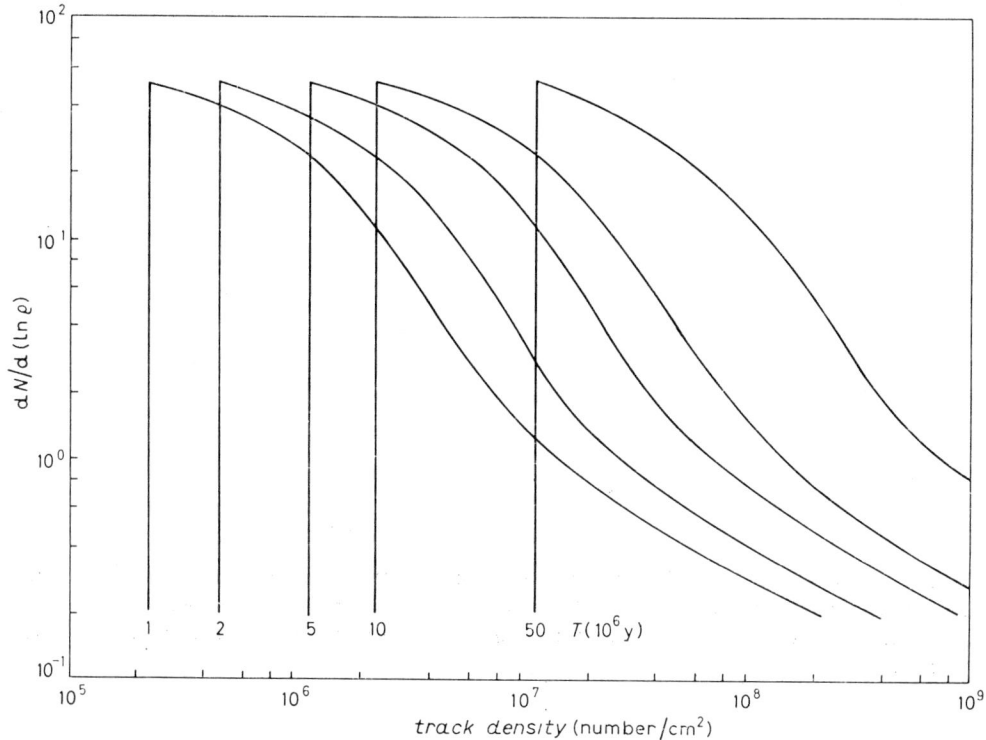

Fig. 4. – Expected track density frequency distributions in the case of static irradiation of a 3 cm thick soil layer for different exposure durations.

to those shown in fig. 1 are found up to depths of 2 metres [31]. The average share in terms of the time of irradiation for grains in a 2 metre column for their exposure in the top $(0 \div 1)$ cm depth would be less than $4 \cdot 10^9$ y $\times$ $\times 1/200 \simeq 20 \cdot 10^6$ y. Time scales of this order in surface regions can indeed generate the high-ϱ grains. However, it would require $\geqslant 1 \cdot 10^9$ y to generate 10^8 cm^{-2} track density for shielding depths exceeding few centimetres.

It is quite easy to show that the track density distributions as observed in the lunar regolith can only be generated by alternately exposing material in layers of thicknesses < 10 cm for periods of $\leqslant 20 \cdot 10^6$ y and blanketing them with a fresh unirradiated layer. If the deposited layers are thinner, the time required to achieve the desired result would be proportionately less.

A number of papers have appeared discussing the nature of accretion/ mixing of the lunar regolith based on observations of the various near-surface processes, besides solar-flare tracks:

    1) solar-wind implantations of rare gases, carbon, nitrogen, etc.;

    2) solar-flare-produced isotopes, e.g. ^{26}Al, ^{22}Na, ^{132}Xe;

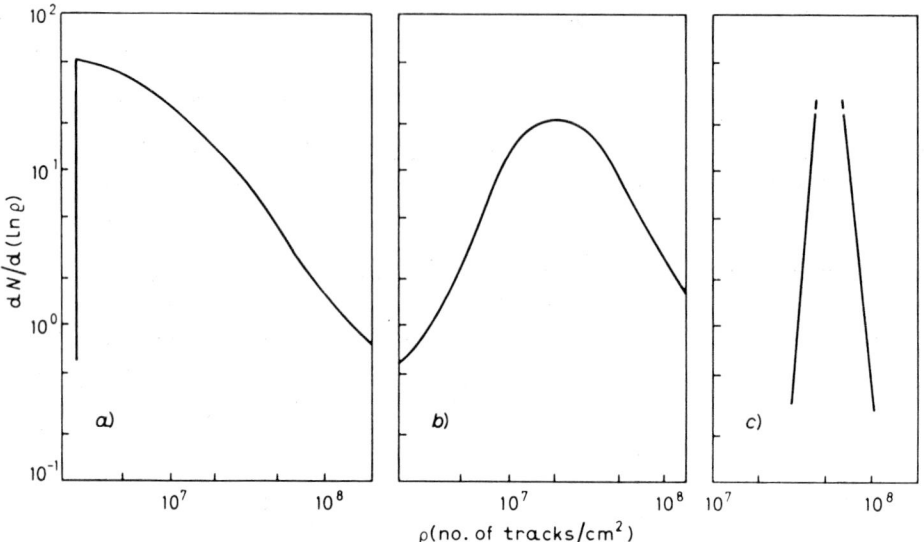

Fig. 5. – Expected track density frequency distribution in the case of static irradiation of a 3 cm thick soil layer for $10 \cdot 10^6$ y (a)). The changes in the frequency distribution pattern due to slow and rapid mixing of a layer are shown in a qualitative way in b) and c), respectively.

    3) agglutinate formation;

    4) microcrater formation.

The above surface-correlated parameters are found to be well correlated with the following track parameters [22, 32]:

ϱ_{med} (median track density): 50% of the total number of grains analysed at a given depth have track densities $< \varrho_{med}$.

ϱ_q (quartile track density): 25% of the total number of grains analysed at a given depth have track densities $< \varrho_q$.

ϱ_{min} (minimum track density): the lowest value of track density observed in a given depth interval.

N_H/N: fraction of grains having track densities $\geqslant 10^8$ cm^{-2}.

As an example, we show in fig. 6 the correlation between the agglutinate content and quartile track density values for Apollo 12, 14, 16 and 17 soils, and in fig. 7 between fraction of grains having $\varrho > 10^8$ cm^{-2} and solar-wind ^{36}Ar concentration. A saturation in agglutinate content sets in at $\varrho_q > (30 \div \div 40) \cdot 10^6$ cm^{-2} and in solar-wind ^{36}Ar for $N_H/N > 0.3$. Note in fig. 6 the group of soils with high agglutinate concentration but low ϱ_q values; these represent mixtures of mature and immature soils.

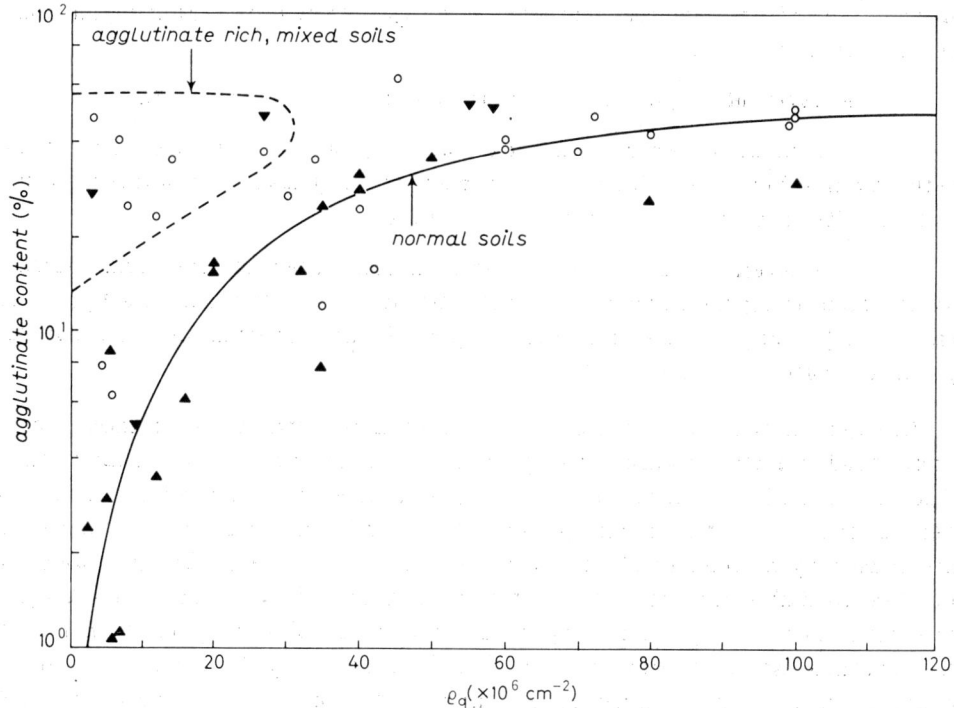

Fig. 6. – Percentage agglutinate content for Apollo 12 (▲), 14 (▼), 16 (●) and 17 (○) soil samples plotted as a function of quartile track densities. After [33].

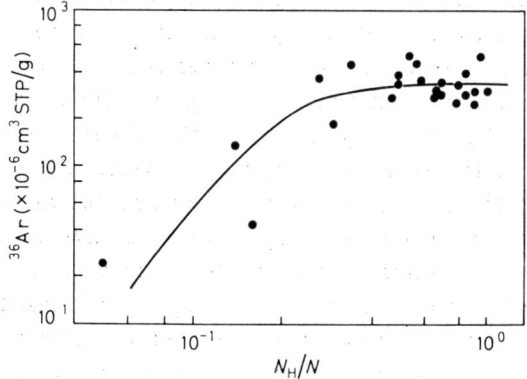

Fig. 7. – ^{36}Ar concentration *vs.* N_H/N values for lunar soils. After [30].

The exposure ages of soils have been estimated on the basis of tracks, agglutinates and solar-flare proton-produced xenon and neon isotopes [22, 34-37]. There exists a general agreement between these independent estimates. Based on these studies, the general features of the evolution of the lunar regolith can be summarized as follows:

1) A layer of $(2 \div 3)$ cm thickness is deposited on the surface.

2) It remains exposed for a period of $(20 \div 50) \cdot 10^6$ y before being blanketed by another layer. During this period, the top layer is reworked by micrometeorite impacts, up to depths of ~ 0.5 cm.

3) The average amount of material accumulated on the lunar surface by the blanketing mechanism is of the order of $(0.5 \div 0.8)$ g·cm$^{-2} \cdot 10^{-6}$ y, based on tracks [32, 35], cosmogenic isotopes [38, 39] and thermal-neutron capture produced isotopic effects [40-42].

The general features of the evolution of lunar regolith discussed above can be understood in terms of gardening by impacts of meteorites on the lunar surface. The Monte Carlo calculations of Arnold, Maurette and their collaborators are very instructive in this regard [43-45]. The gardening and regolith formation are caused by meteorites of sizes varying from 10^{-7} to 10^{10} g. In fig. 8 we show our best available current estimates for the meteorite flux [19, 46]. ARNOLD [45] has developed a simple cratering model based on laboratory experiments on the characteristics of craters formed in hypervelocity impacts. The corresponding mean crater diameters, D_c, and depths, d_c, are also shown in fig. 8. Considering these data, we have shown the intervals in meteorite masses responsible for surface reworking $(m = (10^7 \div 1)$ g$)$, turnover and layering $(m = (1 \div 10^8)$ g$)$ and excavation of deep-seated «fresh» regolith $(m > 10^8$ g$)$.

The total mass of meteoritic material contained in four decades, $(10^{-8} \div 10^{-4})$ g, constitutes an important agent for erosion, melting and evaporation [19]. If the excavation efficiency is assumed to be $\sim 10^8$ g/erg (for assumed impact velocity of 20 km·s^{-1}), independent of mass [47], the fluxes in fig. 8 correspond to an impacting mass of 1.2 g/cm$^2 \cdot 10^9$ y; the amount of matter excavated is $2.4 \cdot 10^4$ g/cm$^2 \cdot 10^9$ y [19]. On the Moon these impacts produce agglutinates, garden the regolith by mixing it vertically and redistributing it horizontally. Note that, in fig. 8, we have considered the mass interval up to 1 g as responsible for reworking of a layer, since the average thickness of layers on the Moon has been determined to be few centimetres.

Impacts will, of course, disturb the layer by layer blanketing sequence. The calculations of Arnold [45] show that the median disturbed depth varies as $T^{0.89}$, with the median disturbed depth being 2.1 m for a period of $3 \cdot 10^9$ y. This result is not inconsistent with observations of solar-flare proton-produced ^{53}Mn activities, which gives an idea of the extent of vertical mixing during the past $5 \cdot 10^6$ y [48, 49].

An important feature of the low-energy irradiation is that, in extended bombardment situations, a saturation value results. This is a result of a balance between production and erosion/attrition/fragmentation of the sample.

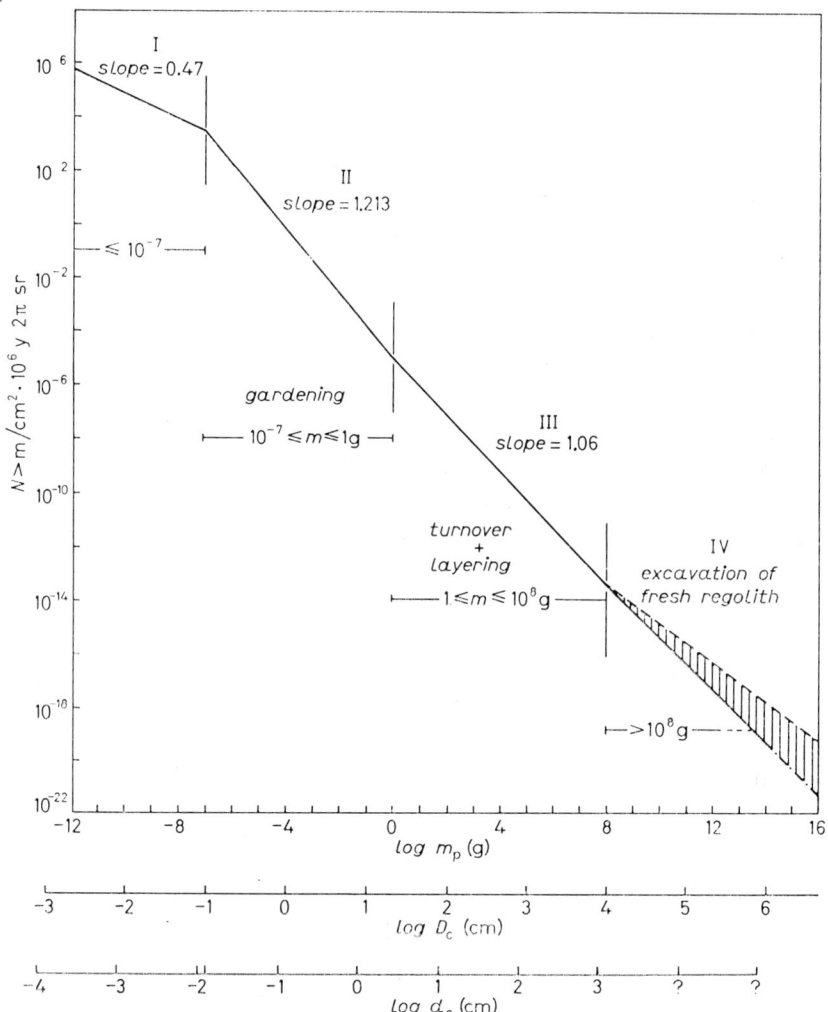

Fig. 8. – Cumulative meteoroid flux plotted as a function of meteoroid mass. The approximate diameter and depth of the craters (D_c and d_c) corresponding to meteoroids of various masses are also indicated. These values hold for meteoroids of density 3.4 g·cm⁻³ and impact velocity of ∼ 20 km/s. The meteoroid mass ranges responsible for gardening, mixing and addition of fresh material in the regolith are also indicated in the figure.

The effective time of irradiation is energy dependent, being larger for higher energies. If the production is $Q(X)$, then for the case of uniform erosion rate, $dX/dt = \varepsilon$, the effective time of irradiation, T_e, at saturation is given by

$$(4) \qquad\qquad C(X)/Q(x) = T_e ,$$

where

(5)
$$C(X) = \int\limits_0^\alpha Q(x + \varepsilon t) \exp\left[-\lambda t\right] \mathrm{d}t \,.$$

Relation (5) holds for a radioactive isotope of disintegration constant λ. The term $\exp\left[-\lambda t\right]$ should be deleted for the case of fossil tracks.

The effective erosion-controlled irradiation periods have been discussed for the case of tracks and radioactive isotopes [50, 51]. If we consider measured erosion rates based on study of lunar rocks, $\sim 5 \cdot 10^{-8}$ cm y^{-1}, the values of T_e for solar flare and galactic cosmic-ray tracks are in the range of $(10^5 \div 10^6)$ y and $(10^6 \div 10^8)$ y, respectively. The corresponding period for solar wind will be of the order of $(10^2 \div 10^3)$ y.

Useful information on the regolith dynamics is obtained from a study of the exposure history of the rocks and rocklets. These studies have been summarized by LAL [30]. They indicate fragmentation as the controlling factor for the exposure age of rocks on the lunar surface—and in general a very complex irradiation history for rocks due to their exposure at different depths within the regolith.

2. – The irradiation pattern of gas-rich meteorites.

Having discussed the nature of irradiation of the lunar regolith, we will now consider the nature of record in gas-rich meteorites. To-date data are available primarily on particle tracks and solar wind. Only limited data are available on microcraters and agglutinates. But they are adequate to discern the main differences between the lunar regolith and gas-rich meteorites.

In fig. 9 we show the track density distributions in irradiated grains from a number of gas-rich meteorites. The percentage of grains irradiated differs by an order of magnitude between gas-rich meteorites, but is never much greater than 20 %. To afford a comparison with the lunar regolith, in fig. 10 we have shown the averaged distribution in soil samples from Apollo 12 and 15 missions. Particle track data and the solar-wind concentrations as indicated by ^{36}Ar are given in fig. 11 alongside with similar data for the lunar-soil samples.

The time of compaction of grains, as determined from the ^{244}Pu fission tracks originating from the matrix of the carbonaceous chondrites, are known for a number of meteorites [7]. They range between 4.2 and $4.6 \cdot 10^9$ y for five meteorites: Nogoya, Murchison, Mighei, Murray and Cold Bokkeveld.

In contrast to the lunar regolith, where the near-surface irradiation has gone on for a long period, the irradiation of meteoritic grains must have occurred in a relatively short period, without any appreciable shielding. There are a number of reasons [15] for this statement, which refers primarily to carbonaceous chondrites:

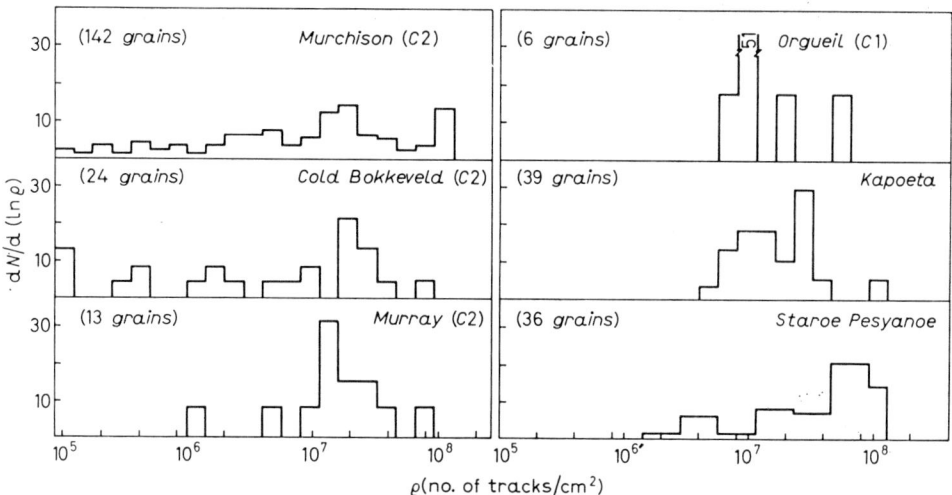

Fig. 9. – Track density frequency distributions in irradiated grains from several carbonaceous chondrites and gas-rich meteorites. These grains must have been irradiated prior to their compaction into the meteorites.

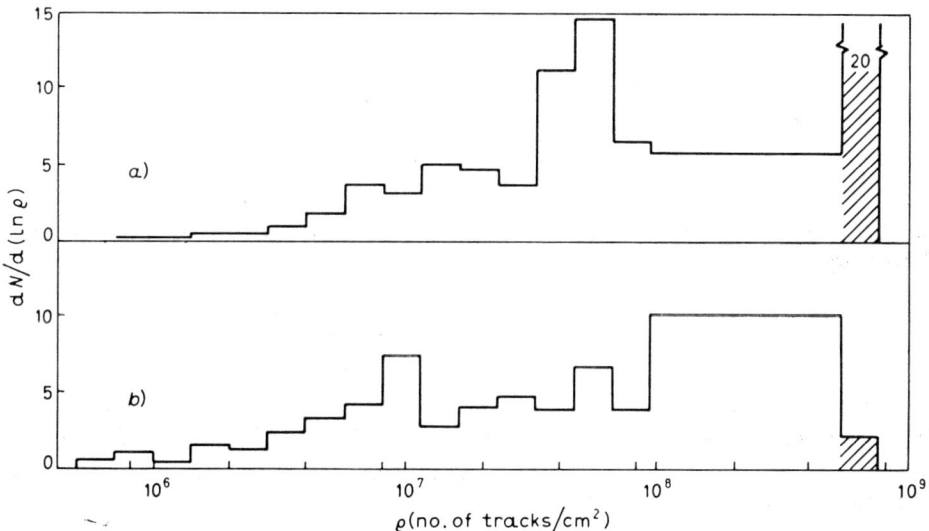

Fig. 10. – Track density frequency distributions for grains taken from the 60 and 240 cm long drill cores collected at the *a)* Apollo 12 (no. of grains $= 954$) and *b)* Apollo 15 (no. of grains $= 1216$) sites, respectively.

1) A population of high-track-density grains with a median track density of $2 \cdot 10^7$ cm^{-2} and a marked underabundance of high-track-density grains, $\geqslant (10^8 \div 10^9)$ cm^{-2}.

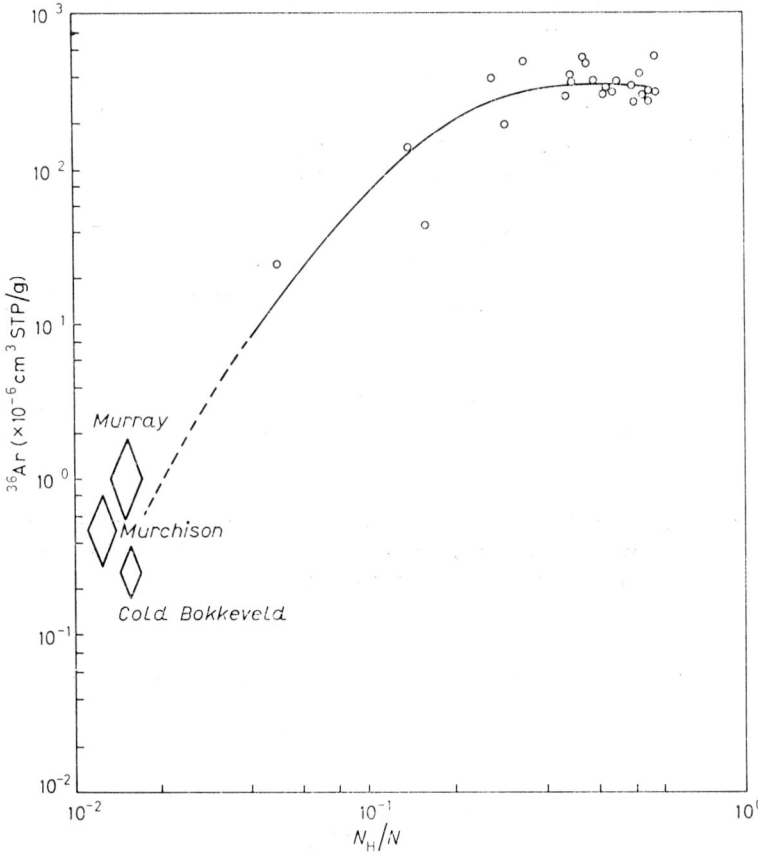

Fig. 11. – ^{36}Ar concentrations in lunar soils (○) and three carbonaceous chondrites (◇) plotted against their N_H/N values.

2) An irradiation depth distribution with 80 % of grains having been irradiated with shielding depths < 0.1 cm; this information is based on gradients in track densities within individual grains and the ratio of tracks due to VVH ($Z \geqslant 30$) and VH ($Z \geqslant 20$) nuclei.

3) The total irradiation time of grains, including the irradiation episode which led to formation of tracks in the irradiated grains as shielded by less than $\sim (100 \div 150)$ g cm^{-2} of matter, was less than $1 \cdot 10^6$ y. Track-rich grains, therefore, have to be irradiated very differently from those in a regolith, where a larger fraction at any time must receive a shielded irradiation. Given a short exposure duration, the only way of generating the high-track-density grains will be to expose them essentially unshielded.

4) The track density depth profiles are considerably steeper in the carbonaceous chondrites, indicating that the track densities were acquired in short time intervals, so that erosion effects were not important.

5) The « total » cosmic-ray exposure ages of the carbonaceous chondrites are generally small [16]. In three cases where the exposure ages are less than $1.5 \cdot 10^6$ y (Cold Bokkeveld: $0.3 \cdot 10^6$ y; Nogoya: $(0.15 \div 0.19) \cdot 10^6$ y; Murchison: $(1.0 \div 1.6) \cdot 10^6$ y), the ^{21}Ne (spallogenic) exposure ages are in agreement with those based on ^{26}Al, indicating a very small exposure time during the ealy irradiation, prior to compaction [6].

It should be noted here that conclusions regarding the irradiation history are based on studies of the olivine grains, which constitute less than 10% of the matrix. Implicit, therefore, in our deductions is the underlying assumption that the olivine grains constituted a part of the matrix during the cosmic-ray irradiation. The measurements of cosmogenic ^{21}Ne in the olivine grains and the matrix of the two low-exposure-age meteorites studied would be consistent with this assumption [18]; the exposure ages are identical for the Cold Bokkeveld and Murchison matrix and olivine grains. Still it could be argued that the irradiated olivine grains had a separate preirradiation history and were later mixed with the matrix and unirradiated olivine grains. It is difficult to check this point experimentally.

The other rather marked differences between the lunar regolith and the meteorite breccias relates to the role played by impacts in the early history, before compaction. Firstly, the size distribution of grains in the interval $(10 \div 100)$ μm [52, 53] is quite different. There is a considerable dearth of grains in this size region in the meteorites compared to the lunar regolith, indicating that meteorite impact did not proceed to any important extent to produce grains in this size range by comminution [52]. Secondly, one finds inclusions of different types of meteorites in several gas-rich meteorites, also in unequilibrated chondrites and mesosiderites [54]. Mixing and processing of matter took place in the early history of the solar system [55]. However, although lunar regolith contains an appreciable $((1 \div 2)\%)$ amount of matter of composition of primitive meteorites, the meteoritic fragments are very rare in the lunar regolith. WILKENING [54] has discussed this in some detail and concluded that the mixing in the case of meteorites must have occurred in a low-relative-velocity environment. Collisions between asteroidal materials would seem to satisfy the observations; however, those between comets and asteroids would be expected to lead to the lunar situation.

On the basis of on the various facts presented above, it seems highly unlikely that the observed particle track and solar-wind irradiation records in the carbonaceous chondrite were implanted in a lunar-regolith-type environment. The material must have been irradiated before the parent bodies formed, as single grains—flying in space as well as on the surface of centimetre to metre sized bodies for periods of $< 1 \cdot 10^6$ y followed by their aggregation into kilometre size or larger bodies, where they were stored, shielded from cosmic-ray effects until their recent excavation and exposure in space [15]. The short time scale involved in the irradiation and accretion process would not be inconsistent with our present understanding of the time scales of formation of planetesimals [56, 57].

One of the pressing reasons for proposing a regolith-type irradiation for gas-rich meteorites was the fact that the majority of the irradiated grains exhibit asymmetry in their track irradiation record, which is not possible in case of single grains exposed in space [9-11, 25, 58]. According to GOSWAMI and LAL [15], the irradiation occurred when material was dispersed in space as single grains as well as aggregates (clumps) of small size, < 1 m. In this situation, an asymmetric irradiation would be expected, at the same time permitting one to reach the levels of solar-wind/charged-particle track implantation as observed in the available short duration of irradiation. It may, however, be mentioned that, if the implantation took place at distances of a few AU, then, unless the galactic cosmic-ray flux was unusually low at that time, we would require to postulate high solar-flare heavy-nuclei fluxes in the early history of the Sun, by one to two orders of magnitude. This hypothesis would be particularly called for in the case of Cold Bokkeveld.

For the case of distant irradiation, solar-wind and solar-flare particle intensity would both be expected to follow an inverse-square-law dependence with distance from the Sun. The solar wind-track correlation plot for the carbonaceous chondrites seems to fall on the linear part of the lunar-soil graph (fig. 11). This would be expected also since in the two cases we are comparing only the near-surface effects—implantation of solar wind and track formation without any appreciable shielding.

It seems important to note that the proposed scenario for the formation of carbonaceous chondrite need not apply to other types gas-rich meteorites (H, L chondrites, howardites, aubrites, etc.), where the precompaction irradiation time scales are larger; also these meteorites are characterized by relatively more complex textural and mineralogical compositions. Further, definite evidences have been obtained for the presence of « young » xenoliths in some of these meteorites (e.g. [13, 14]), which clearly indicate recent mixing of these materials with the parent material of the gas-rich meteorites. One would, therefore, be inclined to believe that the irradiation history of gas-rich meteorites other than the carbonaceous chondrites is much more complex. Some of the irradiation processes may even be on-going today [9, 10, 59].

3. – Summary.

Studies of the fossil nuclear-bombardment records have elucidated the dynamic evolution history of the upper few metres of the lunar surface (the lunar regolith) under the influence of the continuous bombardment of meteoroids. The evolution of the lunar regolith involves discrete deposition of layers followed by their near-surface exposure for a few million years to a few tens of million years before being blanketed by another soil layer. Meteoroid impacts tend to disturb the layer by layer blanketing sequence and the median disturbed depth in the regolith due to meteoroid impact varies almost linearly with time.

The observation of dusty layers on the asteroidal bodies, the inferred chemical composition of asteroids based on reflection spectroscopy and polarimetry and intercomparisons of charged-particle tracks, microcraters and other surface irradiation features between meteoritic breccia and lunar regolith have led to a simple picture for the origin of gas-rich meteorites. A casual study of the literature reveals that there exists a general consensus on this; it is believed that the gas-rich meteorites originated in asteroidal regoliths [6, 10, 11, 13, 14, 60]. The problem of retention of impact ejecta on asteroids has been considered in some detail. It has been estimated that the amount of material returning to the asteroids after impacts for the larger objects Ceres, Pallas and Vesta should be more than 95 % [8]. Even for smaller objects of radius $(50 \div 100)$ km, the retention is expected to be about 50 % [10]. Hence the analogy with lunar regolith seems to be good. Based on these and other considerations, ANDERS [10] has concluded that all gas-rich meteorites originated in an asteroidal regolith. Further, it has been concluded [9, 10] that the gas-rich meteorites received their gas implantation within $(1 \div 8)$ AU from the Sun and that, as evidenced from the presence of young xenoliths, the gas implantation is an on-going process today.

In this lecture we have outlined the irradiation patterns observed in lunar regolith and gas-rich meteorites with particular reference to near-surface processes—charged-particle track formation, solar-wind implantation, formation of microcraters and agglutinates. The evidence in the case of carbonaceous chondrites, which occupy a special status in that they generally have low cosmic-ray exposure ages $((0.15 \div 5) \cdot 10^6$ y$)$, with concordant ^{26}Al, ^{22}Ne exposure ages indicating that their exposure duration in the regolith has been small $< 1 \cdot 10^6$ y [6]. Furthermore, their size during the recent cosmic-ray exposure was generally large enough so that, during their most recent cosmic-ray exposure in space, the charged-particle track production has been very small, allowing a study of the low track densities arising from the early irradiation, i.e. prior to compaction. The analysis [15] outlined here in some details leads one to the conclusion that the irradiation pattern in carbonaceous chondrites at least could not have resulted during their existence as an asteroidal regolith. We believe that the irradiation pattern of carbonaceous chondrites arose before the accumulation of the asteroidal bodies to any appreciable size. Material was either irradiated as individual grains in space or as small clumps of few centimetre radii during the stage when the asteroidal material was gravitationally unstable to form clusters. During this stage the impact ejecta were not retained on these small bodies. The irradiation continued until finally the asteroidal bodies evolved. By a model of this type only can one explain the carbonaceous-chondrite solar-flare and cosmic-ray records. The model implies that the parent bodies of carbonaceous chondrites are made up of the irradiated material. The irradiated material stored at depths exceeding several tens of metres, where the galactic cosmic-ray spallation during the last $4 \cdot 10^9$ y is not important, constitutes the matrix of the carbonaceous chondrites.

In the case of other gas-rich meteorites where young xenoliths have been found [13, 14] clearly « young » material has recently been mixed. Mixing between meteorites has been extensively documented [54]. In addition, the complexity in their structural and mineralogical composition combined with other precompaction irradiation records seems to indicate that their evolution was complex, including a late regolith-type irradiation.

It is hoped that continuing studies of the early irradiation record would provide a reasonable working hypothesis for the evolutionary history of different classes of gas-rich meteorites.

REFERENCES

[1]   H. E. SUESS, H. WANKE and F. WLOTZKA: *Geochim. Cosmochim. Acta*, **28**, 595 (1964).
[2]   P. EBERHARDT, J. GEISS and N. GROGLER: *J. Geophys. Res.*, **70**, 4375 (1965).
[3]   D. LAL and R. S. RAJAN: *Nature (London)*, **223**, 269 (1969).
[4]   P. PELLAS, G. POUPEAU, J. C. LORIN, H. REEVES and J. ADOUZE: *Nature (London)*, **223**, 269 (1969).
[5]   D. E. BROWNLEE and R. S. RAJAN: *Science*, **182**, 1341 (1973).
[6]   J. N. GOSWAMI, I. D. HUTCHEON and J. D. MACDOUGALL: *Proceedings of the VII Lunar Science Conference, Geochim. Cosmochim. Acta, Suppl.* 7, Vol. **1** (1976), p. 543.
[7]   J. D. MACDOUGALL and B. K. KOTHARI: *Earth Planet. Sci. Lett.*, **33**, 36 (1976).
[8]   D. L. MATSON, T. V. JOHNSON and G. J. VEEDER: *Proceedings of the VIII Lunar Science Conference, Geochim. Cosmochim. Acta, Suppl.* 8, Vol. **1** (1977), p. 1001.
[9]   E. ANDERS: *Icarus*, **24**, 363 (1975).
[10] E. ANDERS: *Asteroids: An exploration assessment*, preprint, NASA SP-000 (1978).
[11] R. S. RAJAN: *Geochim. Cosmochim. Acta*, **38**, 777 (1974).
[12] P. B. PRICE, I. D. HUTCHEON, D. BRADDY and D. MACDOUGALL: *Proceedings of the VI Lunar Science Conference, Geochim. Cosmochim. Acta, Suppl.* 6, Vol. **3** (1975), p. 3449.
[13] J. C. HUNEKE, S. P. SMITH, R. S. RAJAN, D. A. PAPANASTASSIOU and G. J. WASSERBURG: *Lunar Science VIII*, Abstr. 485 (1977).
[14] L. SCHULTZ and P. SIGNER: *Earth Planet. Sci. Lett.*, **30**, 363 (1977).
[15] J. N. GOSWAMI and D. LAL: *Icarus*, **40** (1979) (in press).
[16] E. MAZOR, D. HEYMAN and E. ANDERS: *Geochim. Cosmochim. Acta*, **34**, 781 (1970).
[17] D. D. BOGARD, R. S. CLARKE, J. E. KEITH and M. A. REYNOLDS: *J. Geophys. Res.*, **76**, 4076 (1971).

[18] J. D. MACDOUGALL and D. PHINNEY: *Proceedings of the VIII Lunar Science Conference, Geochim. Cosmochim. Acta, Suppl.* 8, Vol. 1 (1977), p. 293.

[19] D. E. GAULT, F. HORZ, D. E. BROWNLEE and J. B. HARTUNG: *Proceedings of the V Lunar Science Conference, Geochim. Cosmochim. Acta, Suppl.* 5, Vol. 3 (1974), p. 2365.

[20] D. LAL, D. MACDOUGALL, L. WILKENING and G. ARRHENIUS: *Proceedings of Apollo 11 Lunar Science Conference, Geochim. Cosmochim. Acta, Suppl.* 1, Vol. 3 (1970), p. 2295.

[21] G. CROZAZ, U. HAACK, M. HAIR, M. MAURETTE, R. WALKER and D. WOOLUM: *Proceedings of the Apollo 11 Lunar Science Conference, Geochim. Cosmochim. Acta, Suppl.* 1, Vol. 3 (1970), p. 2051.

[22] G. ARRHENIUS, S. LIANG, D. MACDOUGALL, L. WILKENING, N. BHANDARI, S. BHAT, D. LAL, G. RAJAGOPALAN, A. S. TAMHANE and V. S. VENKATAVARADAN: *Proceedings of the II Lunar Science Conference, Geochim. Cosmochim. Acta, Suppl.* 2, Vol. 3 (1971), p. 2583.

[23] G. M. COMSTOCK, A. O. EVWARAYE, R. L. FLEISCHER and H. R. HART jr.: *Proceedings of the II Lunar Science Conference, Geochim. Cosmochim. Acta, Suppl.* 2, Vol. 3 (1971), p. 2569.

[24] G. CROZAZ, R. WALKER and D. WOOLUM: *Proceedings of the II Lunar Science Conference, Geochim. Cosmochim. Acta, Suppl.* 2, Vol. 3 (1971), p. 2543.

[25] J. N. GOSWAMI, D. BRADDY and P. B. PRICE: *Proceedings of the VII Lunar Science Conference, Geochim. Cosmochim. Acta, Suppl.* 7, Vol. 1 (1976), p. 55.

[26] N. BHANDARI, J. N. GOSWAMI and D. LAL: *Proceedings of the IV Lunar Science Conference, Geochim. Cosmochim. Acta, Suppl.* 4, Vol. 3 (1973), p. 2275.

[27] G. E. BLANFORD, R. M. FRULAND and D. A. MORRISON: *Proceedings of the VI Lunar Science Conference, Geochim. Cosmochim. Acta, Suppl.* 6, Vol. 3 (1975), p. 3557.

[28] I. D. HUTCHEON, D. MACDOUGALL and P. B. PRICE: *Proceedings of the V Lunar Science Conference, Geochim. Cosmochim. Acta, Suppl.* 5, Vol. 3 (1974), p. 2561.

[29] R. WALKER and D. YUHAS: *Proceedings of the IV Lunar Science Conference, Geochim. Cosmochim. Acta, Suppl.* 4, Vol. 3 (1973), p. 2379.

[30] D. LAL: *Phil. Trans. R. Soc. London Ser. A*, **285**, 69 (1977).

[31] J. N. GOSWAMI and D. LAL: *The Moon and the Planets*, Vol. **18** (1978), p. 371.

[32] R. L. FLEISCHER and H. R. HART: *Earth Planet. Sci. Lett.*, **18**, 420 (1973).

[33] J. N. GOSWAMI and D. LAL: *Proceedings of the VIII Lunar Science Conference, Geochim. Cosmochim. Acta, Suppl.* 8, Vol. 1 (1977), p. 813.

[34] G. CROZAZ: *Phys. Chem. Earth*, **10**, 197 (1977).

[35] N. BHANDARI, J. N. GOSWAMI and D. LAL: *Proceedings of the XIII International Conference on Cosmic Rays*, Vol. 1 (Denver, Col., 1973), p. 287.

[36] K. GOPALAN, J. N. GOSWAMI, M. N. RAO, K. M. SUTHAR and T. R. VENKATESAN: *Proceedings of the VIII Lunar Science Conference, Geochim. Cosmochim. Acta, Suppl.* 8, Vol. 1 (1977), p. 793.

[37] N. B. BHAI, K. GOPALAN, J. N. GOSWAMI, M. N. RAO and T. R. VENKATESAN: *Proceedings of the IX Lunar Science Conference, Geochim. Cosmochim. Acta*, 1629 (1978).

[38] D. D. BOGARD, L. E. NYQUIST, W. C. HIRSCH and D. R. MOORE: *Earth Planet. Sci. Lett.*, **21**, 52 (1973).

[39] R. O. PEPIN, J. R. BASFORD, J. C. DRAGON, M. R. COSICO and V. R. MURTHY: *Proceedings of the V Lunar Science Conference, Geochim. Cosmochim. Acta, Suppl.* 5, Vol. **2** (1974), p. 2149.

[40] G. P. RUSS III, D. S. BURNETT and G. J. WASSERBURG: *Earth Planet. Sci. Lett.*, **15**, 172 (1972).

[41] G. P. Russ III: Ph. D. Thesis (1973).

[42] D. B. Curtis and G. J. Wasserburg: *Lunar Science VI*, Abstr., **1**, 172 (1975).

[43] Y. Langevin and J. R. Arnold: *Annu. Rev. Earth Planet. Sci.*, **5**, 449 (1977).

[44] J. Borg, G. M. Comstock, Y. Langevin, M. Maurette, B. Jouffrey and C. Jouret: *Earth Planet. Sci. Lett.*, **29**, 161 (1971).

[45] J. R. Arnold: *Proceedings of the VI Lunar Science Conference, Geochim. Cosmochim. Acta, Suppl.* 6, Vol. **2** (1975), p. 2375.

[46] G. Neukum and H. Dietzel: *Earth Planet. Sci. Lett.*, **12**, 59 (1971).

[47] J. F. Vedder: *J. Geophys. Res.*, **77**, 4304 (1972).

[48] M. Imamura, K. Nishiizumi, M. Honda, R. C. Finkel, J. R. Arnold and C. P. Kohl: *Proceedings of the V Lunar Science Conference, Geochim. Cosmochim. Acta, Suppl.* 5, Vol. **2** (1974), p. 2093.

[49] K. Nishiizumi, M. Imamura, C. P. Kohl, M. T. Nurrell, J. R. Arnold and G. P. Russ III: *The extent of lunar regolith mixing on a ten million year time scale, Earth Planet. Sci. Lett.*, in press (1979).

[50] D. Lal: *Space Sci. Rev.*, **14**, 3 (1972).

[51] D. Lal: *Phil. Trans. R. Soc. London Ser. A*, **277**, 395 (1974).

[52] S. K. Bhattacharya, J. N. Goswami, D. Lal, P. P. Patel and M. N. Rao: *Proceedings of the VI Lunar Science Conference, Geochim. Cosmochim. Acta, Suppl.* 6, Vol. **3** (1975), p. 3509.

[53] T. V. V. King and E. A. King: *Meteoritics*, **13**, 47 (1978).

[54] L. L. Wilkening: in *Comets, Asteroids, Meteorites*, edited by A. H. Delsemme (Toledo, O., 1977).

[55] J. M. Herndon and L. L. Wilkening: in *Protostars and Planets*, edited by T. Gehrels (Tucson, Ar., 1979), p. 502.

[56] P. Goldreich and W. R. Ward: *Astrophys. J.*, **183**, 1051 (1973).

[57] V. S. Safronov: *Evolution of the Protoplanetary Clouds and Formation of the Earth and the Planets* (Moscow, 1969).

[58] J. D. Macdougall, R. S. Rajan and P. B. Price: *Science*, **183**, 73 (1974).

[59] D. Lal: *Proceedings of the Nobel Symposium 21, From Plasma to Planet*, edited by A. Elvius (New York, N. Y., 1972), p. 49.

[60] G. Poupeau, T. Kirsten, F. Steinbrunn and D. Storzer: *Earth Planet. Sci. Lett.*, **24**, 229 (1974).

Chemical Evolution of the Galaxy.

R. Gallino and E. Gaschino

Istituto di Fisica Generale dell'Università - Torino, Italia
Istituto di Cosmogeofisica del C.N.R. - Torino, Italia

1. – Introduction.

When considering the problem of isotopic anomalies in the solar system, one has first to recognize what kind of difficulties one encounters in interpreting the standard distribution of the solar-system chemical composition.

Heavy elements are nucleosynthesized by short-lived massive stars and are ejected in interstellar space during the final supernova explosion. Consequently, the solar-system chemical composition is the result of a sequence of complex phenomena, involving

    i) the dynamical evolution of the Galaxy from the halo phase up to the development of the disc and spiral arms,

    ii) the rate of star formation and the mass spectrum for each generation of stars,

    iii) the amount of metals contributed by each generation of stars according to stellar evolutionary theories,

    iv) the remixing of stellar processed material with interstellar gas,

    v) the possible infall of unprocessed gas into the disc and the possible outfall in the intergalactic medium of metal-enriched material.

Further, locally, the solar nebula could have been contaminated by one or more exploding supernovae. In this respect, the average chemical composition of interstellar matter at the epoch of formation of the Sun would not be representative of the solar-system distribution.

Each of the above processes has been subjected to a large series of investigations; actually, the main problem depends on several uncertain parameters and one must be very cautious about the information that one can extract. In any case, we are not allowed to use too detailed hypotheses when calculating chemical enrichment.

In this note, we first briefly examine the principal constraints on the theory and the observative evidences. Further, we present a particular model that roughly reproduces the main features of chemical enrichment in the Galaxy.

2. – Chemical observational constraints.

2˙1. *Galactic nucleosynthesis of* ^{4}*He*. – Although ^{4}He is the principal product of stellar nucleosynthesis on the main sequence, only a small fraction is ejected in space at the end of the life of the star. Then, the net yield of He contributed by one generation of stars to interstellar enrichment is not sufficient to explain its high observed abundance [1].

On the other hand, observations of H II regions [2] and of planetary nebulae [3] indicate that the primordial abundance by mass at the beginning of the contraction of the Galaxy was $Y_0 = 0.20 \div 0.23$, which is consistent with the prediction of the cosmological theory [4].

However, a tendency toward a slight increase of Y with galactic time is furnished by observations of population I stars. Indeed, for the Sun we can adopt $Y_\odot = 0.26$ [5], and for the youngest stars $Y_{\text{now}} = 0.28 \div 0.30$ [3].

In this framework, the galactic nucleosynthesis of ^{4}He has to account for an enrichment $\Delta Y \simeq 0.10$.

2˙2. *Galactic nucleosynthesis of heavy elements*. – The initial explosion of the Universe cannot account for the production of heavy elements beyond ^{4}He. Thus, all metals contributed to the interstellar medium must have been synthesized in stellar cores.

Observationally, halo stars are all metal-deficient, in the sense that population II stars with $Z \gtrsim Z_\odot/10$ do not reach high latitudes, above $h \simeq 4$ kpc, while all stars with $Z \gtrsim Z_\odot/3$ are within the disc. Nevertheless, a general correlation between metal deficiency of stars and their height above the galactic plane, as suggested by MORGAN [6], has not been clearly confirmed by a recent analysis of the distribution of RR Lyrae ([7], see also [8]).

We must realize that metal-deficient stars are rare. According to DIXON [9], only about 20 % of all solar-mass stars present in the Galaxy at the solar galactocentric distance have $Z \leqslant Z_\odot/3$. Moreover, only 1 % of these stars have $Z \leqslant Z_\odot/10$. Further, no metal-deficient stars have been observed with Z less than a few times 10^{-5}.

An important investigation about the distribution of metallicity $Z(S)$ among halo globular clusters has been performed by HARTWICK [10], starting from an analysis of the space distribution by WOLTJER [11]. This distribution is shown in fig. 1, plotting Z/Z_1 *vs.* S/S_1, where S is the cumulative number of stars with metal abundance less than or equal to Z, $Z_1 = 0.003$ being the maximum metal abundance observed in globular clusters in the halo. We note that there are relatively more stars with lower metal abundance.

Concerning disc population stars, heavy elements must grow from an initial value of about $Z_\odot/3$, through a solar value $Z_\odot \simeq 0.015$ [5] at the Sun birth, up to a present value $Z_{now} \simeq 0.02$ [12]. Here again, the most important fea-

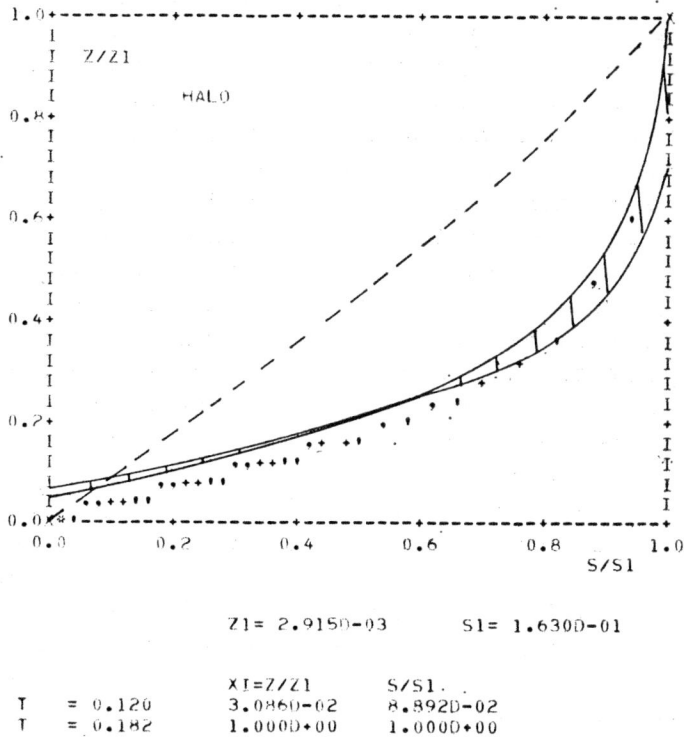

Z1= 2.915D-03   S1= 1.630D-01

	XI=Z/Z1	S/S1
T = 0.120	3.086D-02	8.892D-02
T = 0.182	1.000D+00	1.000D+00

Fig. 1. – Chemical distribution Z/Z_1 *vs.* S/S_1 in the halo phase. The hatched area represents the observed distribution of globular clusters by HARTWICK [10]. The upper curve represents the simple-model solution, in accord with eq. (2) and assuming $Z_1 = 0.003$, $p = 0.01$. The dashed curve represents our computed solution (see sect. **9**).

ture is the metal distribution, related with the so-called F-G dwarf problem [13, 14]: there are very few metal-poor stars, all disc stars showing a Z value within a factor of 2 with respect to $Z_\odot$ [15-17]. The cumulative distribution Z/Z_{now} *vs.* S/S_{now} for the disc phase is shown in fig. 2, based on a statistical analysis by PAGEL and PATCHETT [12]. Here we have exactly the reverse situation than for the halo.

The galactic nucleosynthesis of metals must account for an enrichment $\Delta Z \simeq 0.02$. We notice that, according to HACYAN *et al.* [18], the observed ratio $\Delta Y/\Delta Z \simeq 2.7$ is a powerful constraint on stellar evolutionary prescriptions.

3. – Dynamical prescriptions and star formation rate.

The Galaxy is assumed to have undergone two major evolutionary phases:
i) a rapid halo contraction, ii) a subsequent disc plus bulge phase.

First hydrodynamic calculations for the evolution of typical disc galaxies
have been performed by LARSON [19] and by TINSLEY and LARSON [20] for
our Galaxy. The collapsing protogalaxy was treated as a two-component
system: a gaseous component and a stellar component.

Concerning the galactic halo, we may simply assume that in the early phases
the protogalaxy follows a homologous free-fall contraction, starting from
a homogeneous rotational ellipsoid with height $h_0 = (10 \div 20)$ kpc and radius
$R_0 = (25 \div 50)$ kpc. The contraction proceeds almost freely, on a time scale
of the order of 1 to $2 \cdot 10^8$ y, until the halo reaches a height of a few kpc
above the galactic plane. From then on, the collapse slows down by
frictional effects; a spheroidal bulge with a radius of a few kpc settles in
the central part of the Galaxy, whereas the outer material falls in more slowly
towards the disc. The time scale for this intermediate phase may be evaluated to be of the order of 0.5 to 1 billion years [21]. The overall contraction
of the Galaxy stops when a flat thin disc forms, with a height of about 400 pc
and a radius of the order of 20 kpc. The present age of the Galaxy, T_G, is
somewhat uncertain, depending on the adopted value of the Hubble constant
which is method dependent. We will adopt T_G as a free parameter, ranging
from 10 to 20 billion years. Fortunately, as we shall see, this uncertainty is
not crucial for determining the present chemical distribution, that rather
depends on the present gas mass to total mass ratio, μ_g. For the solar neighbourhood, we can assume $\mu_g = 0.10$, with an uncertainty of a factor 2 (see
for references [12]).

The star formation rate (SFR) appears to be one of the more important
factors characterizing the structure of the model as well as the chemical
features. The SFR can be defined as the galactic mass fraction condensing
in stars per unit time:

$$B(t) = \frac{1}{M_G} \frac{dM_s(t)}{dt} = \frac{dS(t)}{dt},$$

where M_G is the total mass of the Galaxy and $M_s(t)$ is the cumulative mass
condensed in stars up to time t.

The SFR is currently assumed to depend on the local gas density ϱ_g according
to a simple power law: $B(t) = \varrho_g^n$, where $1 \leqslant n \leqslant 2$. ([22], see for references [23]).

With these prescriptions, the SFR increases in the early phases, when the
gas density increases due to the shrinkage of galactic volume. As the halo
collapse slows down, the SFR decreases owing to the consumption of the

available gas, that progressively condenses in stars. Consequently, the SFR must show a pronounced peak near the end of the halo contraction.

According to LARSON [24], the choice $n = 1.85$ provides a satisfactory representation of the observed properties of elliptical galaxies. For disc galaxies, however, the simple power law can only be suitable for the early halo collapsing phase. Indeed, according to LARSON [19], « if a disc component is to be formed, the SFR must decrease more strongly during the later stages of the collapse than is predicted by the power law ». Among other possibilities, as the effect of tidal forces exerted by the central spherical bulge against star formation, this sudden decrease can be accounted for, if one admits that the primeval interstellar medium was not quiet and uniform, but was subjected to strong turbulent motions. In a qualitative way, one can represent the interstellar gas as a two-component structure, with dense high-velocity clouds embedded in a less dense intercloud gas. In this framework, an enhanced SFR driven by cloud-cloud collisions could occur in the halo phase, whereas during later stages of collapse—the efficiency of collisions falling down—the SFR decreases abruptly.

For the disc phase, star formation occurs only in spiral arms. This fact suggests an even more complex situation, with different star formation mechanisms applying in various galactic phases.

4. – Initial mass function.

Let $B(m, t) \, \mathrm{d}m \, \mathrm{d}t$ represent the mass fraction of the Galaxy born as stars of mass m in the mass interval $(m, m + \mathrm{d}m)$ and in time $(t, t + \mathrm{d}t)$. Then

$$B(t) = \int_0^\infty B(m, t) \, \mathrm{d}m .$$

It is generally assumed that $B(m, t)$ is a separable function of mass and time:

$$B(m, t) = \psi(m) B(t) ,$$

where the initial mass function (IMF) $\psi(m)$ is the ratio of galactic mass fraction condensed in stars per unit time per unit mass interval to the total mass fraction born per unit time as stars of all masses. The IMF must satisfy the relation

$$\int_0^\infty \psi(m) \, \mathrm{d}m = 1 .$$

The IMF, assumed to be constant with time, can be deduced from observations, taking into account stellar lifetimes and the luminosity-mass relation.

TABLE I.

M_V	$m/m_\odot$	τ_m (y)	$K(M_V)$	$K(M_V) \times \times m/m_\odot$	q_1	q_2	q_3
— 6	95	3.0D + 06	8.6D — 05	0.8D — 02	0.51	0.42	0.00
— 5	50	4.0D + 06	4.7D — 04	2.4D — 02	0.48	0.38	0.36
— 3	15.5	1.3D + 07	2.1D — 03	3.3D — 02	0.37	0.21	0.09
— 1	6	3.8D + 07	6.5D — 03	3.9D — 02	0.34	0.24	0.24
+ 1	2.8	1.9D + 08	2.2D — 02	6.2D — 02	0.44	0.42	0.42
+ 3	1.65	1.3D + 09	3.3D — 02	5.5D — 02	0.58	0.55	0.55
+ 5	1	1.2D + 10	4.8D — 02	4.8D — 02	0.75	0.72	0.72

In table I we report the stellar lifetimes τ_m vs. absolute magnitude M_V, according to TRURAN, HANSEN and CAMERON [25], together with the relation $M_V(m)$. An approximate analytical expression, giving τ_m as a function of the stellar mass m, was proposed by TINSLEY [26]:

$$\log \tau_m = 10.02 - 3.57 \log m + 0.90 \log m^2 .$$

SALPETER [27] first suggested that the IMF roughly satisfies a simple power law relation

$$\psi(m) = A m^{-b}$$

with $b = 1.35$. Since then, many modifications to the IMF have been proposed, especially for the first evolutionary stages where the IMF is believed to support a much higher number of massive stars [14]. TALBOT and ARNETT [28] suggested that a steeper slope, $b = 1.6$, may apply in the range of massive stars. For a discussion of the problem we may refer to [23].

However, a recent analysis by BURKI [29] of young galactic clusters and associations shows that the global mass spectrum for all clusters is consistent with Salpeter's value.

In conclusion, we think that the adoption of a Salpeter IMF through all galactic ages constitutes perhaps the best choice at the present time.

The IMF must be truncated at a certain value, corresponding to the upper mass limit on the main sequence against vibrational instability. This value ranges between 60 $m_\odot$ and 120 $m_\odot$ [30, 31].

Concerning the lower limit, we can simply pile up all unevolved stars with $m \lesssim 1\ m_\odot$ having lifetimes longer than the age of the Galaxy. From the data by WEISTROP [32], TALBOT and ARNETT [28] assumed that the mass fraction of the IMF consisting of stars with $m \gtrsim 1\ m_\odot$ is

$$\zeta = \int_1^\infty \psi(m)\ dm \simeq 0.25 .$$

5. – Remixing of interstellar gas.

Some attempts have been made in order to construct galactic models that follow the formation of the disc taking account of the effect of an infall of halo material [19, 33, 34], or of an infall of unprocessed intergalactic material [35, 36].

The main interest in these works consists in the possibility of interpreting the observed chemical gradient along the disc in our Galaxy as well as in other galaxies. Indeed, observations indicate a higher metal abundance as one goes towards the centre of the Galaxy [3, 17, 37]. This gradient can be connected with the residual gas fraction at different galactocentric distances; indeed, the spheroidal central bulge is gas poor, the solar neighbourhood contains about 10 % of gas, while the outer regions have as much as 50 % of gas.

The central bulge may then be considered as the result of the rapid homologous contraction of the halo, thus approaching to the conditions encountered in elliptical galaxies, while the disc may grow outward with time owing to a later slow infall of halo material. The delayed infall of unprocessed gas from the halo in the outer disc regions reduces the star formation rate and dilutes the interstellar metal abundance. Actually, many uncertainties remain on the true dynamic evolution of the Galaxy, when halo matter settles into the disc.

6. – Stellar evolution.

In order to evaluate the amount of cosmic ^{4}He and heavy elements supplied by stellar nucleosynthesis, we need to know the various stages of stellar evolution, that are different for different stellar masses and are subjected to large uncertainties, especially concerning the mechanism of the final explosion and the fraction of metals left behind in collapsed remnants.

We do not intend to examine in detail the very complex situation concerning present estimates of chemical yields by stellar activity. For an analysis of the overall problem, we refer to [38, 39]. Here, we simply sketch the adopted picture concerning the final chemical composition of stars of different masses:

a) Lower-mass stars $(m \leqslant 5\ m_\odot)$. All these stars are supposed to evolve from the main sequence up to the giant phase, then suffering a ^{4}He-shell thermal instability [40, 41] or a dynamical instability driven by radiation pressure [42].

These stars are thought to be the progenitors of planetary nebulae and then of white dwarfs. Consequently, concerning chemical processing, they contribute only to ^{4}He enrichment, all metals being frozen in collapsed remnants (white dwarfs) of about 1 $m_\odot$.

b) *Intermediate-mass stars* $(5 \, m_\odot \lesssim m \lesssim 10 \, m_\odot)$. These stars are supposed to follow a common evolutionary pattern [42] and to evolve up to central ignition of ^{12}C (^{16}O) in degenerate conditions. The ultimate fate is somewhat uncertain, depending on the rate of neutrino energy losses through the degenerate URCA-shell process [43]. The existence of pulsars, being currently associated with original stars in this mass range [44], can be assumed as evidence that after the supernova explosion here again all metal cores are left behind as collapsed remnants of about $1.5 \, m_\odot$.

c) *Massive stars* $(m > 10 \, m_\odot)$. Heavy elements are contributed by stars in this mass range. The evolution proceeds up to the central Si-Fe phase. The more advanced phases, beyond central ^{4}He-exhaustion, have been followed carefully by ARNETT ([45] and references therein) and by WEAVER *et al.* [46]. Concerning the ultimate phases, we can roughly distinguish three different mass intervals:

i) $10 \lesssim m/m_\odot \lesssim 20$.

After a ^{28}Si-core forms, the star is likely to meet a dynamical instability driven by β-inverse decays on ^{28}Si. This instability is probably unable to disrupt the entire star, but sufficient to eject all the external C-O zone, leaving a core remnant of $\sim 1.5 \, m_\odot$.

ii) $20 \lesssim m/m_\odot \lesssim 80$.

The star evolves up to the formation of a Fe-core, the dynamical instability being triggered by the so-called Fe $\to$ He transition. The collapse of Fe cores has been followed by hydrodynamical calculations [47], but the ultimate fate remains uncertain, the possibility of a collapse towards a black-hole not being excluded.

iii) $m \gtrsim 80 m_\odot$.

These stars, although rare, give an important contribution to interstellar metal abundances. They evolve up to ^{12}C-ignition, then suffer a collapse driven by the electron-positron pair dynamical instability. According to BARKAT *et al.* [48] and FRALEY [49], the collapse is later followed by an explosion when the ^{16}O-core ignites in degenerate conditions. The star is entirely disrupted, ejecting about half of its original mass in the form of metals, mostly ^{16}O, ^{28}Si and ^{56}Fe. The existence of such high masses is somewhat controversial. If they do not exist, we would have to modify our ideas about metal yields contributed by the less massive stars.

We define the total yield, p, as the ratio of the mass fraction p_Z of newly synthesized metals ejected by one generation of stars to the mass fraction α locked up in long-lived stars or remnants. With the adopted evolutionary picture, we have $p \simeq 0.01$, $p_Z = 0.007$.

Table I gives the adopted values of the mass fractions processed to He, to heavy elements and left as dead remnants for some indicative stellar masses.

7. – The simple model.

In the previous sections we have briefly examined the difficulties connected with the choice of a plausible evolutionary picture of the Galaxy.

A first analytical approach consists in the so-called simple model, which is based on the assumptions of instantaneous-recycling approximation, a constant IMF, a complete homogeneous remixing of the gas with no infall or outfall of matter. This model predicts a heavy-element abundance by mass varying with time according to the following expression [50]:

$$(1) \qquad\qquad Z = p \ln (1/\mu) \,,$$

where p is the metal yield by one generation of stars and μ the gas mass fraction of the Galaxy.

Due to the existence of a chemical gradient along the disc, connected with a different rate of gas consumption at various galactocentric distances, the simple model is suitable for the solar neighbourhood only if we further accept the one-zone hypothesis, $i.e.$ that there are no radial flows of matter and that the interstellar gas is well remixed along a narrow shell centred at the distance of the Sun and perpendicular to the galactic plane.

For $p = 0.02$, as currently assumed in the literature [28, 51, 52], and $\mu_{\mathrm{ncw}} = 0.10$, eq. (1) gives a fairly large present metal abundance: $Z_{\mathrm{now}} = 4.6 \cdot 10^{-2}$. Actually, a more acceptable value could be obtained, with a suitable choice of p and μ in the range of present uncertainties. However, the simple model fails in interpreting the $Z(S)$ distribution. Indeed, from eq. (1) and from the equation of conservation of mass

$$\mu = 1 - \alpha S \,,$$

where α is the mass fraction of each generation of stars that remains locked in long-lived stars or collapsed remnants, one obtains

$$(2) \qquad\qquad \frac{S}{S_1} = \frac{1 - \exp[-Z/p]}{1 - \exp[-Z_1/p]} \,.$$

This relation predicts too many metal-poor stars, as shown in fig. 2.

In order to solve this difficulty, we have first to drop the instantaneous-recycling approximation. Indeed, owing to the large spread of stellar lifetimes, stars of solar mass born at a given time t do not die immediately, but a long time after, then diluting the interstellar gas with unenriched material. Further, we have to distinguish in a more accurate way the halo phase from the disc phase, since, as we have recalled in sect. **2**, the disc evolves starting with a

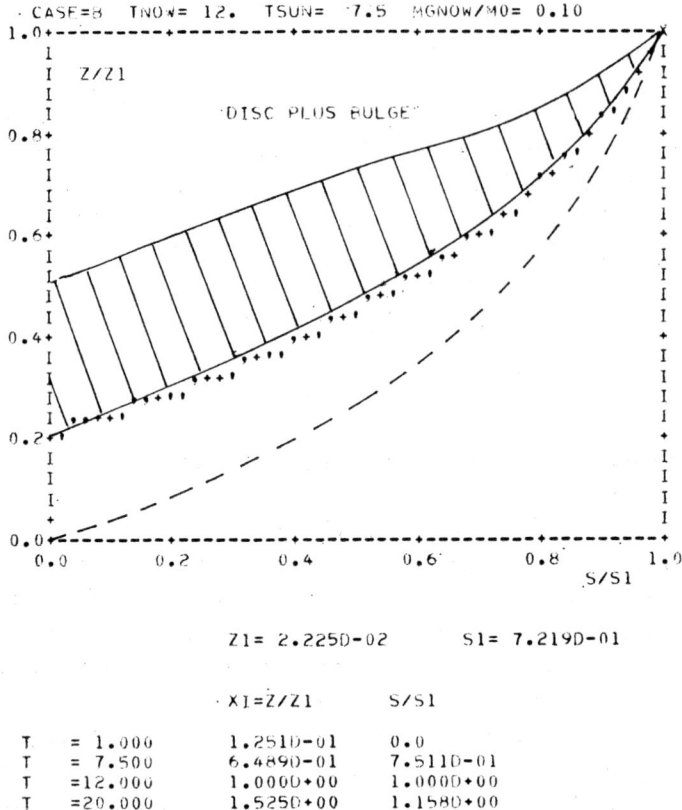

Fig. 2. – Chemical distribution Z/Z_{now} vs. S/S_{now} for the disc phase. The hatched area represents the observed distribution adopted from [12]. The lower curve represents the simple-model solution. The dashed curve, that practically coincides with the bottom edge of the hatched area, represents our computed solution, where $S_{\mathrm{now}} = 0.72$ and $Z_{\mathrm{now}} = 0.022$.

nonzero initial abundance, but with $Z \simeq Z_\odot/3$. In doing so, we must consider the general integral equations governing chemical enrichment.

8. – General equations for the adopted model.

With the above prescriptions, the chemical history of the Galaxy can be described by the following equations.

a) *Equation of conservation of mass:*

$$(3) \qquad \mu(t) = 1 + \mu_{\bullet j}(t) - S(t) \,,$$

where $\mu(t)$ is the ratio of gas mass to total mass, $S(t)$ the cumulative mass condensed in stars of all masses from the beginning up to time t:

$$S(t) = \int\limits_{M_V} \int\limits_0^t K(M_V)\, m(M_V)\, B(t')\, \mathrm{d}t'\, \mathrm{d}M_V \, ,$$

and $\mu_{\mathrm{ej}}(t)$ the cumulative mass ejected in space by dead stars from the beginning up to time t:

$$\mu_{\mathrm{ej}}(t) = \int\limits_{M_V} \int\limits_0^{t-\tau(M_V)} K(M_V)\big(m(M_V) - m_{\mathrm{r}}(M_V)\big) B(t')\, \mathrm{d}t'\, \mathrm{d}M_V \, .$$

Both $S(t)$ and $\mu_{\mathrm{ej}}(t)$ are measured in units of the total mass of the Galaxy. In the preceding relations it is

$m(M_V)$	mass of a star of magnitude M_V,
$\tau(M_V)$	lifetime of a star of magnitude M_V,
$m_{\mathrm{r}}(M_V)$	mass of collapsed remnant,
$B(t)$	rate of star formation in units of the total mass.

Moreover, $K(M_V)$ is related to the IMF by the relation

$$K(M_V)\, m(M_V)\, \mathrm{d}M_V = \psi(m)\, \mathrm{d}m \, .$$

We prefer to integrate over the absolute magnitude M_V, as in [25], instead of integrating over the stellar mass m, owing to the fact that the function $K(M_V)\, m(M_V)$ remains nearly constant over a large range of absolute magnitudes. We underline that, adopting the spectrum by TRURAN *et al.* [25], for massive stars a slightly larger contribution results than that indicated in sect. **4**:

$$\int\limits_1^\infty K(M_V)\, m(M_V)\, \mathrm{d}M_V = 0.47 \, .$$

Concerning the halo, an essential modification must be introduced in eq. (3), in order to reproduce the observed $Z(S)$ distribution of globular clusters. HARTWICK [10] suggested that, at every time t, a certain fraction of the interstellar gas should be removed from the star formation process and chemical remixing. The cumulative removed mass $D(t)$ may be assumed proportional to the cumulative mass condensed in stars:

$$D(t) = d \cdot S(t) \, ,$$

where the constant d must be determined by observative requirements. Assuming instantaneous recycling, HARTWICK found $d = 10$, which implies that at the end of the free-fall contraction almost all the interstellar gas is removed from star processing.

This removal of gas may perhaps be physically justified by the following considerations. If the efficiency of star formation in the halo is governed by cloud-cloud collisions (see sect. **3**), we expect that stars are mainly formed in giant associations. These associations would behave as giant H II regions, with a long lifetime, of the order of a few hundred million years. The recurring explosion of many supernovae heats a large surrounding region through turbulent motions driven by shock waves. A multisupernova remnant would then result.

In these conditions, the metals ejected by the more massive short-lived stars diffuse in the outer cold interstellar medium, whereas this cold medium can hardly penetrate into the hot region. An enhanced metal enrichment of the cold material will then ensue.

At a later time, when all O-B stars present in the association will die, the « frozen » material will dissolve and will again participate to the process of chemical remixing and star formation.

The above considerations strengthen the idea of a temporal removal of gas, though this removal should not be considered as a physical ejection of matter away from the halo, out of the Galaxy or directly in the disc region, as suggested by HARTWICK [10] and by CAIMMI [53].

For the halo phase, the continuity equation must then be modified and becomes

$$(4) \qquad \mu_{\text{eff}}(t) = 1 + \mu_{\text{ej}}(t) - (1 + d)\, S(t)\,,$$

where $\mu_{\text{eff}}(t)$ is now the effective mass participating to chemical processing (fig. 3).

When the halo collapse slows down toward the formation of the disc and all the halo massive stars have died, the giant H II regions will gradually vanish, thus releasing the trapped gas. This process acts in the same sense of the delayed ejection of unenriched matter by evolved long-lived solar-mass stars in later ages and it helps to solve the F-G dwarf problem.

b) Star formation rate. Following the discussion of sect. **3**, we shall distinguish the halo from the disc phase. Concerning the first rapid halo contraction, we adopt a power law relation

$$(5) \qquad B(t) = AV(t)^{-s}\,,$$

where $V(t)$ is galactic volume at time t and $s = 1.85$, according to an analysis of elliptical galaxies by LARSON [24]. Assuming a rotational ellipsoid in free-

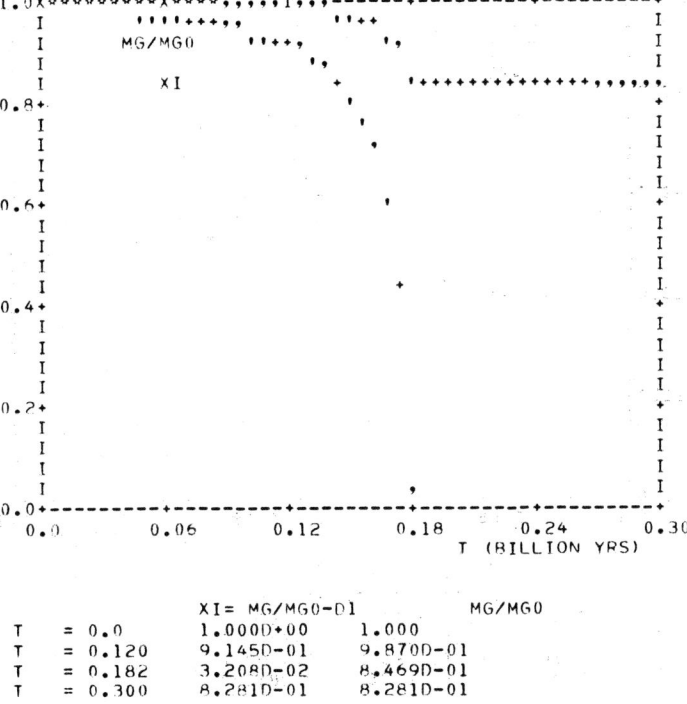

		XI= MG/MG0-D1	MG/MG0
T	= 0.0	1.000D+00	1.000
T	= 0.120	9.145D-01	9.870D-01
T	= 0.182	3.208D-02	8.469D-01
T	= 0.300	8.281D-01	8.281D-01

Fig. 3. – Lower curve: temporal variation of the effective mass $\mu_{\text{eff}}(t) = \mu(t) - D(t)$ in the halo phase participating to chemical remixing. Upper curve: total gas mass fraction.

fall contraction with constant eccentricity, we follow the shrinkage of the galactic volume $V(t)$ as in [54] up to $t_1 \simeq 1.8 \cdot 10^8$ y when $h_1 = 4$ kpc.

In the later slow halo contraction stage and in the subsequent disc phase, according to TRURAN et al. [25], instead of imposing the behaviour of the SFR, we assume that the gas mass fraction decreases exponentially with time:

$$(6) \qquad \mu(t) = \mu(t_1) \exp\left[- a(t - t_1)\right],$$

where the coefficient a is determined if we know the present gas mass fraction in the solar neighbourhood, $\mu(T_{\text{G}})$. We adopt $T_{\text{G}} = 12$ billion years and $\mu(T_{\text{G}}) = 0.10$.

The SFR can then be deduced from the equation of conservation of mass. As represented in fig. 4, the SFR shows a peak near the end of the free-fall contraction and then decreases roughly proportionally to $\mu(t)$.

c) Initial mass function and stellar evolutionary prescriptions. As discussed above, we adopt the visual-magnitude spectrum $K(M_V)$ as in [25], which is in fair agreement with the Salpeter IMF.

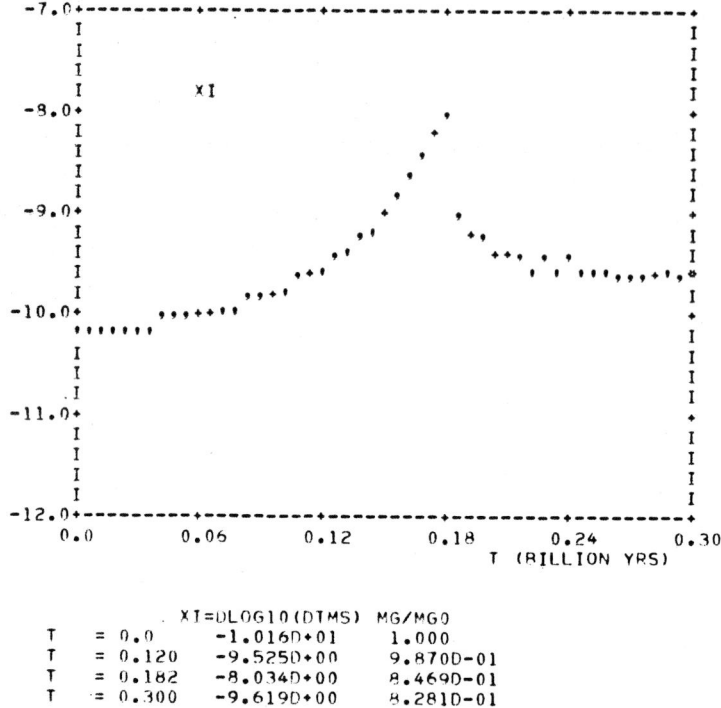

T		XI=DLOG10(DTMS)	MG/MGO
T	= 0.0	-1.016D+01	1.000
T	= 0.120	-9.525D+00	9.870D-01
T	= 0.182	-8.034D+00	8.469D-01
T	= 0.300	-9.619D+00	8.281D-01

Fig. 4. – Star formation rate during the rapid halo contraction, $B(t)$, in units of 10^6 y.

Concerning the yield by one generation of stars, we adopt the chemical production matrix discussed in sect. **6**.

d) Equation of metal enrichment. The behaviour with galactic time of the abundance by mass X_i of any chemical element i is described by the following equation:

$$(7) \qquad X_i(t) = \frac{1}{\mu(t)} \left[X_i(0) + \int_{M_V} \int_0^{t-\tau(M_V)} K(M_V)\, m(M_V)\, \alpha_i(M_V, t')\, B(t')\, \mathrm{d}t'\, \mathrm{d}M_V - \right.$$

$$\left. - \int_{M_V} \int_0^t K(M_V)\, m(M_V)\, X_i(t')\, B(t')\, \mathrm{d}t'\, \mathrm{d}M_V \right],$$

where $\alpha_i(M_V, t')$ is the fractional stellar mass in the form of element i ejected in space at time $t = t' + \tau(M_V)$ by a star of magnitude M_V born at t'. In particular, for He and heavy elements we have

$$(8) \qquad \begin{cases} \alpha_Y(M_V, t') = \big(q_1(M_V) - q_2(M_V)\big)\big(1 - Z(t')\big) + \big(1 - q_1(M_V)\big) Y(t')\,, \\ \alpha_Z(M_V, t') = q_2(M_V) - q_3(M_V) + \big(1 - q_2(M_V)\big) Z(t')\,, \end{cases}$$

q_1, q_2, q_3 being, respectively, the mass fractions processed to He, processed to heavy elements and left as dead remnants. $X_i(0)$ represents the pregalactic mass fraction of element i.

As we have said in sect. **2**, we adopt an initial He abundance $X_{\text{He}}(0) =$ $= 0.20 \div 0.23$, whereas for heavy elements we assume $Z(0) = 0$.

The second integral in eq. (7) must be multiplied by the factor $1 + d$ in the halo phase, in order to account for the removal of gas. Subsequently, when the removed gas is brought back to the interstellar medium, $X_i(t)$ must be suitably averaged.

9. – Numerical results.

We now examine a numerical model that has been obtained by taking into account the various prescriptions discussed in preceding sections.

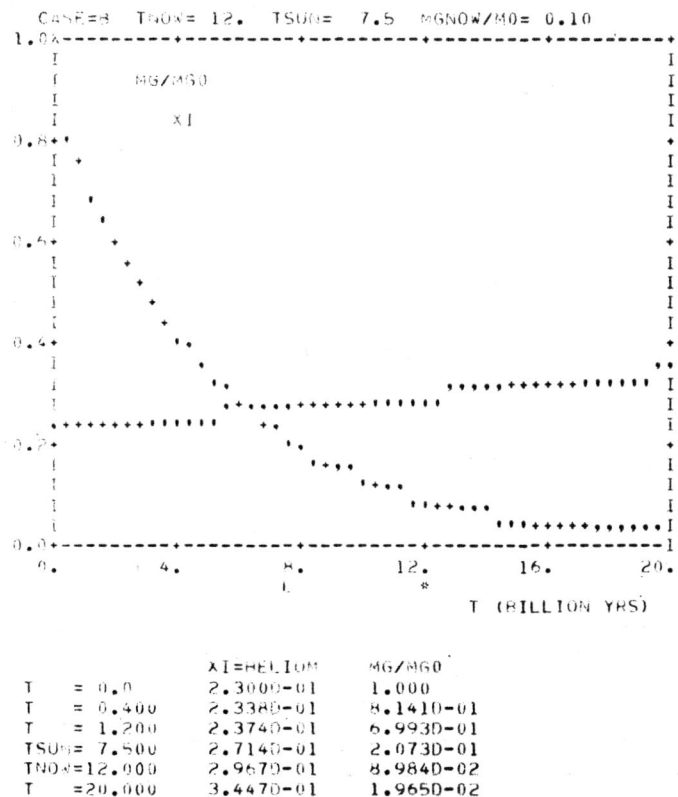

T		XI=HELIUM	MG/MG0
T	= 0.0	2.300D-01	1.000
T	= 0.400	2.338D-01	8.141D-01
T	= 1.200	2.374D-01	6.993D-01
TSUN=	7.500	2.714D-01	2.073D-01
TNOW=	12.000	2.967D-01	8.984D-02
T	= 20.000	3.447D-01	1.965D-02

Fig. 5. – Galactic helium enrichment, starting from a pregalactic abundance $Y(0) = 0.23$, for the choice $T_{\text{G}} = 12 \cdot 10^9$ y, $\mu(T_{\text{G}}) = 0.10$. At solar-system formation we have $Y_{\odot} = 0.2,7$, while presently it is $Y_{\text{now}} = 0.30$. The symbol $*$ indicates present age, while the symbol $\odot$ indicates solar birth. The decreasing curve represents the gas mass fraction $\mu(t)$.

Figure 5 shows the resulting chemical enrichment of He by stellar nucleo-synthesis, with the choice $Y(0) = 0.23$ and $\mu(T_G) = 0.10$. The theoretical curve $Y(t)$ is in fair agreement with observations. It appears that stellar nucleosynthesis gives a moderate, though not negligible contribution to ga-lactic He. When assuming no pregalactic He, stellar activity does not provide the observed abundance [38, 55], unless one adopts a completely different stellar evolutionary picture.

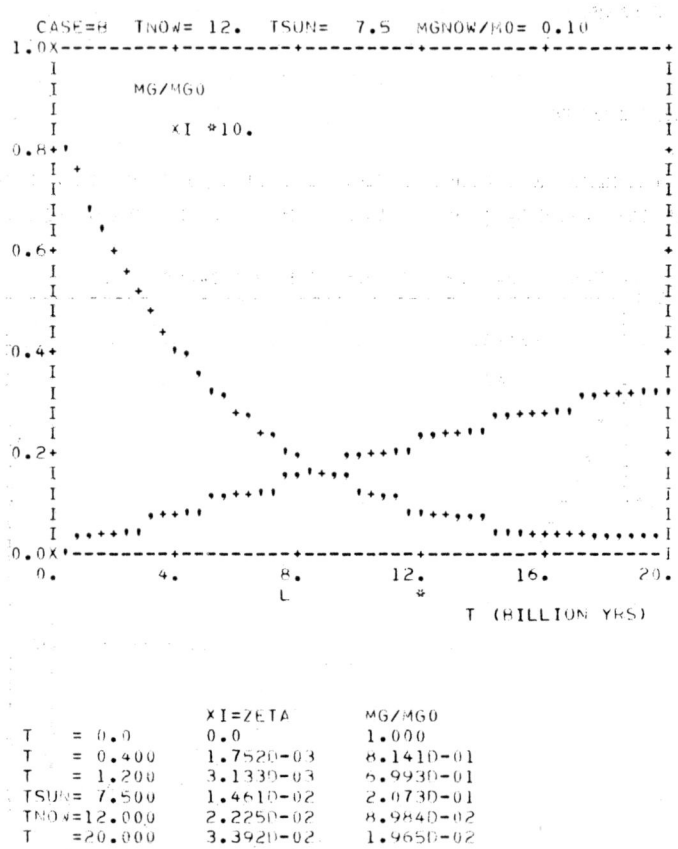

```
CASE=B   TNOW= 12.   TSUN=  7.5   MGNOW/M0= 0.10
1.0X--------------+------------------+----------------+----------+
  I                                                                 I
  I        MG/MG0                                                   I
  I                                                                 I
  I             XI  *10.                                            I
0.8+'                                                               +
  I    +                                                            I
  I'                                                                I
  I        '                                                        I
  I      '                                                          I
0.6+   +                                                            +
  I      +                                                          I
  I         +                                                       I
  I           +                                                     I
  I             +                                                   I
0.4+          ' '                                                   +
  I             '                                                   I
  I              '  '                          '.++++''             I
  I            + '                       ,'++++''                   I
  I               + +             ,,++++'                           I
0.2+                ' '      ,,'+++'                                +
  I                 ,,+'++,,                                        I
  I             ,,++''   '+,,                                       I
  I           +++''           '''++,,,,                            j
  I    ,+++''                        '''++++++,,,,,,                I
0.0X'--------------+------------------+----------------+----------j
  0.              4.             8.            12.           16.           20.
                   L                          *
                                              T (BILLION YRS)
```

		XI=ZETA	MG/MG0
T	= 0.0	0.0	1.000
T	= 0.400	1.752D-03	8.141D-01
T	= 1.200	3.133D-03	5.993D-01
TSUN=	7.500	1.461D-02	2.073D-01
TNOW=	12.000	2.225D-02	8.984D-02
T	= 20.000	3.392D-02	1.965D-02

Fig. 6. – Galactic heavy-element enrichment, $Z(t)$. The curve is multiplied by a scale factor 10. In particular, we obtain $Z_\odot = 0.015$ and $Z_{\text{now}} = 0.022$.

As we said before, the key parameter defining the present chemical com-position is not the galactic age, rather the present gas mass fraction. Indeed, as has been shown by BUSSO *et al.* [38], changing the present value of T_G, but leaving unchanged $\mu(T_G)$, one essentially obtains the same results.

Figure 6 shows the resulting galactic enrichment of heavy elements, $Z(t)$. Again, the observational constraints are satisfactorily reproduced by our model. For the disc phase, Z varies roughly linearly with time

$$\frac{\mathrm{d}Z}{\mathrm{d}t} = 1.7 \cdot 10^{-2}/10^{10}\,\mathrm{y}\,.$$

This gradient is slightly larger than that suggested by observations [16, 17]. However, observational uncertainties are rather large and do not permit to go deeper into the discussion.

Concerning the solar system, we must notice that the data by MAYOR [17]

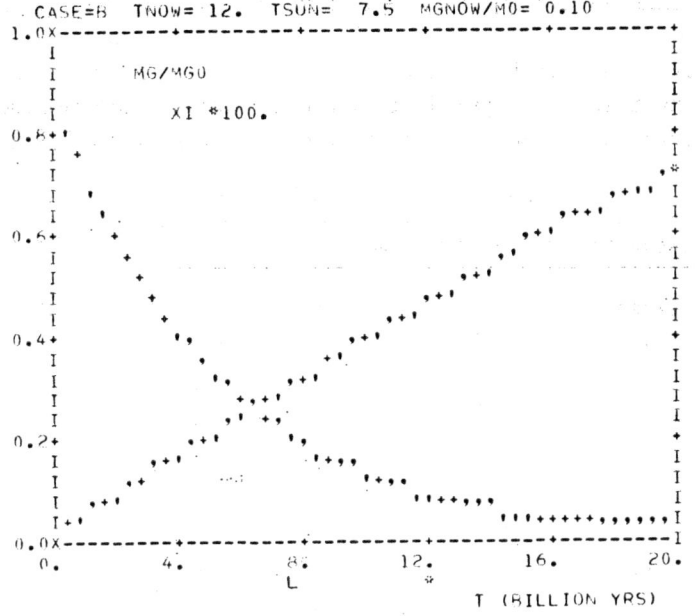

	XI=CARBON	XI=OXYGEN	XI=NEON	XI=MAGNESIUM	XI= SI+FE
T = 0.0	0.0	0.0	0.0	0.0	0.0
T = 0.400	3.700D-04	8.917D-04	1.391D-04	9.815D-05	2.544D-04
T = 1.200	6.598D-04	1.599D-03	2.472D-04	1.743D-04	4.557D-04
TSUN=7.500	3.084D-03	7.499D-03	1.152D-03	8.120D-04	2.140D-03
TNOW=12.000	4.708D-03	1.144D-02	1.760D-03	1.240D-03	3.269D-03
T =20.000	7.218D-03	1.751D-02	2.700D-03	1.903D-03	5.015D-03
XI SOLAR=	3.4D-03	8.3D-03	1.48D-03	4.8D-04	2.0D-03

Z SOLAR=1.850D-02

Fig. 7. – Chemical enrichment of ^{12}C. At the bottom the abundances by mass of C, O, Ne, Mg, Si+Fe are reported for some characteristic galactic ages, together with the observed solar-system abundances adopted by CAMERON [56].

indicate a solar abundance larger by a factor up to 2 with respect to the average value of *F-G* stars with the same age. If this difference should be confirmed, the solar system would be investigated more carefully, taking account of local chemical dishomogeneities.

We now discuss the $Z(S)$ distribution. As shown in fig. 1, in the halo phase we succeed in reproducing the chemical distribution of halo globular clusters, with the right value $Z_1 \simeq 3 \cdot 10^{-3}$, by adopting in eq. (5) $A = 1.04 \cdot 10^{-2}$ (V in kpc³, t in years) and $d = 5$. This last value is a factor of 2 lower than suggested by HARTWICK, the reduction being ascribable to the noninstantaneous-recycling approximation. Indeed, a noticeable fraction of gas is trapped by less massive stars and ejected in space not immediately, but well after the rapid halo collapse.

The cumulative mass condensed in stars of all masses when $Z \simeq 3 \cdot 10^{-3}$ is $S_1 = 0.16$, which constitutes an upper limit to the fraction of halo population stars and is consistent with the analysis of high-velocity stars by EGGEN *et al.* [57].

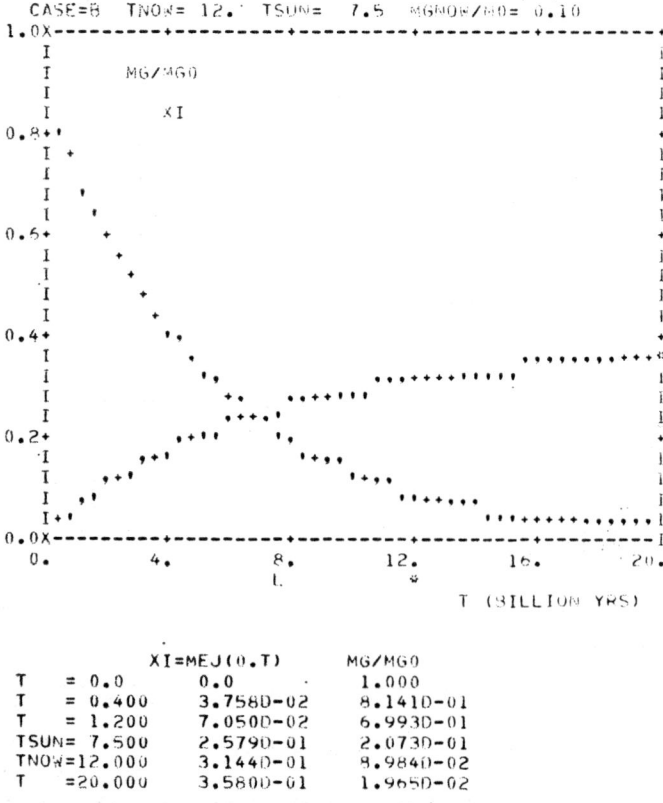

	XI=MEJ(0,T)	MG/MGO
T = 0.0	0.0	1.000
T = 0.400	3.758D-02	8.141D-01
T = 1.200	7.050D-02	6.993D-01
TSUN= 7.500	2.579D-01	2.073D-01
TNOW=12.000	3.144D-01	8.984D-02
T =20.000	3.580D-01	1.965D-02

Fig. 8. – Cumulative galactic fractional mass ejected by evolved stars from the beginning up to time t, $\mu_{\text{ej}}(t)$, as a function of galactic age.

We must notice that the observed distribution of globular clusters appears to contradict the evidence by DIXON [9], following which only about 1/20 of halo population stars have $Z \lesssim Z_\odot/10$. If this evidence is statistically correct, a large fraction of very-metal-deficient stars must be unseen, or has escaped from the Galaxy, or else the solar-mass stars were avoided from condensing in the first generations. A different scenario would be invoked if the chemical distribution of globular clusters will not be representative of the average chemical enrichment of interstellar gas.

For the disc phase, the $Z(S)$ distribution, starting with an initial abundance $Z = Z_\odot/3$, is shown in fig. 2. The calculated distribution lies within observational uncertainties. However, when future observations will involve a definitely higher $Z(S)$ curve, then the theoretical model will have to be modified by taking in more accurate account the process of chemical remixing.

With the stellar evolutionary prescriptions adopted in sect. **6**, we obtain $\Delta Y/\Delta Z \simeq 3.0$, which is fairly consistent with the observed value 2.7 by HACYAN et al. [18]. This fact gives some support to the adopted evolutionary picture.

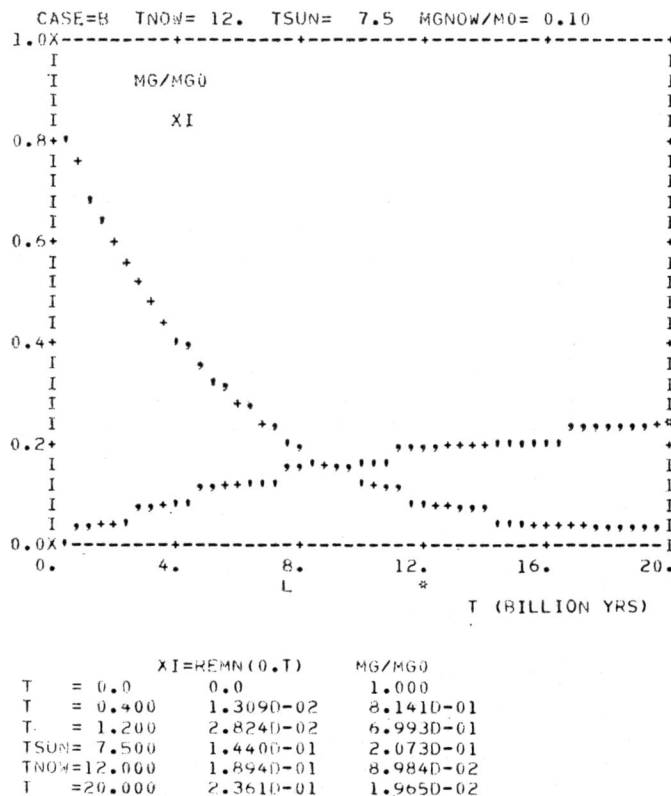

Fig. 9. – Cumulative galactic fractional mass condensed in remnants (white dwarfs, neutron stars, black-holes) from the beginning up to time t. Dashed curve: cumulative galactic fractional mass condensed in white dwarfs only.

The relative contribution to Z by the most important chemical elements, C, O, Ne, Mg, Si+Fe, may be calculated according to the estimates by ARNETT [45] and with our choices about the core remnant fractions. In particular, concerning ^{12}C, we have added the contribution by stars in the mass range $20\,m_\odot \leqslant m \leqslant 80\,m_\odot$. Indeed, for these stars the effect of the ^4He convective shell leads some ^{12}C to be avoided from further burning in the core and to be ejected together with the envelope [58].

Figure 7 shows the chemical enrichment of ^{12}C. Since all metals are contributed by short-lived massive stars, the same behaviour is followed by all other heavy elements, except for the first halo ages. The calculated abundances by mass are indicated at the bottom of the figure for some characterizing epochs, together with the corresponding observed solar-system abundances, according with CAMERON [56]. The agreement is satisfactory, thus giving further support to our model.

Finally, fig. 8 and 9 represent, respectively, the cumulative fractional mass ejected from evolved stars, $\mu_{ej}(t)$, and the cumulative fractional mass condensed in remnants (white dwarfs, neutron stars, black-holes).

REFERENCES

[1]   A. FERRARI, R. GALLINO and A. MASANI: *Astron. Astrophys.*, **10**, 481 (1971).
[2]   M. PEIMBERT and S. TORRES-PEIMBERT: *Astrophys. J.*, **203**, 581 (1976).
[3]   S. D'ODORICO, M. PEIMBERT and F. SABBADIN: *Astron. Astrophys.*, **47**, 341 (1976).
[4]   R. V. WAGONER, W. A. FOWLER and F. HOYLE: *Astrophys. J.*, **148**, 3 (1967).
[5]   D. M. POPPER, M. E. JØRGENSEN, D. C. MORTON and D. S. LECKNONE: *Astrophys. J. Lett.*, **161**, L57 (1970).
[6]   W. W. MORGAN: *Astrophys. J.*, **64**, 432 (1959).
[7]   D. BUTLER, T. D. KINMAN and R. P. KRAFT; *Astron. J.*, **84**, 993 (1979).
[8]   R. CANTERNA and A. SCHOMMER: *Astrophys. J. Lett.*, **219**, L119 (1978).
[9]   M. E. DIXON: *Mon. Not. R. Astron. Soc.*, **131**, 325 (1966).
[10] F. D. A. HARTWICK: *Astrophys. J.*, **209**, 418 (1976).
[11] L. WOLTJER: *Astron. Astrophys.*, **42**, 109 (1975).
[12] B. E. J. PAGEL and B. E. PATCHETT: *Mon. Not. R. Astron. Soc.*, **172**, 13 (1975).
[13] S. VAN DEN BERGH: *Astron. J.*, **67**, 486 (1962).
[14] M. SCHMIDT: *Astrophys. J.*, **137**, 758 (1963).
[15] J. B. HEARNSHAW: *Mem. R. Astron. Soc.*, **77**, 55 (1972).
[16] M. MAYOR: *Astron. Astrophys.*, **32**, 321 (1974).
[17] M. MAYOR: *Astron. Astrophys.*, **48**, 301 (1976).
[18] S. HACYAN, D. DULTZIN-HACYAN, S. TORRES-PEIMBERT and M. PEIMBERT: *Rev. Mex. Astron. Astrophys.*, **1**, 355 (1976).
[19] R. B. LARSON: *Mon. Not. R. Astron. Soc.*, **176**, 31 (1976).
[20] B. M. TINSLEY and R. B. LARSON: *Astrophys. J.*, **221**, 554 (1978).
[21] S. ISOBE, Y. YOSHII and H. SAIO: *XXII Liège Astrophysics Symposium, June 20-22, 1978.*
[22] M. SCHMIDT: *Astrophys. J.*, **129**, 243 (1959).

[23] J. Audouze and B. M. Tinsley: *Ann. Rev. Astron. Astrophys.*, **14**, 43 (1976).

[24] R. B. Larson: *Mon. Not. R. Astron. Soc.*, **173**, 671 (1975).

[25] J. W. Truran, C. J. Hansen and A. G. W. Cameron: *Can. J. Phys.*, **43**, 1616 (1965).

[26] B. M. Tinsley: *Astron. Astrophys.*, **20**, 383 (1972).

[27] E. E. Salpeter: *Astrophys. J.*, **121**, 161 (1955).

[28] R. J. Talbot and W. D. Arnett: *Astrophys. J.*, **186**, 51 (1973).

[29] G. Burki: *Astron. Astrophys.*, **57**, 135 (1977).

[30] R. Stothers and N. R. Simon: *Astrophys. J.*, **160**, 1019 (1970).

[31] K. Ziebarth: *Astrophys. J.*, **162**, 947 (1970).

[32] D. Weistrop: *Astron. J.*, **77**, 849 (1972).

[33] J. B. Ostriker and T. X. Thuan: *Astrophys. J.*, **202**, 353 (1975).

[34] C. Chiosi: in *Chemical and Dynamical Evolution of our Galaxy*, I.A.U. Coll. No. 45, edited by E. Basinska-Grzesik and M. Mayor (1978).

[35] W. J. Quirk and B. M. Tinsley: *Astrophys. J.*, **179**, 69 (1973).

[36] P. Biermann and B. M. Tinsley: *Astron. Astrophys.*, **30**, 1 (1974).

[37] S. M. Faber: in *The Evolution of Galaxies and Stellar Populations*, edited by B. M. Tinsley and R. B. Larson (New Haven, Conn., 1977), p. 157.

[38] M. Busso, R. Gallino and A. Masani: *Astrophys. Space Sci.*, **52**, 479 (1977).

[39] R. Gallino and A. Masani: *Mem. Soc. Astron. Ital.*, **44**, 589 (1977).

[40] M. Schwarzschild and R. Härm: *Astrophys. J.*, **150**, 961 (1967).

[41] I. Iben jr.: *Astrophys. J.*, **196**, 525 (1975).

[42] B. Packzyński: *Acta Astron.*, **20**, 47 (1970).

[43] S. Tsuruta and A. G. W. Cameron: *Astrophys. Space Sci.*, **7**, 374 (1970).

[44] J. E. Gunn and J. P. Ostriker: *Astrophys. J.*, **160**, 979 (1970).

[45] W. D. Arnett: *Astrophys. J.*, **219**, 1008 (1978).

[46] T. A. Weaver, G. B. Zimmerman and S. E. Woosley: *Astrophys. J.*, **225**, 1021 (1978).

[47] J. R. Wilson: *Phys. Rev. Lett.*, **32**, 849 (1974).

[48] Z. Barkat, G. Rakavy and N. Sack: *Phys. Rev. Lett.*, **18**, 379 (1967).

[49] G. S. Fraley: *Astrophys. Space Sci.*, **2**, 96 (1968).

[50] L. Searle and W. L. W. Sargent: *Astrophys. J.*, **173**, 25 (1972).

[51] R. J. Talbot and W. D. Arnett: *Astrophys. J.*, **170**, 409 (1971).

[52] B. M. Tinsley: *Astrophys. J.*, **192**, 629 (1974).

[53] R. Caimmi: *Astrophys. Space Sci.*, **54**, 453 (1978).

[54] M. Kaufman: *Astrophys. Space Sci.*, **33**, 625 (1975).

[55] F. Hoyle and R. J. Tayler: *Nature (London)*, **203**, 1108 (1964).

[56] A. G. W. Cameron: *Space Sci. Rev.*, **15**, 121 (1973).

[57] O. J. Eggen, D. Lynden-Bell and A. R. Sandage: *Astrophys. J.*, **136**, 748 (1962).

[58] R. Gallino and E. Gaschino: preprint (1979).

The Motion of Particles in Stefan Flow.

O. Vittori

FISBAT-C.N.R. Laboratory - Bologna, Italia

I was pleased to receive your invitation and am as well pleased to be here. The researches in which I am involved aim at understanding the role played in the scavenging of airborne particles by some small-scale processes, namely those occurring in the vicinity of condensing and evaporating bodies.

I am aware that analogous processes occurring in a cosmic fluid are governed by a larger variety of physical parameters and require a somewhat different vision. However, I hope that a brief summary of the motion of airborne particles in Stefan flow, as we study it, might be of some interest. There is some evidence that it plays some role in the atmosphere.

Stefan flow brings particles towards condensing surfaces and carries particles away from evaporating surfaces.

In a binary gas mixture (composed of a vapour of concentration C_v and a carrier gas of concentration C_c) the mass flux of the vapour leaving an evaporating surface is given by

$$\dot{m}_v = - D_{vc} \operatorname{grad} C_v ,$$

where D_{vc} is the mutual diffusion coefficient.

The contemporary mass flux of the carrier gas towards the surface is

$$\dot{m}_c = - D_{vc} \operatorname{grad} C_c .$$

Since the total pressure $P_v + P_c = P = \text{const}$, STEFAN [1] states that the carrier gas cannot accumulate in the vicinity of the evaporating surface so that there must be a hydrodynamic flow, $|C_c U_s|$, such that no net migration of carrier gas molecules takes place.

$$U_s = \frac{D_{vc} \operatorname{grad} C_c}{C_c}$$

is the so-called Stefan velocity acting on particles present in the vapour-carrier gas mixture.

The actual velocity U_p of nonvolatile particles (of size comparable to the molecule mean free path) is the sum of the Stefan velocity and the diffusiophoretic velocity [2]. The diffusiophoretic motion of the particle is caused by the diffusion of the two gases into each other and its direction coincides with that of the diffusion of the heavier gas. This is so because a greater proportion of the heavier-gas molecules bombard the particle on one side with respect to the other.

There have been experimental checks of the particle Stefan flow [3, 4]. Let me present that performed by us using water vapour and air at S.T.P.

In the figure you can see two plates (the sink and the source of water vapour, respectively) establishing the water vapour gradient field. The experimental arrangement is such that the field is practically constant in the chamber.

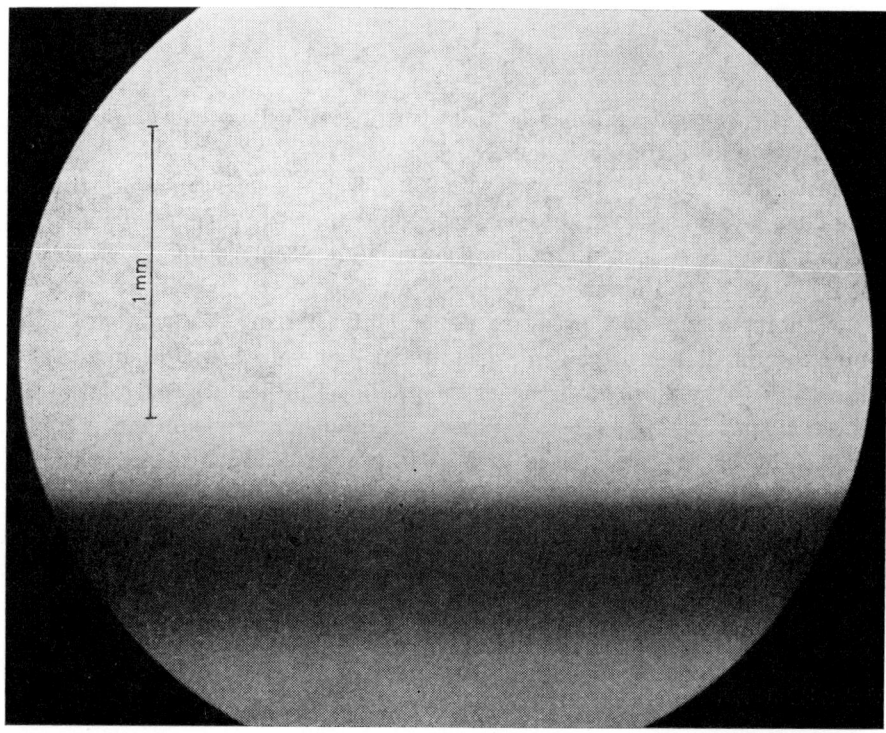

Fig. 1. – The instantaneous boundary during the run. Actually the boundary is even sharper, as, during the exposure time, it moves. The lower image is that due to reflection of the boundary itself in the water surface below.

Particles are injected in the field. Several runs are needed for a good evaluation of U_p. At the very beginning of each run the chamber is full of particles.

Since there is no appreciable convection in the chamber, you can see the plane-parallel sharp boundary of the particle population moving upwards from the vapour source to the sink.

If we take into account the diffusiophoresis, the actual particle velocity in the chamber is expected to be $U_p = 0.8\,U_s$. However, while the boundary moves, Brownian motion takes place, so that particles diffuse appreciably into the dark space (the average particle size is much less than the molecule mean free path).

If the ideal boundary moves with the theoretical Stefan velocity, the observed « slow down » can be computed by assuming the ideal boundary to be that of an aerosol-filled semi-space. The particle Brownian transfer through the ideal boundary is, therefore, the solution of the transport differential equation with the above initial conditions, that is

$$n(x,\,t) = \frac{n_0}{2}\left(1 + \operatorname{erf}\frac{x}{\sqrt{4D_p\,t}}\right),$$

where n is the particle number concentration (initially $n = n_0$) and D_p is the average Brownian diffusion coefficient.

By measuring U_s by the water mass increment of the sink, the Stefan velocity was evaluated with satisfactory accuracy.

To give you at least one number, the particle Stefan velocity is, in our experiment, close to $2 \cdot 10^{-2}$ cm s^{-1}.

U_p is temperature and pressure dependent. To my knowledge there has been no experiment performed in a gas mixture at very low pressure. It would be interesting to perform it in order to predict the importance of the Stefan flow in transporting particles dispersed in another environment. The molecular weight of the two gases is also important. With water vapour diffusing into hydrogen instead of air U_p is approximately doubled.

Turbulence in the fluid, thermophoretic and other fields and some physico-chemical features of the particles change this simple picture.

REFERENCES

[1] J. STEFAN: *Wien. Ber.*, **83**, 843 (1881).
[2] L. WALDMANN: *Z. Naturforsch.*, **14** a, 589 (1959).
[3] P. GOLDSMITH, H. J. DELAFIELD and L. C. COX: *Q. J. R. Meteorol. Soc.*, **89**, 43 (1963).
[4] O. VITTORI and V. PRODI: in *Precipitation Scavenging*, edited by U.S. Atomic Energy Commission (1970).

PROCEEDINGS OF THE INTERNATIONAL SCHOOL OF PHYSICS
« ENRICO FERMI »

TIPOGRAFIA COMPOSITORI - **BOLOGNA**